Handbook of Nanoencapsulation

Nutraceutical encapsulation involves protection of products from oxidative damage, controlled delivery of nanoencapsulated nutraceuticals, and improved nutraceutical bioavailability as well as biological action. It is a promising technique to ensure the stabilization of such labile compounds and to protect the core ingredients from premature reactions and interactions. In a comprehensive manner, the *Handbook of Nanoencapsulation: Preparation, Characterization, Delivery, and Safety of Nutraceutical Nanocomposites* presents various nanosystems/nanocarriers, physical and chemical techniques used in encapsulation of various nutraceuticals, and the targeted delivery of various significant nutraceuticals.

This book bridges the gap between academia and research, as it encompasses the ubiquitous applications of nanoencapsulation technique used on significant nutraceuticals derived from plants, animals, and microalgae.

Key Features:

- Provides a quick and easy access to major plant-, animal-, and microalgae-derived nutraceutical ingredients.
- Discusses nanoencapsulation techniques for protection and targeted release of various food bioactive ingredients.
- Covers safety, bioaccessibility, and multiple applications of nanoencapsulated nutraceuticals in the food industry.

Unveiling pivotal aspects of nanoencapsulation of significant nutraceuticals, this book is a valuable resource for researchers, food toxicologists, food scientists, nutritionists, and scientists in medicinal research.

Handbook of Nanoencapsulation

Preparation, Characterization, Delivery, and Safety of Nutraceutical Nanocomposites

Edited by
Jasmeet Kour, Raees Ul Haq, Sajad Ahmad Wani, and Bhaskar Jyoti

CRC Press
Taylor & Francis Group
Boca Raton London New York

CRC Press is an imprint of the
Taylor & Francis Group, an **informa** business

First edition published 2023
by CRC Press
6000 Broken Sound Parkway NW, Suite 300, Boca Raton, FL 33487-2742

and by CRC Press
4 Park Square, Milton Park, Abingdon, Oxon, OX14 4RN

CRC Press is an imprint of Taylor & Francis Group, LLC

Library of Congress Cataloging-in-Publication Data
Names: Kour, Jasmeet, editor. | Haq, Raees ul, editor. | Wani, Sajad Ahmad, editor. |
Jyoti, Bhaskar, editor.
Title: Handbook of nanoencapsulation : preparation, characterization, delivery, and safety of nutraceutical nanocomposites / edited by Jasmeet Kour, Raees Ul Haq, Sajad Ahmad Wani, Bhaskar Jyoti.
Description: First edition. | Boca Raton : CRC Press, 2023. |
"The ingredients commonly encapsulated are vitamins, minerals, flavours, fragrances, salts, enzymes, acidulants, coloring and leavening agents, essential oils and other bio active compounds like carotenoids, antioxidants, dietary fibres etc."–Introduction. | Includes bibliographical references and index.
Identifiers: LCCN 2022038659 (print) | LCCN 2022038660 (ebook) |
ISBN 9781032194387 (hbk) | ISBN 9781032194424 (pbk) | ISBN 9781003259183 (ebk)
Subjects: LCSH: Food additives. | Microencapsulation. | Functional foods. |
Bioactive compounds. | Nanocomposites (Materials)
Classification: LCC TX553.A3 H35 2023 (print) | LCC TX553.A3 (ebook) |
DDC 664/.06—dc23/eng/20221108
LC record available at https://lccn.loc.gov/2022038659
LC ebook record available at https://lccn.loc.gov/2022038660

ISBN: 978-1-032-19438-7 (hbk)
ISBN: 978-1-032-19442-4 (pbk)
ISBN: 978-1-003-25918-3 (ebk)

DOI: 10.1201/9781003259183

Typeset in Times
by codeMantra

Contents

Preface

Food is a storehouse of several nutrients, and nutrients in turn lead to energy which is sufficient to perform pivotal functions of the body in addition to strengthening the immune system. In today's world, consumers are giving utmost importance to their health, which has actually paved the way for the generation of foods blessed with bioactives, thereby enhancing the nutritional status of food. Food manufacturers are facing manifold challenges related to safety and appearance of food in the form of physical hazards such as heat, light, oxygen, chemical-based contaminants, microbiological hazards, and retention of organoleptic and physicochemical properties. Amid these limitations, encapsulation technique has been a boon in overcoming the aforementioned problems by giving utmost protection to the core materials, leading to promoted bioavailability/bioaccessibility and increased functionality. This technique involves enclosure of a bioactive compound in its liquid, solid, or gaseous state within an inert material in the form of liposomes, polymeric and lipo-nano particles, biodegradable microspheres, cyclodextrin, and hydrogels.

In the food sector, nanoencapsulation offers immense protection to bioactive or nutraceutical ingredients from external stimuli, elevates retention of their flavor, inhibits loss of their flavor, enhances their shelf life, and ensures their targeted delivery. Nanotechnology is one of those evolving areas associated with future industrial productions generating new applications in the form of food packaging and delivery systems for several bioactives. Nowadays, nanotechnology is gaining importance in the food and health care sectors owing to its enormous proven health benefits. Primarily, the focus is on modifying the textural attributes of food products and improving the bioaccessibility/bioavailability of significant dietary factors.

One of the major highlighting features of this book is to draw the attention of food production professionals, food safety personnel, and food scientists. The book has unveiled pivotal aspects of nanoencapsulation of significant nutraceuticals of plant origin and from the animal world and microalgae apart from providing an elaborate knowledge about various nanocarriers and techniques/instruments used in encapsulation, applications in food and other sectors, bioavailability limitations, and safety as well. The entire material presented in this work is the collaboration of several experts in their distinguished fields. This piece of work will cater to researchers, academicians, food toxicologists, food scientists, nutritionists, and people in medicinal research in terms of uniqueness and usefulness. This work will indeed prove to be very informative to the general public for efficiently analyzing the significance of nanoencapsulation of health-friendly, food-based ingredients.

Dr. Jasmeet Kour
Dr. Raees Ul Haq
Dr. Sajad Ahmad Wani
Dr. Bhaskar Jyoti

Editors

Dr. Jasmeet Kour completed her Master's degree in Food Science and Technology from Government College for Women, Gandhi Nagar, Jammu University, Jammu and Kashmir, India. She earned a Doctoral degree at Sant Longowal Institute of Engineering and Technology, Longowal, Sangrur, Punjab, India from the Department of Food Engineering and Technology in 2019. She has been serving as an Assistant Professor (academic arrangement basis) at the Department of Food Science and Technology, Padma Shri Padma Sachdev, Government PG College for Women, Gandhi Nagar, Jammu, Jammu and Kashmir since 2009. She has conducted research in various prominent nutraceuticals of plant origin, and her work has been published in reputed journals with high-impact factors in eminent publishing houses in the field of food science. She has presented her research and review papers in various national and international conferences. Dr. Kour has authored and co-authored numerous book chapters and scientific articles published in international books with prestigious publishers such as Elsevier and Springer. She is also a part of various international projects. As her research work left a great imprint, she was also cordially invited to share her newest research findings at Tokyo University of Agriculture, Japan at the 6th International Conference on Agricultural and Biological Sciences (ABS 2020). She is currently working as an editorial board member and peer reviewer of various journals of repute. Dr. Kour is also credited with editorship of various mega book projects under publishers such as Elsevier and CRC Press / Taylor & Francis Group.

Dr. Raees ul Haq earned his Ph.D. from the Department of Food Engineering and Technology at Sant Longowal Institute of Engineering and Technology, Longowal, Sangrur, Punjab, India. He completed his Master's degree in Food Technology in the Islamic University of Science and Technology, Awantipora, Jammu and Kashmir, India. He has authored and co-authored chapters of many books published by prominent international publishers. Dr. Raees has presented his research work at various national and international conferences, seminars, and workshops. He is actively engaged with the scientific community by publishing his work as well as working as an editorial board member and peer reviewer of reputed international journals. He is the recipient of Maulana Azad National Fellowship awarded by the University Grants Commission (UGC) and has published a good number of research and review articles in reputed journals.

Dr. Sajad Ahmad Wani is currently a Postdoctoral Researcher in the Department of Food Science & Technology, Islamic University of Science & Technology, Awantipura, Jammu and Kashmir (Union Territory), India. He has completed his Master's degree in Food Technology in IUST, Awantipura, Jammu and Kashmir, India and PhD from Sant Longowal Institute of Engineering and Technology, Punjab, India. He has published more than 35 research/review articles, 10 book chapters, and 1 book. He is also writing articles for the popular magazine *Food & Beverage News* (India's first magazine for the food & beverage industry). He has attended 45 international/national conferences, seminars, and workshops throughout the world. Dr. Wani serves as editorial board member of many national and international journals. He is potential peer reviewer of reputed international journals related to food science and technology that belong to popular publishing houses, viz. Elsevier, Taylor & Francis, Wiley, Springer, etc. He is also member of various associations such as IFT, IFERP, AFSTI, and Asian Council of Science Editors. Dr. Wani is the recipient of Maulana Azad National Fellowship (MANF-2013–14) from the University Grants Commission, New Delhi, India.

Dr. Bhaskar Jyoti has completed her Master's degree in Food Science & Technology in Chaudhary Charan Singh University, Meerut, and as a part of her curriculum, she was selected for her first research project at CSIR-CFTRI, Mysore in the Department of HRD, Lab of Computational Fluid Dynamics in collaboration with the FMBCT Department in the topic of Study on Bun & Dough Rheological Properties after Incorporation of Microencapsulated *Garcinia cowa* Fruits Extract. During this time, she proved herself as one of the competent, motivated, and talented team members. Dr. Jyoti is an ethical researcher with bench work experience in the research areas of food science and technology. She is adaptive and proficient in scientific writing and poster presentations. Dr. Jyoti has seven publications (including five peer-reviewed research articles) in well-reputed journals as the first author and overall 12 research articles. She also has six Indian patents in her name and two international German patents (Granted). During her graduation and post-graduation, she has worked with Haldiram's Foods International Ltd. (Noida), Uttar Pradesh (2007) on Food Processing of Indian Snacks and Development of New Products and with Calpro Foods Pvt. Ltd. (Delhi) 2010 on Food Enzymes and Emulsifiers in Bakery Products. She has been awarded the Young Scientist Award at ABRF Sage University, Bhopal in an international conference and has received many awards for her research writing and presentations.

Contributors

Mohammed Shafiq Alam
Department of Processing and Food Engineering
Punjab Agricultural University
Ludhiana, India

Furheen Amin
Department of Food Science and Technology
University of Kashmir
Jammu and Kashmir, India

Mohamad Anis
Department of Grain Science and Technology
CSIR-Central Food Technological Research Institute
Mysore, India

Francisca Aba Ansah
University of Energy and Natural Resources (UENR)
Sunyani, Ghana

Rosy Bansal
Department of Food Processing and Engineering
GSSDGS Khalsa College Patiala
Patiala, India

Vikas Bansal
Department of Food Technology
School of Engineering & Technology
Jaipur National University
Jagatpura, India

Garima Bhardwaj
Department of Chemistry
Sant Longowal Institute of Engineering and Technology
Punjab, India

Hitesh Chopra
Chitkara College of Pharmacy
Chitkara University
Rajpura, India

Monika Choudhary
Department of Food and Nutrition
Punjab Agricultural University
Ludhiana, India

Gagandeep Garg
Department of Food Processing
Maharaja Ranjit Singh Punjab Technical University
Bathinda, India

Prerna Gupta
Department of Food Technology & Nutrition
Lovely Professional University
Jalandhar, India

Rhythm Kalsi
Department of Food Technology & Nutrition
Lovely Professional University
Jalandhar, India

Amarjeet Kaur
Department of Food and Nutrition
Punjab Agricultural University
Ludhiana, India

Jasleen Kaur
Department of Food Technology & Nutrition
Lovely Professional University
Jalandhar, India

Francisca Aba Ansah
Council for Scientific and Industrial Research-Food Research Institute
Department of Agro-processing Technology and Food Biosciences
CSIR-College of Science and Technology
Accra, Ghana

Kawaljeet Kaur
Department of Food Science and Technology
Govt. College for Women Gandhi Nagar
Jammu, India

Mahaldeep Kaur
Department of Microbial Biotechnology
Panjab University
Chandigarh, India

Maninder Kaur
Punjab Agricultural University
Ludhiana, India

Prabhjot Kaur
Department of Food and Nutrition
Punjab Agricultural University
Ludhiana, India

Pushpendra Koli
College of Science, Health, Engineering and Education
Murdoch University
Perth, Western Australia

Mukul Kolish
Department of Food Technology and Nutrition
Lovely Professional University
Phagwara, India

Varun Kumar
Department of Home Science
Bhupendra Narayan Mandal University
Madhepura, Bihar

Aditi Kundu
Division of Agricultural Chemicals
Indian Agricultural Research Institute
New Delhi, India

Emmanuel Kyereh
Council for Scientific and Industrial Research-Food Research Institute
Department of Agro-processing Technology and Food Biosciences
CSIR-College of Science and Technology
Accra, Ghana

Tanu Malik
Department of Food Technology & Nutrition
Lovely Professional University
Jalandhar, India

Mudasir Bashir Mir
Department of Food Science and Technology
Govt. Degree College
Baramulla Jammu and Kashmir, India

Bharti Mittu
National Institute of Pharmaceutical Education and Research (NIPER)
S.A.S. Nagar, Mohali

Vishal Mutreja
Department of Chemistry
Chandigarh University
Punjab, India

Jessica Pandohee
Telethon Kids Institute
Nedlands, Australia

Anindita Paul
Crop Chemistry and Soil Science
ICAR-Central Tobacco Research Institute
Rajahmundry, India

Monica Premi
Department of Food & Nutrition Science
Manipal Academy of Higher Education
Dubai, United Arab Emirates

Reshu Rajput
University Institute of Agricultural Sciences
Chandigarh University
Gharuan, India

Malti Rajput
Department of Chemistry
Padma Shri Padma Sachdev Govt PG College for Women
Gandhi Nagar
Jammu & Kashmir, India

Supradip Saha
Division of Agricultural Chemicals
Indian Agricultural Research Institute: New Delhi
Delhi, India

Sangeeta
Department of Food Processing
Guru Nanak College
Budhlada, India

Rohan Sarkar
ICAR-Directorate of Medicinal Aromatic Plants Research
Anand, India

Jadala Shankaraswamy
Department of Fruit Science
Sri Konda Laxman Telangana State Horticultural University
Wanaparthy, India

Ajay Sharma
Department of Chemistry
Chandigarh University
Ajitgarh, India

Renu Sharma
Department of Chemistry
Akal Degree College Mastuana Sahib
Sangrur, India

Seema Sharma
Department of Food Technology, School of Engineering & Technology
Jaipur National University
Jagatpura, India

Vishal Sharma
Department of Food Technology
Amity University
Jaipur, India

Sakshi Thakur
Department of Chemistry
Chandigarh University
Ajitgarh, India

Lisa F. M. Lee Nen That
School of Science
RMIT University
Bundoora, Australia

Yakindra Prasad Timilsena
Agriculture and Food
Commonwealth Scientific and Industrial Research Organisation (CSIRO)
Werribee, Australia

Dinesh Kumar Yadav
Division of Environmental Soil Science
ICAR-Indian Institute of Soil Science
Bhopal, India

Nargis Yousf
Department of Food Science and Technology
Govt. Degree College
Baramulla Jammu and Kashmir, India

Syeda Saniya Zahra
Department of Pharmacognosy
Shifa Tameer-e-Millat University
Islamabad, Pakistan

CHAPTER 1

An Overview of Various Nanosystems for Encapsulating Nutraceuticals

Mohammed Shafiq Alam, Maninder Kaur, and Jasmeet Kour

CONTENTS

DOI: 10.1201/9781003259183-1

1.1 INTRODUCTION

The food processing sector is developing at a fast pace and is contributing toward the development of new processing techniques to prepare various food products that are more edible, safe, and palatable to consumers. Also, it makes the food available during off-season, thereby offering a huge variety of foods in different processed forms. The processing techniques commonly used are heating, canning, drying, blanching, pasteurization, cooling, dehydration, and evaporation. Nowadays, more focus is placed on novel processing techniques that offer various benefits as compared to traditional abovementioned techniques. Encapsulation of food ingredients is one of the flourishing processing techniques in the food sector, as it encapsulates and protects the food material from exposure to moisture, heat, and other extreme damaging conditions, thereby enhancing the stability and viability of encapsulated food materials.

Encapsulation can be done for those food ingredients that have health benefits and prevent the occurrence of diseases (DeFelice, 1995). The incorporation of these bioactive compounds in the encapsulated form in neutraceuticals is essential as these alone without encapsulation can be easily subjected to decomposition during processing and storage. The need for encapsulation arises from here to deliver safe, stable, and vital bioactive compounds and neutraceuticals to the body. Hence, an appropriate carrier system is required to make these bioactive compounds available without any damage and undesirable changes while going through processing (Fuchs et al., 2010; Katouzian and Jafari, 2016).

Before proceeding further with the discussion on systems for encapsulation, it is necessary to have a look at the basic terms associated with the encapsulation of neutraceuticals. A brief review of the various terms and definitions associated with encapsulation along with typical encapsulation techniques will be covered in this chapter.

1.2 ENCAPSULATION: DEFINITION AND TYPES

Encapsulation can be defined as the process of packing or entrapping sensitive food ingredients within another substance (i.e. coating or wall material), thereby protecting them from adverse reactions, facilitating production of particles with reduced diameters, and ensuring controlled release of nutrients. It may include encasing of solids, liquids, or gases, which makes them easy to handle. The encapsulated material is called a fill, core material, active agent, payload phase, or internal phase. The encapsulating material is called carrier material, shell, external phase, membrane, coating, or matrix. The major requirement of carrier material in the food sector is that it should be food grade and capable of protecting ingredients from degradation.

Requirements to design an effective encapsulation system:

1. The capsule dimensions and final physical state are of prime importance while designing an encapsulation system as they affect the sensory properties of food.
2. The design of wall material should be according to the desired function of the entrapped material.
3. The environmental release conditions should be taken care of.
4. To increase the encapsulation efficiency, the solubility of organic solvent should be high and polymer solubility should be low with high concentration of polymer.
5. The encapsulating material should be economical with high capacity to encapsulate the core material and should be resistant to high shear and mechanical stresses.

Encapsulation can be classified into two types:

1. **Microencapsulation:** Microencapsulation is the process of entrapping the tiny particles of solid, liquids, gases, or droplets of bioactive ingredients in a coating of 1–1000 μm to prepare capsules ranging in size from submicrons to several millimeters with desirable properties. Microencapsulation is done for entraining the natural food compounds, fish and essential oils, natural food colorants, flavors, and phenolic compounds with antimicrobial properties (Jafari and McClements, 2017). Thus, microencapsulation presents a huge potential to be utilized by food industries to develop packaging systems by trying out new safe materials to enclose and prolong the shelf life of foods. It can be potentially used to e enhancing the availability of ntrap flavors and odor and can act as a barrier between the environment and sensitive bioactive compounds. It is of great use for entrapping EOs with antimicrobial properties in different food matrices, which usually have unpleasant odor and poor volatility and water solubility. The commonly encapsulated bioactive food components are given in Table 1.1.
2. **Nanoencapsulation:** The term "Nano" is derived from a Greek word which means dwarf and describes the dimension of the order of 100 nanometers (nm) or less, i.e., 10^{-9}m (Chaudhry et al., 2008). Later, the idea of nanotechnology was put forth in 1959 by Richard Feyman, and in 1974, the term nanotechnology was first used by Norio Taniguchi to denote the ability to process materials at a nanometer scale. The concept of nanotechnology in the food sector dates back to the ages of pasteurization which is done to kill the spoilage bacteria existing in 1000nm (Chellaram et al., 2014). Applications of nanotechnology in food science include alteration of the texture of food components, encapsulation of food components, development of new tastes, controlled release of flavors, and enhancing the availability of nutritional and bioactive components. All these applications form a part of encapsulation, and thus nanoencapsulation plays a major role toward maintaining the nutritional characteristics, safety, and packaging of foods.

Table 1.1 Application of Microencapsulated Bioactive Food Ingredients

Bioactive Ingredient	Example	Purpose of Encapsulation
Vitamins	Vitamin A, B1, B2, B12, C, D, E, K, niacin, and folic acid	• Reduction of off-flavors • Enhancing stability against oxidation and extremity of temperature
Lipids	Fish oil, rice bran oil, linolenic acid, and other essential oils (EOs)	• Stabilization and prevention of oxidative degradation during processing • Controlled release
Flavoring agents	Citrus oil, onion and garlic oils, and spice oleoresins	• Reduction of volatility • Transformation into stable powders • Taste masking
Antioxidants and phenolic compounds	Curcumin, carotenoids, etc.	• Stabilization and protection
Microorganisms and enzymes	Lipase, glucose oxidase, invertase, lactic acid bacteria	• Increase in stability during storage • Controlled release
Acidulants	Vitamin C, acetic acid, lactic acid, etc.	• Production of stable acids for baking industry • Controlled release of CO_2 during baking and processing
Sweeteners	Sugars, aspartame, or artificial sugars	• Reduction in hygroscopicity • Improvement in flow characteristics • Prolonged sweetening perception

Nanoencapsulation can be defined as a process of encapsulation or packing of substances with a protective layer of films, coverings, and layers at the nanometer scale (Paredes et al., 2016). Nanoencapsulation technologies have the potential to meet the challenges faced by food industries related to effective delivery of functional ingredients concerning health, controlled release of flavor, and aroma compounds. Techniques like nanocomposite, nanoemulsification, and nanostructuration are used as carrier vehicles to carry functional ingredients to the final action site. Thus, we can say that application of nanotechnology in the food sector is an emerging field and is expected to grow at a fast rate in the future.

1.3 PRESENT TRENDS OF NANOENCAPSULATION IN THE FOOD SECTOR

Several scientists and researchers are working on the application of nanoencapsulation technology in the food sector so that it can contribute toward public health, ensuring safe food and providing essential nutrients and bio-components (Huang et al., 2010; Sozer and Kokini, 2009). The beverage industry is also flourishing globally and the sale of nanotechnology packed beverage products has increased. Thus, there is a need for the stabilization of payload in dealing with aqueous solutions. Encapsulation of neutraceuticals in the liquid matrix is still a cumbersome operation, and there is a need to carry out focused research to develop safe food-grade delivery systems for nanoencapsulation. A number of scientific publications have been reported by various researchers (Chaudhry et al., 2008; Cushen et al., 2012; Nazzaro et al., 2012; Neethirajan and Jayas, 2011; Han et al., 2015; Mohammadi et al., 2015). At present, 400 companies are involved in developing nanoencapsulated food products and packaging materials (Neethirajan and Jayas, 2011). Reports suggest that the nano food and beverage industry is growing at a fast rate, and there is going to be an intense competition among the companies to survive in the nanoencapsulation market.

1.4 NANOENCAPSULATION PRODUCTION TECHNIQUES

Nanoencapsulation techniques according to the classical concept are categorized into "Top-Down" and "Bottom-Up" techniques. As the name indicates, the "Top-Down" technique involves the reduction in the size of ingredients/particles during the process of encapsulation. Energy in the form of mechanical forces is required to carry out nanonization in the above technique. In the "Bottom-Down" process, in contrast to the above, the particle size is decreased. Nanonization is achieved by clustering the ions, monomers, molecules, and atoms which are controlled with the application of several physical and chemical techniques to form nanocapsules. The production techniques and their details are given in Figure 1.1. Nanocarrier systems are lipid-based nanosystems, protein-based nanosystems, and polymeric-type nanosystems that are used to develop nanocapsules.

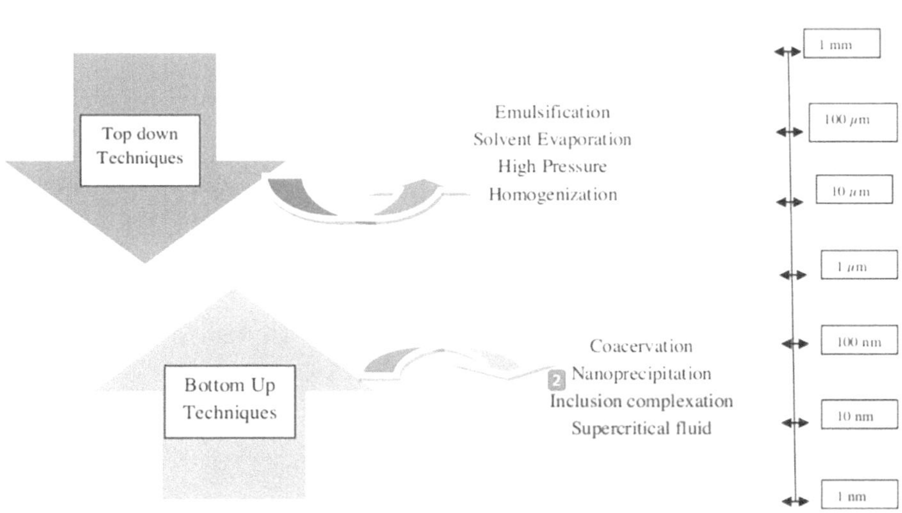

Figure 1.1 Top-down and bottom-up production techniques for nanoencapsulation.

Before moving into the nanocarrier systems, first we will discuss in detail the different production techniques. These are emulsification, solvent evaporation, high-pressure homogenization, microfluidization, ultrasonication, supercritical precipitation, inclusion complexion, coacervation, and nanoprecipitation.

1.4.1 Nanoemulsion

Nanoemulsions are formed by incorporating lipids into an aqueous medium or vice versa by dispersing droplets in the nanometric scale (20–200 nm) called as the dispersed phase in a continuous phase. They are also called ultrafine emulsions, miniemulsions, submicron emulsions, etc. Nanoemulsions can be achieved by two methods:

a. **High-energy emulsification or dispersion method**: This is the most common method used to ↔ 10 nm prepare nanoemulsions by using mechanical devices such as high-pressure homogenizers, microfluidizers, ↔ 1 nm ultrasound generators, and high shear stirrers to generate massive disruptive forces. All oil-in-water (O/W) and water–in-oil (W/O) nanoemulsions are produced by the dispersion method.

Advantages

- Simplified method of nanoemulsion formation with the development of smaller size of droplets with high energy input.
- Higher oil-to-surfactant ratio nanoemulsions can be prepared by this method.

Disadvantage

- Energy requirement to produce nanoscale droplets is high, and thus it is a cost-inefficient method.

b. **Low-energy emulsification or condensation method:** This method has garnered great attention from the researchers, as it is an energy-efficient method to produce nanoemulsions. It utilizes the internal chemical energy of the system, either by changing the temperature or by changing the composition of the oil–water system, and thus needs only simple stirring to produce droplets (Sole et al., 2006).

Advantages

- Droplets of emulsions are smaller in size as compared to high-energy methods, irrespective of composition and type of the system.
- Energy generation is minimum in this method, and thus the degradation of heat-sensitive compounds is minimum.

Disadvantage

- These methods depend upon physicochemical properties such as temperature, composition, and solubility.

1.4.2 Solvent Evaporation Technique

In the solvent extraction method, a drug is prepared by mixing with an organic solvent and emulsifying the polymer into an aqueous solution. Then, the organic solvent is subjected to vacuum or atmospheric conditions or heating to have polymer precipitation as nano- or microspheres (Pinto Reis et al., 2006) followed by centrifugation. Stirring rate, type and quantity of the dispersing agent, and viscosity of organic and aqueous phases along with temperature determine the size of capsules formed (Tice and Gilley, 1985). To obtain capsules of a smaller size, high-speed homogenization and ultrasonication can be performed.

Advantage

- This is an efficient technique of producing nanoencapsules to achieve particles of size below 250 nm.

Application

- It can be used for nanoencapsulation of different ingredients like curcumin, quercetin, coenzyme Q10, and α-tocopherol.

1.4.3 High-Pressure Homogenization

This technique makes the use of high-pressure induced homogenization with the application of device consisting of positive displacement pump and pressure valves homogenization chambers. It involves the application of pressure in the range of 500–5,000 psi in the homogenization chamber. The displacement pump during the delivery stroke causes the coarse emulsion to pass through a small inlet orifice of micrometer range. Disruption of coarse emulsion later on into tiny droplets is influenced by various factors like turbulence, shear stress, and cavitation (Tesch et al., 2003; Schultz et al., 2004). Thus, the whole process of high-pressure emulsification is divided into two phases, i.e., the disruption phase and stabilization phase.

Advantages

- This method efficiently breaks down the droplets and increases their stability.
- It is easy to scale up this process, easy to apply, and easy to reproduce. and it gives high throughput making the high-pressure homogenization technique the most suitable for producing nanoemulsions in the food sector.
- Better and efficient method to produce high-quality nanoemulsions.

Disadvantage

- Requires high energy and temperature to produce emulsions.

Applications

- It is used to prepare nanoemulsions of mandarin, lemon, bergamot, etc.
- In the dairy industry, it has found application in the disruption of fat globules.
- It is used in the food processing sector to provide supplementary nutrients and also acts a preservation technique.

1.4.4 Microfluidization

This method of emulsification is somewhat similar to high-pressure homogenization, except that it uses a microfluidizer to generate high pressure in the range of 500–20,000 psi, and the interaction chamber consists of a microchannel through which the coarse emulsion is allowed to pass through an impingement area to form uniform droplets in the size range of 50–300 μm followed by filtration. The coarse emulsion is delivered at high pressures and velocities of 400 m/s at the inlet of microchannels, which gets broken down into two streams forming Y or T junction. These junctions further in the interaction chamber strike with each other at very high velocity, impulsive force, and shear rate causing disruption and formation of very fine emulsions (Souto et al., 2005).

Advantage

- This technique is efficient in forming droplets by disruption and results in the formation of uniform-sized fine droplets.

Disadvantages

- It involves the utilization of high pressure to form nanoemulsions, which increases their temperature.
- It takes longer time to form emulsions by this technique, thereby resulting in the formation of droplets of increased size.

Applications

- Nanoemulsions of thyme, lemongrass, and sage oil are obtained by the microfluidization technique.
- Nanoemulsions with active ingredients to form edible films can be prepared by the process of microfluidization with different functional and physical properties.
- High nutritional edible oils can also be encapsulated using this technique.

1.4.5 Ultrasonication Method

This technique is based on the principle of application of ultrasonic energy with a frequency of 20 kHz by which the coarse droplets/emulsions break down to form nanoemulsions. The application of sound waves increases the cavitation threshold along with an increase in mechanical vibrations. The generation of high pressure and turbulence by the acoustic and shock waves collapses the droplets to form fine nanoparticles. The entire process of ultrasonication emulsion formation takes place in the ultrasonic chamber (Kamogawa et al., 2004; Jafari et al., 2007; Kentish et al., 2008). Various factors that affect the formation of nanoemulsions are viscosity ratio of continuous and dispersed phase, concentration of emulsifier, and the amplitude of waves applied (Nakabayashi et al., 2011).

1.4.6 Supercritical Precipitation

Supercritical fluids like carbon dioxide have been used as an important solvent for the encapsulation of food ingredients as they exhibit the properties of both gas and liquid and are also inexpensive. At room temperature and above its vapor pressure, CO_2 exists as liquid with density comparable to solvents and has excellent wetting properties. Above its critical temperature and pressure, it exists in the supercritical state and has liquid-like density and gas-like viscosity. It acts as a versatile and selective solvent as a small variation in pressure or temperature can cause drastic change in its density, viscosity, and dielectric properties.

Application:

- It is most commonly used for encapsulation of thermally sensitive components and bioactive compounds.

1.4.7 Coacervation

In this technique, newly formed coacervate phase is by phase separation of single or a mixture of polyelectrolyte from a solution. The coacervation process is classified into two types depending upon the number of types of polymers used. Simple coacervation involves the use of only one type of polymer, whereas complex coacervation involves the use of two or more than two types of polymers. Achievement of high payloads and controlled release–based mechanical stress, temperature, or sustained release makes this encapsulation technology distinctive and promising (Gouin, 2004). The only limitation posed by this encapsulation technology is the lack of commercialization as the enzyme used during cross linking has to be utilized carefully according to the legislations set.

1.4.8 Inclusion Complexation

Inclusion complexation is the phenomenon of encapsulation of or intermolecular interaction of encapsulated ingredient and shell material which is bearing a substrate into the cavity leading to the penetration of the guest molecule partly or completely into the cavity. This technique is highly preferred for encapsulating organic volatile molecules, EOs, and vitamins; for masking odors and flavors; and for preserving aromas.

Advantages

- The major advantage of this technique is higher encapsulation efficiency as compared to other bottom-up techniques.
- Higher stability of the core component.

Disadvantage

- The only limitation of this technique is that only few molecular compounds are suitable for this type of encapsulation.

1.4.9 Nanoprecipitation

This technique is based on the spontaneous emulsification technique in which biodegradable polymers are precipitated from an organic solution and the organic solvent is diffused in an aqueous medium (Galindo-rodriguez et al., 2004). It is also called the solvent displacement method.

1.5 ENCAPSULATION TECHNIQUES BASED ON CARRIER SYSTEMS

Classification of nanoencapsulation technologies on the basis of carrier system is an advanced concept and according to the wall material used, these are classified as lipid-based nanosystems, polymeric-type nanosystems, and protein-based nanosystems. Among these systems, lipid-based nanosystems are widely used and have found possibility in industrial production as these have higher encapsulation efficiency and low toxicity effect (Fathi et al., 2012). On the contrary, carbohydrate- and protein-based nanocarrier systems cannot be scaled up due to the requirement of various heat treatments and chemicals.

1.5.1 Lipid-Formulated Nanosystems

Lipid-based nanosystems have been utilized for pharmaceutical preparations and in food industries for encapsulating the various food ingredients and neutraceuticals. These systems have the capability to improve the performance of various antioxidants by improving their stability and bioavailability. These are further classified as nanoemulsions, nanoliposomes, nanochelates, and colloidosomes (Figure 1.2).

1.5.1.1 Nanoemulsions

Emulsions with a droplet size of 100–500 nm and formed by high-pressure homogenizers are termed as nanoemulsions. These are also utilized in the dried form as developed by the spray drying technique. The lipophilic bioactive compounds can be delivered with higher bioavailability and stability in powdered form (Weiss et al., 2006). O/W nanoemuslions are preferred for encapsulating lipophilic

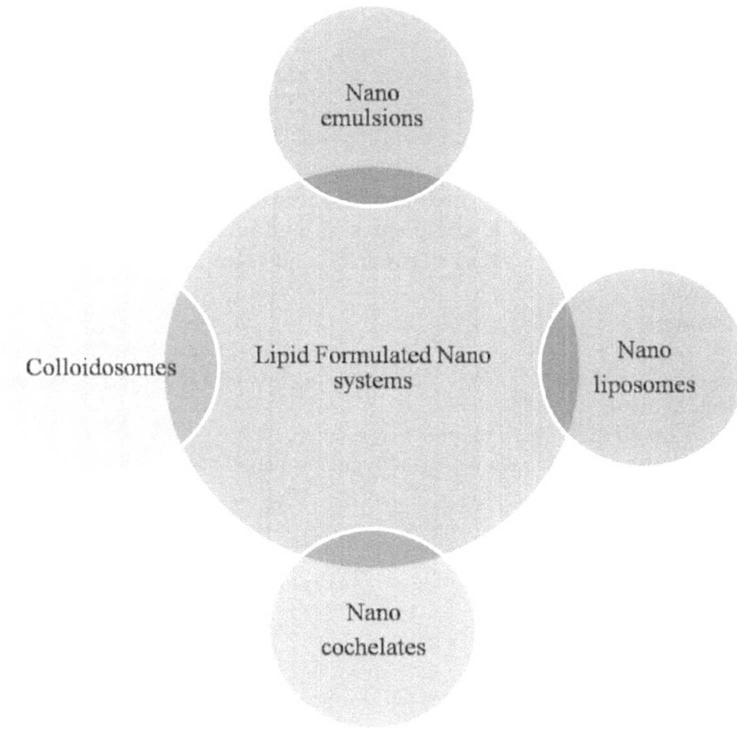

Figure 1.2 Lipid-based wall materials for nanoencapsulation.

bioactive agents, whereas W/O emulsions are suitable for encapsulating polyphenols (Mohammadi et al., 2016a).

Advantages

- This nanoencapsulation technique contains more content of the lipid phase.
- Safety from a toxicological point of view is another important feature of nanoemulsions.
- This can also be scaled up for large-scale production by using the high-pressure homogenization method.

Disadvantage

- It is a little bit difficult to have a controlled release of drugs from nanoemulsions.

Applications of Nanoemulsions in the Food Sector

- Used for decontamination of food packaging equipment and in the packaging of food.
- Highly suitable for encapsulating poorly water-soluble ingredients such as fish oil and lipophilic vitamins.

1.5.1.2 Nanoliposomes

The version of liposome at a nanoscale is called nanoliposome. Liposomes are tiny spherical-shaped artificial vesicles that are formed from natural nontoxic phospholipids. These are promising carriers for safe drug delivery due to their size and hydrophobic and hydrophilic character (besides biocompatibility). These liposomes have a unique feature of encapsulating and stabilizing aqueous compounds in their aqueous core and hydrophobic compounds in lipid bilayers, respectively. The mean diameter of nanoliposomes varies between 50 and 150 nm.

Advantages

- Nanoliposomes are chemically more stable and are able to protect bioactive ingredients at higher water activity.
- They have all the benefits of other nanocarriers.
- More fresh surface areas are created by the formation of nanoliposomes.

Disadvantage

- Energy requirement for generating nanoliposomes is high.

Applications

- These are used as carrier vehicles for neutraceuticals, enzymes, food additives, and food antimicrobials.
- These are also used for flavor release of ready meals (Kirby, 1993; Fathi et al., 2012; Mozafari et al., 2008).

1.5.1.3 Nanocochelates

Nanocochelates are formed by wrapping the nano-coiled particles around the micronutrients. These are basically a series of lipid bilayers with cylindrical, cigar-like microstructures. Nanocochelate delivery systems are stable phospholipid-cation precipitates composed of simple, naturally occurring materials, generally phosphatidylserine and calcium. Their structure is made up of multilayers of solid, sheet of lipid bilayer rolled up in a spiral or in stacked sheets, with little or no internal aqueous space (Bhosale et al., 2014).

Advantages

- Nanocochelates are stable than liposomes and less susceptible to oxidation.
- Biological molecules can be efficiently incorporated into the lipid bilayers.
- Easy and safe to produce is another promising factor of nanocochelates.
- They have the capability to encapsulate or entrap the active drug within a crystal matrix rather than chemically bonding with the drug.

Disadvantages

- They require specific storage conditions.
- Their cost of production is high.

Applications

- These encapsulation techniques are used for protecting an extended range of micro nutrients.
- They have the potential to enhance the nutritional value of processed foods.

1.5.1.4 Colloidosomes

Colloidosomes are the colloids self-assembled at interfaces of emulsion droplets to form microcapsules called colloidosomes. A hollow shell is fabricated and molecules of various substances such as fat blockers, vitamins, and medicines are placed inside this shell. The solid shell is formed to form a closed packed layer of colloidal particles, which forms an integrated part of the capsule surface and the spaces in between the particles form an array of uniform pores. This technique is utilized in the encapsulation of vitamins, drugs, proteins, flavors, fragrances, and controlled release of drugs in cosmetic, biomedical, food, and pharmaceutical industries.

1.5.1.5 Solid Lipid Nanoparticles

Solid lipid nanoparticles are formed of solid lipid core matrix, which is spherical in shape with a diameter between 10 and 1,000 nm and can solubilize lipophilic molecules (Awad et al., 2008). The formation process of solid lipid nanoparticles is represented. The solid lipid core or shell is stabilized using emulsifiers. There are basically two techniques to produce solid lipid nanoparticles on a large scale, namely, hot homogenization and cold homogenization, and on the laboratory scale, they can be prepared by the emulsification-evaporation method followed by sonification (Varshosaz et al., 2010). The major advantage of solid lipid nanoparticles as compared to nanoliposomes and nanoemulsions is that their production does not involve the utilization of organic solvents; also they have long-term stability and a tendency to retain high concentration of bioactive compounds.

1.5.2 Polymeric-Formulated Nanoparticles

Polymeric nanoparticles are formed using biodegradable polymers by their aggregation and self-association to have a controlled release and safe delivery of bioactive compounds. Generally, they are composed of a polymer and a surfactant. Various processes such as emulsion polymerization, polycondensation, emulsion solvent evaporation, nanoprecipitation, and spray drying are involved in making polymeric nanoparticles. These formulated nanoparticles are used to entrap various active ingredients like beta-carotene, antimicrobial agents, vitamins, and other functional compounds (Guadarrama-Lezama et al., 2012; Pereira et al., 2015; Ramsden, 2005).

1.5.2.1 Nanocomposites

Polymer nanocomposites are made of fine nanoparticles of size <100 nm, which are incorporated into plastics so as to improve their properties. Materials like polymeric resins, nanoscale metals and oxides, nanoclays and carbon nanoparticles with an inclusion in the range of 2%–8% by weight are used to make thermoplastic polymer nanocomposites. Organic clay fillers and dispersed clay layers have been widely studied in polymeric clay nanocomposites. Such nanocomposites have the advantage of high mechanical strength making them applicable for making food packaging films and for imparting various antioxidants and antimicrobial characteristics. Thus, these polymeric nanoclay composites have a huge potential to be utilized as a novel food packaging material with advanced thermal, mechanical, and barrier properties.

1.5.2.2 Nanofibers

Nanofibers are the fibers lying in the nano range with diameters <100 nm. Different polymers are used to generate nanofibers, and thus they have different physical properties and are subjected to different applications. The diameter of nanofibers generated also depends upon the type of polymer used and also on the method of production employed. The natural polymeric nanofibers include cellulose, silk fibroin, gelatin, collagen, keratin, and polysaccharides such as alginate and chitosan. Examples of synthetic polymers are polyurethane (PU), poly(lactic-co-glycolic acid) (PLGA), poly(lacticacid) (PLA), poly(3-hydroxybutyrate-co-3-hydroxyvalerate) (PHBV), polycaprolactone (PCL), and poly(ethylene-co-vinylacetate) (PEVA). Electrospinning, self-assembly, etc. are the methods used to prepare nanofibers out of which electrospinning is the most common method employed as it has the capability to generate ultrathin nanofibers with mass continuous production, compositions, controllable diameters, and orientations from various polymers. In the food sector, generally synthetic polymers are used and the utilization of fibers consisting of biopolymers still needs to be explored (Weiss et al., 2006).

1.5.2.3 Carbohydrate-Based Nanoparticles

For manufacturing carbohydrate nanoparticles, polysaccharides are most commonly used as a nanocarrier material due to their unique and massive molecular structure, higher ability to entrap bioactive compounds, abundance, and biodegradability (Fathi et al., 2014). Among the various polysaccharides, the most commonly used are starch, guar gum, cellulose, chitosan, dextran, alginate, cyclodextrins, and pectin. Researchers have shown great interest in developing nanoparticles from starch as it can be obtained from various sources plus it is abundant in nature (Simi and Emilia Abrahim, 2007).

Starch: Starch is the most common biodegradable, digestible, and biocompatible polymer used to encapsulate flavors, flax seeds, insulin, and unsaturated fatty acids (Li et al., 2010; Chung et al., 2008; Fathi et al., 2014). Starch in its natural form is hydrophobic and thus cannot be employed to encapsulate hydrophobic food components. Also, another limitation to its use is that starch is subject to amylase hydrolysis and acid attack due to its higher sensitivity. Low cost with fewer requirements of purification processes makes it a widely acceptable polysaccharide to be used to prepare carbohydrate-based nanoparticles (Scheller and O'Sullivan, 2011).

Guar gum: Guar gum is a water-soluble polysaccharide and is being utilized as an emulsifying and thickening agent in food products. It is a water-soluble biopolymer with good solubility both in cold and hot water; it forms a gel-like structure.

Cellulose: Cellulose is also an abundant polysaccharide similar to starch, but due to its low solubility in water, it is not widely utilized as an encapsulating agent. Some modifications are required in the chemical, physical, and biochemical characteristics of cellulose to make it a suitable encapsulating agent (Ozeki et al., 2011; Alonso et al., 2010). Furthermore, the modified form of celluloses includes cellulose esters, which are classified into enteric and nonenteric esters (Bagheri, 2014; Ozeki et al., 2011; Jin et al., 2009; Alonso et al., 2010). Enteric cellulose esters are insoluble in acidic solutions, thus they can be utilized as encapsulating agents but nonenteric cellulose esters are water insoluble and hence are not worthy of being used as an encapsulating agent.

Chitosan: Chitosan is obtained by alkaline deacetylation of chitin and is a natural biodegradable polymer exhibiting various antioxidant and antimicrobial properties (Hsieh et al., 2006; Avila-Sosa et al., 2012; Alonso et al., 2010). Chitosan is soluble at pH range of <6.5 but becomes insoluble at higher pH values; thus, it needs to modify physically and chemically to enhance its functional characteristics.

Dextran: Dextran is a linear bacterial polysaccharide consisting of glucose molecules and hydroxyl groups. Hydrophobic compounds are covalently bonded with different organic functional groups. These can be water soluble or insoluble depending upon the degree of substitution and thus the biopolymer can be utilized for the self-assembly and entrapping of bioactive ingredients with variable hydrophobicities. The work done on encapsulating food bioactive components using dextran is limited.

Alginate: Alginate is a linear hydrophilic polysaccharide biopolymer with properties to encapsulate lipase, EOs, hydrophilic food bioactives, and lipid nanoparticles (Rojas-Graü et al., 2007; Belščak-Cvitanović et al., 2011; Yeh et al., 2011). The major limitation for its use is that it gets promptly dissolved in intestinal pH or in sodium ions and has low encapsulation efficiency.

Cyclodextrins: Cyclodextrins are used to form inclusion complexes with hydrophobic food bioactives as they have central lipophilic cavity and outer hydrophilic surface. They have also been used to entrap food bioactive components, which are chemically labile, less soluble, and temperature sensitive such as fish oil, antioxidants, EOs, and antimicrobials.

Pectin: Pectin is a linear anionic polysaccharide which is suitable for acid-sensitive food bioactive components and is formed by the esterification process. The degree of esterification classifies them into two classes, i.e., low methoxyl pectin capable of forming gels in the presence of divalent calcium ions and high methoxyl pectin capable of forming gels in high sugar contents under acidic conditions. The major disadvantage of using pectinate carriers is low entrapping efficiency and quick release of entrapped bioactive components due to their porous structure (Polavarapu et al., 2011; Esfanjani et al., 2015; Fathi et al., 2014).

1.5.3 Protein-Formulated Nanoparticles

Protein-formulated nanoparticles are prepared using various techniques out of which the desolvation technique is the most commonly used. This technique works on the principle of replacement of water molecules surrounding the protein molecules using a desolvating agent, thereby leading to change in structural characteristics and precipitation of proteins in the form of nanoparticles. Examples of such nanoparticles include collagen, albumin, milk proteins, and gelatin. The major advantage of using the desolvation technique is simple procedure and applied materials used along with the ability to utilize different solvents and desolvating agents. Also, they exhibit traits like nontoxicity, biodegradability, stability, and nonantigenicity (Hilty et al., 2009).

1.6 ENCAPSULATION TECHNOLOGIES ON THE BASIS OF SPECIFIED EQUIPMENT

1. **Nanoencapsulation by electrospinning:** Electrospinning is the process of creating polymers in the range of 100 nm by the electrohydrodynamic method with the application of high voltage to spin a polymer solution (Wu and Clark, 2008). The instrument for electrospinning consists of mainly a spinneret, high voltage supply, and a target collector plate. It is a rapid method to produce ultrathin electrospuns by injecting polymer solution from the spinneret to the target collector plate. The voltage used is in the range of 1–30 kV, which results in the formation of thin fibers due to decrease in their surface area. This method is especially suitable for producing submicron and nano-sized fibers. Encapsulation efficiency and effectiveness are high and provide stable encapsulated bioactive food-grade polymers. Also, it helps in controlled release of drugs and their safe delivery.
2. **Nanoencapsulation by electrospraying:** This method utilizes the concept of atomization and is also called electrohydrodynamic atomization where the application of high voltages is responsible for causing atomization of liquids into fine droplets. The major advantages of this method include high encapsulation efficiency along with single production process. Various factors affecting the formation of nanoparticles include voltage, collector distance, volume velocity, temperature, and pressure (Hao et al., 2013; Kawakami, 2010).
3. **Nanoencapsulation by nano spray dryer:** This technique utilizes the spray drier to extend its application for the encapsulation of functional foods, cosmetics, and pharmaceuticals by converting their solutions into fine powders efficiently and rapidly (Esfanjani and Jafari, 2016). The concept was introduced in 2009, and since then, it has been making rapid progress with major benefits of production of ultrafine powders having narrow size distributions and high throughput.

1.7 ADVANCEMENTS IN NANOENCAPSULATION TECHNOLOGY

Application of the double-emulsion technique has gained a huge importance in the encapsulation of food bioactive compounds (Assadpour et al., 2016; Esfanjani et al., 2015, 2017; Mehrnia et al., 2016, 2017; Mohammadi et al., 2016a, b). The trend has also emerged toward the use of multiple-emulsion delivery systems. The oil-in-water-in-oil (O/W/O) and water-in-oil-in-water (W/O/W) are examples of novel encapsulation systems that are used to segregate the different aqueous components, which can adversely react with each other in the same aqueous phase (Garti and Benichou, 2004).

Development of nanosensors, also known as nanodevices, is another advancement in this field, which is basically used to monitor and control foodborne pathogenic bacteria and other pollutants during storage and transportation employed along with various polymers (Baeumner, 2004; Lerner et al., 2011). These nanosensors form part of a smart packaging system where they can be used to indicate the food's freshness as well as authenticity by tracing the temperature, time, and expiry date of the food product enclosed in the package. These have also been employed in packaging so as to detect the toxins and food pathogens (Li and Sheng, 2014; Yang et al., 2011). Analysis of food in terms of flavor along with drinking water analysis can be done with the use of nanosensors (Plexus Institute, 2006; Vo-Dinh et al., 2001). A low-cost nano-bioluminescent spray has been developed using this technology, which produces visible glow when in contact with bacteria (Gfeller et al., 2005). Thus, nanosensors not only help in monitoring the environmental parameters like humidity and temperature but also control the deterioration of food products by means of microbial contamination (Bouwmeester et al., 2009). Therefore, the application of nanotechnology and nanoscience in the food sector can help in the detection of various viruses, toxins, and pesticides, thereby facilitating in monitoring the food quality.

Utilization of nanocoatings to create smart surfaces is the advancement in nanotechnology where nanostructured materials (NSMs) can be mixed with various biopolymers, which enhances the characteristics of the polymer. These NSMs enhance the functional characteristics of active and intelligent packaging material and thus, the same packaging material can be used for a variety of purposes such as improved, intelligent, and active packaging (Hannon et al., 2015; Hu and Huang 2013).

Nano-sized SiO_2 is used as an anticaking agent in powdered products and to thicken the pastes for transporting the flavors and fragrances to food and nonfood products (Dekkers et al., 2011).

1.8 SAFETY ANALYSIS AND RISK MANAGEMENT IN NANOTECHNOLOGY

The application of nanoscience and nanotechnology is expanding day by day in the food sector side by side with the increasing risk of exposure of nanomaterials to humans, and studies have been conducted to examine the toxic effects of nanomaterials present in food and their interaction with the packaging material (Giubilato et al., 2020). Chewing gum contains nanoparticles of TiO_2 and can be released, absorbed, and accumulated in the human body (Avella et al., 2005; Chen et al., 2013). Nano-sized particles of SiO_2 being utilized as flavor and aroma carriers also have the issue of long-term safety in the case of nanoencapsulation and thus needs risk assessment (Borel and

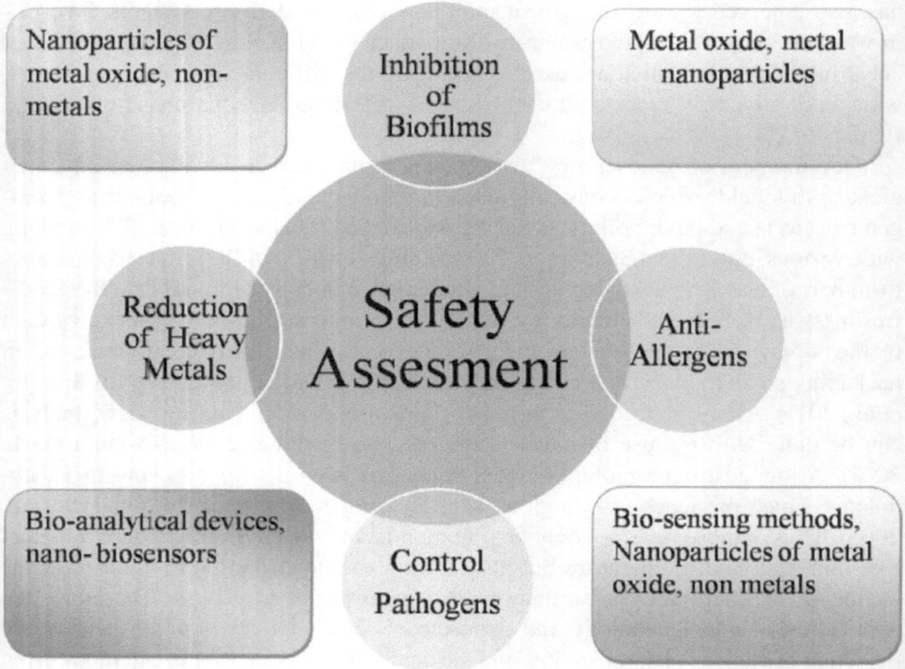

Figure 1.3 Safety assessment under risk analysis for nanotechnology in the food sector.

Sabliov, 2014; Ezhilarasi et al., 2013; Lizundia et al., 2016; McCracken et al., 2013). Thus, there is the need to have extensive characterization of nanoparticles as the toxicity and fate of nanomaterials are guided by their physicochemical properties along with dosage in food packaging (Alfadul and Elneshwy, 2010). The safety assessment of nanoparticles in the food sector can be done by the following methods as shown in Figure 1.3.

- Reduction of heavy metals like metal and nonmetal oxides.
- Safety against various pathogens.
- Safer in terms of antiallergens.
- Inhibition in the formation of biofilms.

1.9 CONCLUSIONS

The application of nanoencapsulation and nanocarriers for entrapping the bioactive ingredients is an emerging field in the food sector and industry. The practical application lies behind the fact that it helps the food technologists to enhance the functional properties of various food formulations, thereby ensuring better public health and nutrition. Also, encapsulation helps in masking flavors, aromas, and protecting the major food ingredients from unfavorable conditions and thus helps in their targeted delivery. Various nanocarrier systems have already been explored by the scientists in the field of food as well as drug application, but various engineering techniques are still under progress and need to be implemented and tested as these are consumed by humans. Important consideration should be given such that nanoencapsulation systems are safe, food grade, nontoxic, and economical to use. Thus, optimization of nanoencapsulation techniques along with various nanocarrier systems is required for the successful implementation of this technique to ensure food quality and safety.

REFERENCES

Alfadul, S. M., & Elneshwy, A. A. (2010). Use of nanotechnology in food processing, packaging and safety–review. *African Journal of Food, Agriculture, Nutrition and Development, 10*(6), 2719–2732.

Alonso, D., Gimeno, M., Sepúlveda-Sánchez, J. D., & Shirai, K. (2010). Chitosan-based microcapsules containing grapefruit seed extract grafted onto cellulose fibers by a nontoxic procedure. *Carbohydrate Research, 345*(6), 854–859.

Assadpour, E., Maghsoudlou, Y., Jafari, S. M., Ghorbani, M., & Aalami, M. (2016). Evaluation of folic acid nano-encapsulation by double emulsions. *Food and Bioprocess Technology, 9*(12), 2024–2032.

Avella, M., De Vlieger, J. J., Errico, M. E., Fischer, S., Vacca, P., & Volpe, M. G. (2005). Biodegradable starch/clay nanocomposite films for food packaging applications. *Food Chemistry, 93*(3), 467–474.

Avila-Sosa, R., Palou, E., Munguía, M. T. J., Nevárez-Moorillón, G. V., Cruz, A. R. N., & López-Malo, A. (2012). Antifungal activity by vapor contact of essential oils added to amaranth, chitosan, or starch edible films. *International Journal of Food Microbiology, 153*(1–2), 66–72.

Awad, T. S., Helgason, T., Kristbergsson, K., Decker, E. A., Weiss, J., & McClements, D. J. (2008). Effect of cooling and heating rates on polymorphic transformations and gelation of tripalmitin solid lipid nanoparticle (SLN) suspensions. *Food Biophysics, 3*(2), 155–162.

Baeumner, A. (2004). Nanosensors identify pathogens in food. *Food Technology (Chicago), 58*(8), 51–55.

Bagheri, L., (2014). What is nanotechnology? *Food Chemistry*, 17(1), 272.

Belščak-Cvitanović, A., Stojanović, R., Manojlović, V., Komes, D., Cindrić, I. J., Nedović, V., & Bugarski, B. (2011). Encapsulation of polyphenolic antioxidants from medicinal plant extracts in alginate–chitosan system enhanced with ascorbic acid by electrostatic extrusion. *Food Research International, 44*(4), 1094–1101.

Bhosale, R. R., Osmani, R. A., Ghodake, P. P., Shaikh, S. M., & Chavan, S. R. (2014). Mastitis: An intensive crisis in veterinary science. *International Journal of Pharma Research and Health Sciences, 2*(2), 96–103.

Borel, T., & Sabliov, C. M. (2014). Nanodelivery of bioactive components for food applications: Types of delivery systems, properties, and their effect on ADME profiles and toxicity of nanoparticles. *Annual Review of Food Science and Technology, 5*, 197–213.

Bouwmeester, H., Dekkers, S., Noordam, M. Y., Hagens, W. I., Bulder, A. S., De Heer, C., … & Sips, A. J. (2009). Review of health safety aspects of nanotechnologies in food production. *Regulatory Toxicology and Pharmacology, 53*(1), 52–62.

Chaudhry, Q., Scotter, M., Blackburn, J., Ross, B., Boxall, A., Castle, L., … & Watkins, R. (2008). Applications and implications of nanotechnologies for the food sector. *Food Additives and Contaminants, 25*(3), 241–258.

Chellaram, C., Murugaboopathi, G., John, A. A., Sivakumar, R., Ganesan, S., Krithika, S., & Priya, G. (2014). Significance of nanotechnology in food industry. *APCBEE Procedia, 8*, 109–113.

Chen, X. X., Cheng, B., Yang, Y. X., Cao, A., Liu, J. H., Du, L. J., … & Wang, H. (2013). Characterization and preliminary toxicity assay of nano-titanium dioxide additive in sugar-coated chewing gum. *Small, 9*(9–10), 1765–1774.

Chung, C., Sanguansri, L., & Augustin, M. A. (2008). Effects of modification of encapsulant materials on the susceptibility of fish oil microcapsules to lipolysis. *Food Biophysics, 3*(2), 140–145.

Cushen, M., Kerry, J., Morris, M., Cruz-Romero, M., & Cummins, E. (2012). Nanotechnologies in the food industry: Recent developments, risks and regulation. *Trends in Food Science & Technology, 24*(1), 30–46.

DeFelice, S. L. (1995). The nutraceutical revolution: its impact on food industry R&D. *Trends in Food Science & Technology, 6*(2), 59–61.

Dekkers, S., Krystek, P., Peters, R. J., Lankveld, D. P., Bokkers, B. G., van Hoeven-Arentzen, P. H., … & Oomen, A. G. (2011). Presence and risks of nanosilica in foodproducts. *Nanotoxicology, 5*(3), 393–405.

Esfanjani, A. F., & Jafari, S. M. (2016). Biopolymer nano-particles and natural nano-carriers for nano-encapsulation of phenolic compounds. *Colloids and Surfaces B: Biointerfaces*, *146*, 532–543.

Esfanjani, A. F., Jafari, S. M., Assadpoor, E., & Mohammadi, A. (2015). Nano-encapsulation of saffron extract through double-layered multiple emulsions of pectin and whey protein concentrate. *Journal of Food Engineering, 165*, 149–155.

Esfanjani, A. F., Assadpour, E., & Jafari, S. M. (2018). Improving the bioavailability of phenolic compounds by loading them within lipid-based nanocarriers. *Trends in Food Science & Technology*, *76*, 56–66.

Ezhilarasi, P. N., Karthik, P., Chhanwal, N., & Anandharamakrishnan, C. (2013). Nanoencapsulation techniques for food bioactive components: a review. *Food and Bioprocess Technology, 6*(3), 628–647.

Fathi, M., Mozafari, M. R., & Mohebbi, M. (2012). Nanoencapsulation of food ingredients using lipid based delivery systems. *Trends in Food Science & Technology, 23*(1), 13–27.

Fathi, M., Martín, Á., & McClements, D. J. (2014). Nanoencapsulation of food ingredients using carbohydrate based delivery systems. *Trends in Food Science & Technology*, *39*(1), 18–39.

Fuchs, S., Kutscher, M., Hertel, T., Winter, G., Pietzsch, M., & Coester, C. (2010). Transglutaminase: New insights into gelatin nanoparticle cross-linking. *Journal of Microencapsulation, 27*(8), 747–754.

Galindo-Rodriguez, S., Allemann, E., Fessi, H., & Doelker, E. (2004). Physicochemical parameters associated with nanoparticle formation in the salting-out, emulsification-diffusion, and nanoprecipitation methods. *Pharmaceutical Research, 21*(8), 1428–1439.

Garti, N., & Benichou, A. (2004). Recent developments in double emulsions for food applications. *Food Emulsions*, *35,* 353–412.

Gfeller, K. Y., Nugaeva, N., & Hegner, M. (2005). Micromechanical oscillators as rapid biosensor for the detection of active growth of Escherichia coli. *Biosensors and Bioelectronics, 21*(3), 528–533.

Giubilato, E., Cazzagon, V., Amorim, M. J., Blosi, M., Bouillard, J., Bouwmeester, H., Costa, A. L., Fadeel, B., Fernandes, T. F., Fito, C., Hauser, M., Marcomini, A., Nowack, B., Pizzol, L., Powell, L., Prina-Mello, A., Sarimveis, H., Scott-Fordsmand, J. J., Semenzin, E., Stahlmecke, B., Stone, V., Vignes, A., Wilkins, T., Zabeo, A., Tran, L., & Hristozov, D. (2020). Risk management framework for nano-biomaterials used in medical devices and advanced therapy medicinal products. *Materials, 13*(20), 4532.

Gouin, S. (2004). Microencapsulation: industrial appraisal of existing technologies and trends. *Trends in Food Science & Technology, 15*(7–8), 330–347.

Guadarrama-Lezama, A. Y., Dorantes-Alvarez, L., Jaramillo-Flores, M. E., Pérez-Alonso, C., Niranjan, K., Gutiérrez-López, G. F., & Alamilla-Beltrán, L. (2012). Preparation and characterization of non-aqueous extracts from chilli (Capsicum annuum L.) and their microencapsulates obtained by spray-drying. *Journal of Food Engineering, 112*(1–2), 29–37.

Han, H. J., Lee, J. S., Park, S. A., Ahn, J. B., & Lee, H. G. (2015). Extraction optimization and nanoencapsulation of jujube pulp and seed for enhancing antioxidant activity. *Colloids and Surfaces B: Biointerfaces, 130*, 93–100.

Hannon, J. C., Kerry, J., Cruz-Romero, M., Morris, M., & Cummins, E. (2015). Advances and challenges for the use of engineered nanoparticles in food contact materials. *Trends in Food Science & Technology, 43*(1), 43–62.

Hao, S., Wang, Y., Wang, B., Deng, J., Liu, X., & Liu, J. (2013). Rapid preparation of pH-sensitive polymeric nanoparticle with high loading capacity using electrospray for oral drug delivery. *Materials Science and Engineering: C, 33*(8), 4562–4567.

Hilty, F. M., Teleki, A., Krumeich, F., Büchel, R., Hurrell, R. F., Pratsinis, S. E., & Zimmermann, M. B. (2009). Development and optimization of iron-and zinc-containing nanostructured powders for nutritional applications. *Nanotechnology, 20*(47), 475101.

Hsieh, W. C., Chang, C. P., & Gao, Y. L. (2006). Controlled release properties of chitosan encapsulated volatile citronella oil microcapsules by thermal treatments. *Colloids and Surfaces B: Biointerfaces, 53*(2), 209–214.

Hu, B., & Huang, Q. R. (2013). Biopolymer based nano-delivery systems for enhancing bioavailability of nutraceuticals. *Chinese Journal of Polymer Science, 31*(9), 1190–1203.

Huang, Q., Yu, H., & Ru, Q. (2010). Bioavailability and delivery of nutraceuticals using nanotechnology. *Journal of Food Science, 75*(1), R50–R57.

Jafari, S. M., & McClements, D. J. (2017). Nanotechnology approaches for increasing nutrient bioavailability. *Advances in Food and Nutrition Research, 81*, 1–30.

Jafari, S. M., He, Y., & Bhandari, B. (2007). Optimization of nano-emulsions production by microfluidization. *European Food Research and Technology, 225*(5), 733–741.

Kamogawa, K., Okudaira, G., Matsumoto, M., Sakai, T., Sakai, H., & Abe, M. (2004). Preparation of oleic acid/water emulsions in surfactant-free condition by sequential processing using midsonic: Megasonic waves. *Langmuir, 20*(6), 2043–2047.

Katouzian, I., & Jafari, S. M. (2016). Nano-encapsulation as a promising approach for targeted delivery and controlled release of vitamins. *Trends in Food Science & Technology, 53*, 34–48.

Kentish, S., Wooster, T. J., Ashokkumar, M., Balachandran, S., Mawson, R., & Simons, L. (2008). The use of ultrasonics for nanoemulsion preparation. *Innovative Food Science and Emerging Technology, 9*(2), 170–175.

Kirby, C. J. (1993). Controlled delivery of functional food ingredients: Opportunities for liposomes in the food industry. *Liposome Technology, 2*, 215–232.

Lerner, M. B., Goldsmith, B. R., McMillon, R., Dailey, J., Pillai, S., Singh, S. R., & Johnson, A. T. C. (2011). A carbon nanotube immunosensor for Salmonella. *AIP Advances*, *1*(4), 042127. doi: 10.1063/1.3658573.

Li, Z., & Sheng, C. (2014). Nanosensors for food safety. *Journal of Nanoscience and Nanotechnology, 14*(1), 905–912.

Li, X., Anton, N., Arpagaus, C., Belleteix, F., & Vandamme, T. F. (2010). Nanoparticles by spray drying using innovative new technology: The Büchi Nano Spray Dryer B-90. *Journal of Controlled Release, 147*(2), 304–310.

Lizundia, E., Ruiz-Rubio, L., Vilas, J. L., & León, L. M. (2016). Poly (l-lactide)/zno nanocomposites as efficient UV-shielding coatings for packaging applications. *Journal of Applied Polymer Science, 133*(2). DOI: 10.1002/APP.42426.

McCracken, C., Zane, A., Knight, D. A., Dutta, P. K., & Waldman, W. J. (2013). Minimal intestinal epithelial cell toxicity in response to short-and long-term food-relevant inorganic nanoparticle exposure. *Chemical Research in Toxicology, 26*(10), 1514–1525.

Mehrnia, M. A., Jafari, S. M., Makhmal-Zadeh, B. S., & Maghsoudlou, Y. (2016). Crocin loaded nano-emulsions: Factors affecting emulsion properties in spontaneous emulsification. *International Journal of Biological Macromolecules, 84*, 261–267.

Mehrnia, M. A., Jafari, S. M., Makhmal-Zadeh, B. S., & Maghsoudlou, Y. (2017). Rheological and release properties of double nano-emulsions containing crocin prepared with Angum gum, Arabic gum and whey protein. *Food Hydrocolloids, 66*, 259–267.

Mohammadi, A., Hashemi, M., & Hosseini, S. M. (2015). Nanoencapsulation of Zataria multiflora essential oil preparation and characterization with enhanced antifungal activity for controlling Botrytis cinerea, the causal agent of gray mould disease. *Innovative Food Science & Emerging Technologies*, *28*, 73–80.

Mohammadi, A., Jafari, S. M., Assadpour, E., & Esfanjani, A. F. (2016a). Nano-encapsulation of olive leaf phenolic compounds through WPC–pectin complexes and evaluating their release rate. *International Journal of Biological Macromolecules, 82*, 816–822.

Mohammadi, A., Jafari, S. M., Esfanjani, A. F., & Akhavan, S. (2016b). Application of nanoencapsulated olive leaf extract in controlling the oxidative stability of soybean oil. *Food Chemistry, 190*, 513–519.

Mozafari, M. R., Flanagan, J., Matia-Merino, L., Awati, A., Omri, A., Suntres, Z. E., & Singh, H. (2008). Recent trends in the lipid-based nanoencapsulation of antioxidants and their role in foods. *Journal of the Science of Food and Agriculture, 86*(13), 2038–2045.

Nakabayashi, K., Amemiya, F., Fuchigami, T., Machida, K., Takeda, S., Tamamitsu, K., & Atobe, M. (2011). Highly clear and transparent nanoemulsion preparation under surfactant-free conditions using tandem acoustic emulsification. *Chemical Communications, 47*(20), 5765–5767.

Nazzaro, F., Orlando, P., Fratianni, F., & Coppola, R. (2012). Microencapsulation in food science and biotechnology. *Current Opinion in Biotechnology, 23*(2), 182–186.

Neethirajan, S., & Jayas, D. S. (2011). Nanotechnology for the food and bioprocessing industries. *Food and Bioprocess Technology, 4*(1), 39–47.

Ozeki, T., Takashima, Y., Nakano, T., Yuasa, H., Kataoka, M., Yamashita, S., … & Okada, H. (2011). Preparation of spray-dried microparticles using Gelucire 44/14 and porous calcium silicate or spherical microcrystalline cellulose to enhance transport of water-insoluble pranlukast hemihydrate across Caco-2 monolayers. *Advanced Powder Technology, 22*(5), 623–628.

Paredes, A. J., Asensio, C. M., Llabot, J. M., Allemandi, D. A., & Palma, S. D. (2016). Nanoencapsulation in the food industry: manufacture, applications and characterization. *Journal of Food Bioengineering and Nanoprocessing*, *1*(1), 56–79.

Pereira, M. C., Hill, L. E., Zambiazi, R. C., Mertens-Talcott, S., Talcott, S., & Gomes, C. L. (2015). Nanoencapsulation of hydrophobic phytochemicals using poly (dl-lactide-co-glycolide) (PLGA) for antioxidant and antimicrobial delivery applications: Guabiroba fruit (Campomanesia xanthocarpa O. Berg) study. *LWT-Food Science and Technology, 63*(1), 100–107.

Plexus Institute (2006). New nanotechnology food research-if it glows don't eat it. https://www.plexusinstitute.org/news-events/show_news.cfm.id1/4164.

Polavarapu, S., Oliver, C. M., Ajlouni, S., & Augustin, M. A. (2011). Physicochemical characterisation and oxidative stability of fish oil and fish oil–extra virgin olive oil microencapsulated by sugar beet pectin. *Food Chemistry, 127*(4), 1694–1705.

Ramsden, J. J. (2005). What is nanotechnology? *Nanotechnology Perceptions*, 1, 3–17.

Reis, C. P., Neufeld, R. J., Ribeiro, A. J., & Veiga, F. (2006). Nanoencapsulation I. Methods for preparation of drug-loaded polymeric nanoparticles. *Nanomedicine: Nanotechnology, Biology and Medicine, 2*(1), 8–21.

Rojas-Graü, M. A., Avena-Bustillos, R. J., Olsen, C., Friedman, M., Henika, P. R., Martín-Belloso, O., ... & McHugh, T. H. (2007). Effects of plant essential oils and oil compounds on mechanical, barrier and antimicrobial properties of alginate–apple puree edible films. *Journal of Food Engineering, 81*(3), 634–641.

Scheller, M., & O'Sullivan, D. J. (2011). Comparative analysis of an intestinal strain of Bifidobacterium longum and a strain of Bifidobacterium animalis subspecies lactis in Cheddar cheese. *Journal of Dairy Science, 94*(3), 1122–1131.

Schultz, S., Wagner, G., Urban, K., & Ulrich, J. (2004). High-pressure homogenization as a process for emulsion formation. *Chemical Engineering & Technology: Industrial Chemistry-Plant Equipment-Process Engineering-Biotechnology, 27*(4), 361–368.

Simi, C. K., & Emilia Abraham, T. (2007). Hydrophobic grafted and cross-linked starch nanoparticles for drug delivery. *Bioprocess and Biosystems Engineering, 30*(3), 173–180.

Sole, I., Maestro, A., González, C., Solans, C., & Gutiérrez, J. M. (2006). Optimization of nano-emulsion preparation by low-energy methods in an ionic surfactant system. *Langmuir, 22*(20), 8326–8332.

Souto, E. B., Anselmi, C., Centini, M., & Müller, R. H. (2005). Preparation and characterization of n-dodecyl-ferulate-loaded solid lipid nanoparticles (SLN®). *International Journal of Pharmaceutics, 295*(1–2), 261–268.

Sozer, N., & Kokini, J. L. (2009). Nanotechnology and its applications in the food sector. *Trends in Biotechnology, 27*(2), 82–89.

Tesch, S., Freudig, B., & Schubert, H. (2003). Production of emulsions in high-pressure homogenizers–part I: Disruption and stabilization of droplets. *Chemical Engineering & Technology: Industrial Chemistry-Plant Equipment-Process Engineering-Biotechnology, 26*(5), 569–573.

Tice, T. R., & Gilley, R. M. (1985). Preparation of injectable controlled-release microcapsules by a solvent-evaporation process. *Journal of Controlled Release, 2*, 343–352.

Varshosaz, J., Minayian, M., & Moazen, E. (2010). Enhancement of oral bioavailability of pentoxifylline by solid lipid nanoparticles. *Journal of Liposome Research, 20*(2), 115–123.

Vo-Dinh, T., Cullum, B. M., & Stokes, D. L. (2001). Nanosensors and biochips: Frontiers in biomolecular diagnostics. *Sensors and Actuators B: Chemical, 74*(1–3), 2–11.

Weiss, J., Takhistov, P., & McClements, D. J. (2006). Functional materials in food nanotechnology. *Journal of Food Science, 71*(9), R107–R116.

Wu, Y., & Clark, R.L. 2008. Electrohydrodynamic atomization: a versatile process for preparing materials for biomedical applications. Journal of Biomaterials Science, Polymer Edition, 19(5), 573–601.

Yang, J. Y., Li, Y., Chen, S. M., & Lin, K. C. (2011). Fabrication of a cholesterol biosensor based on cholesterol oxidase and multiwall carbon nanotubes hybrid composites. *International Journal of Electrochemical Science, 6*, 2223–2234.

Yeh, K. W., Chang, C. P., Yamamoto, T., & Dobashi, T. (2011). Release model of alginate microcapsules containing volatile tea-tree oil. *Colloids and Surfaces A: Physicochemical and Engineering Aspects*, *380*(1–3), 152–155.

Zhang, S., & Kawakami, K. (2010). One-step preparation of chitosan solid nanoparticles by electrospray deposition. *International Journal of Pharmaceutics*, *397*(1–2), 211–217.

CHAPTER 2

Technical Approaches for Encapsulation of Nutraceuticals

Mechanical, Physical, and Chemical

Vikas Bansal, Monica Premi, Mukul Kolish, Vishal Sharma, and Seema Sharma

CONTENTS

DOI: 10.1201/9781003259183-2

2.1 INTRODUCTION

Due to the health benefits of different nutraceutical/bioactive compounds, their addition in various food products is increasing day by day, and consumer awareness and research are going on as how to make this process simple, easy, and cost effective. Stability of nutraceutical/bioactive compounds during processing and storage is the biggest challenge faced by the food manufacturers (Khalid et al., 2017). Many technologies such as encapsulation have been reported as efficient and successful to retain these (heat-sensitive) nutraceutical compounds by covering them in a polymeric matrix known as wall material. The process aids in masking the undesirable color, odor, and taste apart from improving the shelf life (Ballesteros et al., 2017).

The objective of this chapter is to review the encapsulation system specifically focusing on various encapsulation techniques used for storage of nutraceutical compounds for a longer period of time and to protect them from the adverse conditions of environment, processing, and storage.

2.2 OVERVIEW OF ENCAPSULATION METHODS AND MATRICES

There are various ways to encapsulate the nutraceutical compounds (core material) into a coating material (wall material), and this aspect has been investigated and modified time to time by various researchers as per the requirement. All these methods find applications in food, pharmaceuticals, textiles, agriculture, paper, and other allied industries (Dubey et al., 2009; Mozafari et al., 2008; Okuro et al., 2013; Whelehan and Marison, 2011; Yonekura et al., 2014). Various encapsulation techniques (spray drying, spray cooling, fluidized bed coating, extrusion coating, liposomal entrapment, lyophilization, coacervation, centrifugal suspension separation, co-crystallization, and inclusion complexation) (Gibbs et al., 1999) are summarized in Table 2.1, which outlines the basic principle, nature of core material, particle size, related studies, etc.

The method/technique of encapsulation may be physical, chemical, or physicochemical, and the selection of technique is based on different properties of core and wall material and application of food ingredients (Poshadri and Kuna, 2010).

Mechanical stress, enzymatic activity, pH change, time, temperature, osmotic force, etc. are some methods that help in the release of these nutraceutical/bioactive compounds from the wall material when required.

Table 2.1 Summary of Encapsulation Techniques

Technique of Encapsulation	Mechanism/Principle	Type of Core Material	Particle Size	Advantages	Disadvantages	Related Studies
Spray drying	Spraying of dispersion of the nutraceutical compound in a chamber and evaporation of solvent after coming into contact with hot air	Solid or liquid	10–500 μm	Continuous process, low cost, flexible, better flowability and yield, good sensory characteristics	Can't be used for heat-sensitive compounds, applicable only in the case of bulk quantity of products, difficult to manage particle size, limited to suspension where wall material is soluble in water	Souza et al. (2018), Pieczykolan and Kurek (2019), and Guo et al. (2020)
Centrifugal extrusion	Core and wall materials are passed through nozzles and then split into droplets	Solid and liquid	400–2,000 μm	Low temperature and high yield	Limited to liquid, difficult to isolate in viscous medium, too large particle size	Chang et al. (2019) and Nami et al. (2020)
Fluidized bed drying	Core material is mixed with dry wall material and heated for proper solidification	Solid or liquid adsorbed on support	100–5 mm	Low cost, low porosity, high efficiency	Not suitable for heat-sensitive core material	Coronel-Aguilera and San Martín-González (2015)
Air suspension coating	Coating material is sprayed on the core and spread uniformly in the presence of air	Solid	35–5,000 μm	Variety and enhanced capacity	Agglomeration and limited to solid core only	Zaghari et al. (2020) and Subasi et al. (2021)
Pan coating	Wall material is applied on the core material tumbling in a pan	Solid or liquid adsorbed on support	≥600 μm	Suitable for large particles and controlled release of the core material	High loss and time taking	

(Continued)

Table 2.1 (*Continued*) Summary of Encapsulation Techniques

Technique of Encapsulation	Mechanism/Principle	Type of Core Material	Particle Size	Advantages	Disadvantages	Related Studies
Coacervation	Desolvation of hydrophilic polymer, separation of phases in colloidal solution, and then entrapment of the core material	Soluble in hydrophilic and hydrophobic liquids	1–500 μm	Applicable to heat-sensitive core material, small particle size, rigid shell, high payload	High cost, complex mechanism, toxicity of cross-linking material, presence of residual solvent on the surface of the product	Bosnea et al. (2017) and Mihalcea et al. (2017)
Ionic gelation	Formation of encapsulates by electrostatic interaction between two ionic species	Solid and liquid		No organic solvent and no extreme variation in temperature and pH conditions	Laboratory-level process and high porosity	Cutrim et al. (2019) and Naranjo-Durán et al. (2021)
Liposome entrapment	Vesicles entrap an aqueous phase in phospolipid membrane		Few nanometers to few micrometers	Enhances taste, can bind with mucous layer in our gut, can pack hydrophobic and hydrophilic molecules in the same pack	High cost, complexity, requirement of aqueous solution, not stable at extreme temperature, and low yield	Shukla et al. (2017), Esposto et al. (2021), and Venugopalan et al. (2021)

Source: Farheen et al. (2017) and Choudhury et al. (2021).

2.3 ENCAPSULATION TECHNIQUES

2.3.1 Spray Drying

Spray drying, the most commonly used encapsulating technique, accidentally discovered by A. Boake Roberts in 1937, helped to retain the color and flavor of tomato powder when acetone was accidentally added to the tomato puree. Thereafter, spray drying has become the most favorite commercial process for encapsulation of dry flavoring or nutraceutical ingredients. Vitamins, colorants, aroma compounds, fats and oil flavors, and oleoresins and enzymes are most commonly encapsulated by using spray drying. Carbohydrates (maltodextrin, modified starch, and gum Arabic) and proteins (whey protein concentrate and whey protein isolate) are common carrier agents used for spray drying of nutraceutical ingredients and they help in retaining the stability of compounds like anthocyanin (Jafari, 2017). Before applying the feed to the automizer, the carrier agent is homogenized with the core material usually in the ratio of 4:1.

Gum Arabic, obtained from the trunk and branches of Acacia is a traditional and the oldest carrier agent used in spray drying; however, its low production (300 g per plant per year) and high cost prevent its extensive use. Therefore, alternative carrier agents are being used for drying the flavoring. Natural starch has no emulsification properties but the esterification process helps in inculcating these properties and makes modified starches a better option than gum Arabic in terms of flavor retention during spray drying (Varavinit et al., 2001).

Hydrolyzed starch, due to low emulsification properties, used with other carrier agents like gum Arabic and proteins (whey protein concentrate and whey protein isolate), is low in viscosity at high total solid content. Kenyon and Anderson (1998) had used maltodextrin and corn syrup as carrier agents along with the incorporation of other wall materials like sodium caseinate, whey protein, lecithin, and gelatine agents for improving the emulsifying properties (Lin et al., 1995). Whey protein is used along with starch and other surface-active agents, and Kenyon and Anderson (1998) reported its effectiveness in retaining the stability of flavor.

Some important parameters such as inlet/outlet air temperature, wall-to-core ratio, and feed flow rate aid the optimization of spray drying process and to obtain the product with high quality and stability. Many researchers have used the technique of response surface methodology to optimize these parameters keeping the stability of phenolic acid and anthocyanin from sour cherry juice, betalains from beetroot, and total phenolic compound from Nigella sativa oil as the highest priority.

Various nutraceutical oils containing polyunsaturated fatty acids, have potential nutraceutical capacity, and are recommended for daily intake of up to 650 mg (Coste et al., 2007; Freemantle et al., 2006; Garg et al., 2006). A low intake in various epidemiological studies (Bauch et al., 2006; Kris-Etherton et al., 2000) drives the requirement of supplementation of these oils in the food system and therefore encapsulation by spray drying is considered a suitable technique to protect this oil from oxidation.

Spray granulation (Anwar and Kunz, 2011) also limits the inclusion of air, but high process time and thus high thermal impact limit its uses. Further processing such as coating or agglomeration also affects the efficiency of encapsulation negatively, as it requires rewetting of the surface (Domian et al., 2014; Anwar et al., 2010).

Spray chilling and spray cooling: In these processes, the core material is mixed with carrier agent and atomized by chilled air (Risch, 1995). Generally, the carrier agents used are vegetable oil for spray cooling and hydrogenated or fractionated vegetable oil for spray chilling. This process becomes important in the encapsulation of flavor and other functional ingredients, enzymes, and textural ingredients to improve the heat stability and convert the hydrophilic ingredients into free-flowing powder. In spray cooling, a strong binding of the core material to the fat matrix delays the release of the core material even if the fat matrix is melted or damaged during processing.

2.3.2 Coating

Coating is the easiest of the traditional methods of encapsulation of nutraceutical compounds. This method is applied to granulated, pelleted, and tablet material by tumbling in a pan with a solution of wall material and is used widely in food and pharmaceutical industries for coating of different functional and nutraceutical compounds.

Two main coating methods used are spray and fluidized bed coating. Fluidized bed coating, also termed as the Wurster process, was developed by D. E. Wurster in 1950. This process can be further categorized into three forms: top spray, bottom spray, and hot melt fluidized bed.

In this technique, high-velocity air is blown in a chamber with controlled conditions of temperature and humidity where the solid material is suspended in the wall material. Appreciable results have been reported with a particle size range of 50–500 μm. This technique is widely applied to hydrogenated vegetable oils, fatty acids, emulsifiers, and waxes, and solvent-based coatings such as gums, starches, etc. (Gouin, 2004).

This method does not require high temperature and hence saves time and energy. However, it is used for secondary coating of already-coated material to enhance the stability, e.g., corn starch coating, as a secondary coating, is used in the case of caseinate-encapsulated fish oil.

2.3.3 Extrusion Coating

Extrusion-based encapsulation is applied to produce highly dense encapsulated products. In this method, core material and wall material are pumped separately through a concentric orifice located on the outer circumference of the rotating cylinder. The coating material flows through the outer tube and the core material flows through the concentric orifice as the cylinder rotates. During this process, centrifugal force breaks the wall material into tiny particles, and then, it quickly moves to the surface of the core to cover it completely. Various types of flavorings, vitamins, and seasonings are encapsulated by using this technique of extrusion coating. The most widely used extrusion techniques include resonance and jet cutting technique and electrostatic droplet generation (Petrović et al., 2007; Nedovic et al., 2001).

The wall material used in this technique includes carrageenan, starch, gum acacia, fatty acids, waxes, sodium alginate, and polyethylene glycol (Schlameus, 1995).

2.3.4 Liposome Entrapment

This method was firstly used in the year 1970, but it became popular after 1990. The effectiveness of this method in drug delivery and the effectiveness of encapsulation of nutraceutical compounds has attracted researchers in recent years (Timilsena et al., 2020) and gained popularity not only in encapsulation of nutraceutical ingredients (Gulzar and Benjakul, 2020; Khanniri et al., 2016; Ajeeshkumar et al., 2021) but also in medical and cosmetic industry, where it is used for different purposes such as delivery of vaccines, enzymes, hormones, vitamins, etc. (Shade, 2016).

Liposomes are self-assembled lipids based on one or more than one membrane vesicle along with an aqueous core for protecting the lipophilic and hydrophilic molecules. The major part of liposomes consists of phospholipids. Most commonly used phospholipids include natural (egg lecithin, soy lecithin, and marine lecithin) and synthetic or semisynthetic ones such as Di-palmitoyl phosphatidyl choline. To provide structural stability, cholesterol is also used in formulation along with phospholipids (Olusanya et al., 2018; Imran et al., 2015).

This method is useful for encapsulation of both hydrophobic and hydrophilic substances due to the arrangement of the hydrophilic group toward aqueous phase and the hydrophobic groups toward the inner phospholipid membrane. This happens due to the mechanical treatment (heating and stirring) given during the preparation of liposomes (Alam et al., 2017; Alavi et al., 2017; Lila and Ishida, 2017). Lipid drying, addition of lipid to aqueous media, purification, and analysis are our important stages of liposome preparation (Ajeeshkumar et al., 2021). On the basis of size and layer, liposomes can be divided into various categories. The size varies from 0.025 to 5 μm, while unilamellar vesicle and multilamellar vesicle are the two categories on the basis of layer. On the basis of diameter, unilamellar can be further divided into small unilamellar (diameter $< 200\,\mu m$) and large unilamellar (diameter $> 200\,\mu m$) (McClements, 2014; Trucillo et al., 2018).

High encapsulation efficiency, good storage stability, and simple production methods are few characteristics of a large unilamellar vesicle due to which it is considered as the most appropriate liposome for encapsulation of nutraceutical ingredients. Phase transition temperature plays an important role in liposomes. Low-phase transition temperature and gel state of phospholipids instead of the crystalline state create problems in the formation of liposomes (Ajeeshkumar et al., 2021; Subramani and Ganapathyswamy, 2020).

This technique is used in the case of encapsulation of nutraceutical ingredients having poor water solubility and low chemical stability. Liposome entrapment method has higher efficiency and retains the bioavailability of nutraceutical ingredients.

However, this technique has some limitations such as high cost, irregular particle size, more susceptible to lipid oxidation, poor physical and chemical properties, and requirement of post-processing (Ozkan et al., 2019; Trucillo et al., 2018).

2.3.5 Coacervation

Coacervation, also termed as "phase separation," derived from the latin word "acervus," meaning "heap," is a true encapsulation technique because in this technique, the nutraceutical compound is completely surrounded by the wall material. It was the first technique used at the industrial level to produce encapsulated products. This technique consists of precipitation or separation of colloidal solution into two phases, viz., a colloid rich phase (coacervate) and a colloid poor phase (equilibrium phase) (Jyothi et al., 2010). According to phase separation methodology, the coacervation technique can be divided into two categories: simple and complex coacervation. The mechanism of both techniques remains the same except the step of the phase separation. In simple coacervation, the phase separation agent is added for the purpose of desolvation, and this is achieved by enhancing the interaction between macromolecules. This is done by controlling the conditions of temperature or pH along with the addition of solvents like ethanol and electrolytes such as sodium sulfate (Castro-Rosas et al., 2017). Pectin, milk protein, and alginates are common wall materials used to complete the simple coacervation process.

In the case of complex coacervation, ionic interaction of two oppositely charged colloidal polymers takes place. The whole process consists of the following steps:

i. Preparation of the solution of coacervate and equilibrium phase.
ii. Formation of emulsion by mixing of the lipophilic core with the above prepared solution.
iii. Manipulating the conditions of temperature and pH to form the two immiscible phases.
iv. Coating/deposition of wall material/coacervate around the core/nutraceutical ingredient.
v. Enhancing the solidification of deposition by the application of heat.

In this process of complex coacervation, the particle size ranges between 1 and 100 μm, so this method is capable of producing small-size particles. Simplicity, high loading capacity, no need of advanced or complex equipment, low agitation, and low temperature requirement are some advantages of this process (Castro-Rosas et al., 2017), while very high cost, a narrow pH range of stability, ionic strength, and use of glutaraldehyde (cross-linking agent) are the main hurdles for the commercialization of this technique (Mehran et al., 2020; Belšcak-Cvitanovic et al., 2011; Chan et al., 2010; Ozkan et al., 2019).

The coacervation technique is used as a secondary process after application of the primary process of spray drying (Timilsena et al., 2019). A combination of positively charged protein (gelatin) and negatively charged carbohydrate (gum) is used as the wall material. This is the most commonly applied combination in the production of microcapsules at the industrial level. The benefits of this system include high efficiency with controlled release of the core material.

2.3.6 Freeze Drying

Freeze drying, also termed as "Lyophilization," is a better technique for encapsulation than spray drying, specifically for heat-sensitive food products and aromatic compounds such as essential oils, proteins, and pigments. Except for long dehydration time, this technique is simple and needs no use of oxygen in the drying chamber (as in the case of spray drying) and therefore limits the browning or oxidation reactions (Agudelo et al., 2017; Park et al., 2016). Beside these, the chemical nature of the system also plays an important role in the retention of volatile compounds present in the food material (Kopelman et al., 1977). In their research, Ballesteros et al. (2017) found better retention of anthocyanin, phenolic compounds, and antioxidants in coffee powder, encapsulated in maltodextrin through freeze drying. Similarly, retention of polyphenols in bayberry was observed by Cheng et al. (2017).

The size and nature of the encapsulated particle play an important role in defining the properties such as solubility, bioavailability, and shelf life. Alternatively, these properties can be further controlled by following the techniques of nanoencapsulation in spite of microencapsulation. The size of microencapsulated particles ranges between 1 and 800 μm, while the size of nanoencapsulated particles ranges between 1 nm and 1 μm. The nanoparticle size is achieved by following either bottom-up approaches, such as coacervation, nanoprecipitation, inclusion complexation, and supercritical fluid techniques (Ezhilarasi et al., 2013), or top-down approaches, such as emulsification and emulsification-solvent evaporation (Quintanilla-Carvajal et al., 2010).

2.3.7 Inclusion Complexation

Inclusion complexation, another method of achieving the purpose of encapsulation, is a technique that takes place at the molecular level where β-cyclodextrin (derivative of starch, consisting of seven glucopyranose units, and prepared by enzymatic reaction with partially hydrolyzed starch, e.g., maltodextrin) is used as the wall material. β-cyclodextrin consists of two parts: outer part, hydrophilic in nature, and inner part, hydrophobic in nature. The nutraceutical ingredient can be inserted into the internal cavity through the hydrophobic interaction (Pagington, 1986). The formed particle size of inclusion complexes remains around 0.65 nm, which makes them fit easily into their central cavity (Dziezak, 1998). This characteristic allows the encapsulation of various aromatic compounds (e.g., essential oils) by entrapping them inside the hollow center of β-cyclodextrin. There are three processes used for the encapsulation by using β-cyclodextrin as the wall material, and all these three processes are differentiated on the basis of the quantity of water used for formation of molecular inclusion complexes.

In the first method, an aqueous solution is formed by mixing β-cyclodextrin and a desired quantity of water. After this, nutraceutical extract is obtained in crystalline form followed by heating to complete the encapsulation process. In the second method, lesser quantity of water is used than the first method and a concentrated suspension is formed in place of the aqueous solution. Thereafter, the core material is added to form the inclusion complexes in the form of crystals. Free-flowing particles are obtained after following the recovery and heat treatment steps.

In the third method, a paste is formed by using much lower quantity of water and β-cyclodextrin. The paste is mixed thoroughly to form the inclusion complex. This method is better than the above two methods, as this does not require recovery and drying steps (Pagington, 1986). This type of encapsulation used specifically for flavoring and aromatic compounds plays an important role in protecting their volatilization during various processes such as extrusion. The only limitation of this technique is the limited use of β-cyclodextrin in food application due to the regulatory ban in a number of countries (Yuliani et al., 2004).

2.3.8 Ionic Gelation

Ionic gelation or ionotropic gelation is a process of formation of complexation and cross-linking between the anionic salt and a charged polymer. The product formed after this mixing is termed as gelisphere. These gelisphere particles have the capacity of swelling at an extensive rate in biological fluids and are spherical hydrophilic in nature. This technique of encapsulation is widely applicable in nutraceutical and pharmaceutical industries. The gelation process can be divided into two processes: external and internal (Petraityte and Šipailiene, 2019). Both processes are considered good, as they do not have any major impact on encapsulation efficiency.

In external gelation, core material and charged biopolymer particles are added into anionic salt solution uniformly by using different methods such as electrospinning, extrusion, or a syringe needle (Petraityte and Šipailiene, 2019; Pitigraisorn et al., 2017; Chew and Nyam, 2016). In this process, the outer layer of the encapsulated product is solidified with the help of an anionic agent, which acts as a protection medium for the inner core material due to the formation of cross links. In the case of internal gelation, biopolymer, anionic salt, and core material are mixed uniformly with the emulsion to get the desired size of the product (Petraityte and Šipailiene, 2019).

Similarly, some of the micronutrients (ascorbic acid, β carotene, and ferrous fumarate) were encapsulated by using alginate and examined for their stability at different levels of temperature and humidity for 6 months (Han et al., 2008). The resistance toward the enzymatic and acid attacks was reported in a gastrointestinal study. According to the authors, this technique is capable of producing a high yield and a stable product with no or little impact of heat, acid, and moisture.

The external gelation technique was successfully applied in the encapsulation of betalain by using sodium alginate and bovine serum albumin as wall materials and reported the retention of the antioxidant effect for more than 25 days (Otálora et al., 2015).

2.3.9 Supercritical Fluid–Based Technique

It is an energy-efficient encapsulation technology, and the most widely used supercritical fluid used in this technique is carbon dioxide due to its various characteristics, which make its use possible in pharmaceutical and food industries for encapsulation of various nutraceutical compounds (Cittadini et al., 2022). Although the requirement of high pressure and complex equipment is considered as its limitation and is the reason for less popularization of this technique, still the easy utilization of the supercritical fluid, with characteristics of both liquids (high density and enhanced solvent characteristics) and gases (high mass transfer and diffusion rate and low viscosity), makes this technology a favorite over the traditional technique, specifically in the case of heat-sensitive nutraceutical compounds (Al Khawli et al., 2019). The critical temperature and pressure above which carbon dioxide is used is 31.1°C and 7.38 MPa, respectively. According to compatibility and other properties, some other supercritical fluids such as nitrogen, propane, and water are also used. All of these fluids possess different critical temperature and pressure above which they exhibit the supercritical state.

Carbon dioxide is the most preferred supercritical fluid for the purpose of encapsulation of nutraceutical compounds due to its low cost, easy separation after depressurization, nontoxic nature, and low critical temperature. Coprecipitation, entrapment of compounds, chemical binding, and finally encapsulation are the steps followed to complete the process of this technique (Ranjit and Baquee, 2013). Like other techniques, various types of wall materials are used to complete the encapsulation process. Some of the commonly used wall materials include dextran, cyclodextrin, inulin, aliphatic polyesters, polylactide aliphatic copolymers, polycaprolactones, and starch polymer blends.

Starch as a wall material was successfully used by Almeida et al. (2013) in the encapsulation of oregano essential oil and reported a higher antioxidative activity of oregano essential oil due to higher diffusivity of the supercritical fluid carbon dioxide, which enhances the penetration of essential oil. Similarly, soybean lecithin and pluronic L64 were used for the encapsulation of quercetin (Jhan et al., 2020) and α-tocopherol and lecithin were used for the encapsulation of lycopene (Cheng et al., 2017); enhanced antioxidative and storage stability effects were observed in both cases.

2.3.10 Centrifugal Suspension Separation

Centrifugal suspension separation, also termed as rotational suspension separation, is another encapsulation technique which is based on the rotation of a disc. This technique was first used by Sparks in 1987. In this technique, a suspension of core material and coating material is prepared and the prepared suspension is passed through a rotating disc automizer. Now, due to the effect of centrifugal force, the suspension is divided into two different types based on the size of particles:

i. Larger core particles with a layer of wall material.
ii. Smaller particles of pure wall material.

Both of these particles formed at the edge of the rotating disc automizer, situated at the top of the drying/cooling tower. These particles can be separated by the rotational movement of the disc and they get solidified after falling from the tower. Smaller size particles get separated by using sieve or cyclone for recycling purposes (Sparks et al., 1995). In comparison with the air suspension coating method, this method has the advantage of high production rate in less period of time and less coating material usage, as the remaining material is transferred to the recycling process.

Beside this, a range of particles can be subjected to coating treatment. The particle size and thickness can vary between 30 μm to several millimeters and 30 μm to 2 mm, respectively (Sparks et al., 1995). Centrifugal suspension separation can be used for encapsulation of spray-dried aromatic or flavoring compounds. The coating process gives good results specifically in the case of large-size particles (up to 100 μm) because small-size particles may include pure coating material.

2.4 CONCLUDING REMARKS

The range of applications of the above-discussed techniques is very much wide and may overlap. For example, one may follow a certain technique and exclude others based on certain parameters like characteristics of the core material, health and safety issues, release methods, and economic concerns. Among the aforesaid concerns, health and safety issue is very much important, so US Food and Drug Administration (FDA)-approved material should be used and the "Generally Recognized As Safe" (GRAS) list should be followed.

As research is ongoing on new wall material and its process of application along with innovation in the application of this technique, it is expected that newer encapsulation techniques will be developed in few years, which will be more economic, easy to apply, and ultimately enhance the growth of encapsulation market at the global level.

REFERENCES

Agudelo, C., Barros, L., Santos-Buelga, C., Martínez-Navarrete, N., Ferreira, I.C.F.R. (2017). Phytochemical content and antioxidant activity of grapefruit (Star Ruby): A comparison between fresh freeze-dried fruits and different powder formulations. *LWT - Food Science & Technology*, 80, 106–112.

Ajeeshkumar, K.K., Aneesh, P.A., Raju, N., Suseela, M., Ravishankar, C.N., Benjakul, S. (2021). Advancements in liposome technology: Preparation techniques and applications in food, functional foods, and bioactive delivery: A review. *Comprehensive Reviews in Food Science and Food Safety*, 20, 1280–1306.

Al Khawli, F., Pateiro, M., Domínguez, R., Lorenzo, J.M., Gullón, P., Kousoulaki, K., Ferrer, E., Berrada, H., Barba, F.J. (2019). Innovative green technologies of intensification for valorisation of seafood and their by-products. *Marine Drugs*, 17, 689–709.

Alam, S., Mattern-Schain, S.I., Best, M.D. (2017). Targeting and triggered release using lipid-based supramolecular assemblies as medicinal nanocarriers. In: Atwood, J. (Ed.), *Comprehensive Supramolecular Chemistry II*. Elsevier Inc., Amsterdam, The Netherlands, 5, 329–364, ISBN 9780128031988.

Alavi, M., Karimi, N., Safaei, M. (2017). Advanced pharmaceutical bulletin application of various types of liposomes in drug delivery systems. *Advanced Pharmaceutical Bulletin*, 7, 3–9.

Almeida, A.P., Rodríguez-Rojo, S., Serra, A.T., Vila-Real, H., Simplicio, A.L., Delgadilho, I., Beirão Da Costa, S., Beirão Da Costa, L., Nogueira, I.D., Duarte, C.M.M. (2013). Microencapsulation of oregano essential oil in starch-based materials using supercritical fluid technology. *Innovative Food Science & Emerging Technologies*, 20, 140–145.

Anwar, S.H., Kunz, B. (2011). The influence of drying methods on the stabilization of fish oil microcapsules: Comparison of spray granulation, spray drying, and freeze drying. *Journal of Food Engineering,* 105, 367–378.

Anwar, S.H., Weissbrodt, J., Kunz, B. (2010). Microencapsulation of fish oil by spray-granulation and fluid bed film coating. *Journal of Food Science,* 75, E359–E371.

Ballesteros, L.F., Ramirez, M.J., Orrego, C.E., Teixeira· J.A., Mussatto, S.I. (2017). Encapsulation of antioxidant phenolic compounds extracted from spent coffee grounds by freeze-drying and spray-drying using different coating materials. *Food Chemistry,* 237, 623–631.

Bauch, A., Lindtner, O., Mensink, G.B.M., Niemann, B. (2006). Dietary intake and sources of long-chain n-3 PUFAs in German adults. *European Journal of Nutrition,* 60, 810–812.

Belšcak-Cvitanovic, A., Stojanovi´c, R., Manojlovi´c, V., Komes, D., Cindri´c, I.J., Nedovi´c, V., Bugarski, B. (2011). Encapsulation of polyphenolic antioxidants from medicinal plant extracts in alginate-chitosan system enhanced with ascorbic acid by electrostatic extrusion. *Food Research International*, 44, 1094–1101.

Bosnea, L.A., Moschakis, T., Nigam, P.S., and Biliaderis, C.G. (2017). Growth adaptation of probiotics in biopolymer basedo; coacervate structures to enhance cell viability. *LWT - Food Science & Technology*, 77, 282–289.

Castro-Rosas, J., Ferreira-Grosso, C.R., Gómez-Aldapa, C.A., Rangel-Vargas, E., Rodríguez-Marín, M.L., Guzmán-Ortiz, F.A., Falfan-Cortes, R.N. (2017). Recent advances in microencapsulation of natural sources of antimicrobial compounds used in food: A review. *Food Research International*, 102, 575–587.

Chan, E.S., Yim, Z.H., Phan, S.H., Mansa, R.F., Ravindra, P. (2010). Encapsulation of herbal aqueous extract through absorption with Ca-alginate hydrogel beads. *Food and Bioproducts Processing*, 88, 195–201.

Chang, D., Hayat, K., Abbas, S., Zhang, X. (2019). Ascorbic acid encapsulation in a glassy carbohydrate matrix via hot melt extrusion: Preparation and characterization. *Journal of Food Science and Technology*, 39, 660–666.

Cheng, A.-W., Xie, H.-X., Qi, Y., Liu, C., Guo, X., Sun, J., Liu, L.-N. (2017). Effects of storage time and temperature on polyphenolic content and qualitative characteristics of freeze-dried and spray-dried bayberry powder. *LWT - Food Science and Technology*, 78, 235–240.

Chew, S.C., Nyam, K.L. (2016). Microencapsulation of Kenaf seed oil by co-extrusion technology. *Journal of Food Engineering*, 175, 43–50.

Choudhury, N., Meghwal, M., Das, K. (2021). Microencapsulation: An overview on concepts, methods, properties and applications in foods. *Food Frontiers,* 2, 1–17.

Cittadini, A., Munekata, P.E.S., Pateiro, M., Sarriés, M.V., Domínguez, R., Lorenzo, J.M. (2022). Encapsulation techniques to increase lipid stability. In: Lorenzo, J.M. (Ed.), *Food Lipids Sources, Health Implications, and Future Trends*. Academic Press, Oxford, UK, pp. 1–46. ISBN 9780128233719.

Coronel-Aguilera, C.P., San Martín-González, M.F. (2015). Encapsulation of spray dried b-carotene emulsion by fluidized bed coating technology. *LWT—Food Science & Technology*, 62, 187–193.

Coste, T.C., Armand, M., Lebacq, J. (2007). An overview of monitoring and supplementation of omega 3 fatty acids in cystic fibrosis. *Clinical Biochemistry,* 40, 511–520.

Cutrim, C.S., Alvim, I.D., Cortez, M. (2019). Microencapsulation of green tea polyphenols by ionic gelation and spray chilling methods. *Journal of Food Science & Technology*, 56, 3561–3570.

Domian, E., Sulek, A., Cenkier, J., Kerschke, A. (2014). Influence of agglomeration on physical characteristics and oxidative stability of spray-dried oil powder with milk protein and trehalose wall material. *Journal of Food Engineering,* 125, 34–43.

Dubey, R., Shami, T.C., Bhasker Rao, K.U. (2009). Microencapsulation technology and applications. *Defence Science Journal*, 59, 82–95.

Dziezak, J.D. (1998). Microencapsulation and encapsulated food ingredients. *Food Technology*, 42, 136–151.

Esposto, B.S., Juregi, P., Tapia-Blacido, D.R., Martelli-Tosi, M. (2021). Liposomes vs. chitosomes: Encapsulating food bioactive. *Trends in Food Science & Technology*, 108, 40–48.

Ezhilarasi, P.N., Karthik, P., Chhanwal, N. (2013). Nanoencapsulation techniques for food bioactive components: A review. *Food Bioprocess Technology,* 6, 628–647.

Farheen, T., Shaikh, A., Shahi, S. (2017). A review on a process: Microencapsulation. *International Journal of Pharma Research and Health Sciences*, 5, 1823–1830.

Freemantle, E., Vandal, M., Tremblay-Mercier, J. (2006). Omega-3 fatty acids, energy substrates, and brain function during aging. *Prostaglandins, Leukotrienes and Essential Fatty Acids,* 75, 213–220.

Garg, M.L., Wood, L.G., Singh, H., Moughan, P.J. (2006). Means of delivering recommended levels of long chain n-3 polyunsaturated fatty acids in human diets. *Journal of Food Science,* 71, R66–R71.

Gibbs, B.F., Kermasha, S., Alli, I., Mulligan, C.N. (1999). Encapsulation in the food industry: A review. *International Journal of Food Sciences and Nutrition*, 50, 213–224.

Gouin, S. (2004). Microencapsulation: Industrial appraisal of existing technologies and trends. *Trends in Food Science Technology*, 15, 330–347.

Gulzar, S., Benjakul, S. (2020). Characteristics and storage stability of nanoliposomes loaded with shrimp oil as affected by ultrasonication and micro fluidization. *Food Chemistry*, 310, 125916.

Guo, J., Li, P., Kong, L., Xu, B. (2020). Microencapsulation of curcumin by spray drying and freeze drying. *LWT-Food Science & Technology*, 132, 109892.

Han, J., Guenier, A.-S., Salmieri, S., Lacroix, M. (2008). Alginate and chitosan functionalization for micronutrient encapsulation. *Journal of Agricultural and Food Chemistry*, 56, 2528–2535.

Imran, M., Revol-Junelles, A.M., Paris, C., Guedon, E., Linder, M., Desobry, S. (2015). Liposomal nano delivery systems using soy and marine lecithin to encapsulate food bio preservative Nisin. *LWT-Food Science & Technology*, 62, 341–349.

Jafari, S.M., Ghalegi, G.M., Dehnad, D. (2017). Influence of spray drying on water solubility index, apparent density, and anthocyanin content of pomegranate juice powder. *Powder Technology*, 311, 59–65.

Jhan, S., Pethe, A. M. (2020). Double-loaded liposomes encapsulating lycopene β-cyclodextrin complexes: Preparation, optimization, and evaluation. *Journal of Liposome Research*, *30*(1), 80–92.

Jyothi, N.V.N., Prasanna, P.M., Sakarkar, S.N., Prabha, K.S., Ramaiah, P.S., Srawan, G.Y. (2010). Microencapsulation techniques, factors influencing encapsulation efficiency. *Journal of Microencapsulation*, 27, 187–197.

Kenyon, M.M., Anderson, R.J. (1998). Maltodextrin and low dextrose equivalence corn syrup solids: Production and technology for the flavour industry. In: *Flavour Encapsulation. ACS Symposium Series 370 American Chemical Society*, Washington, DC, 7–11.

Khalid, N., Shu, G., Kobayashi, I., Nakajima, M., Barrow, C.J. (2017). Formulation and characterization of monodisperse O/W emulsions encapsulating astaxanthin extracts using microchannel emulsification: insights of formulation and stability evaluation. *Colloids and Surfaces B: Bio interfaces,* 157, 355–365.

Khanniri, E., Bagheripoor-Fallah, N., Sohrabvandi, S., Mortazavian, A.M., Khosravi-Darani, K., Mohammad, R. (2016). Application of liposomes in some dairy products. *Critical Reviews in Food Science and Nutrition*, 56, 484–493.

Kopelman, I.J., Meydav, S., Wilmersdorf, P. (1977). Storage studies of freeze-dried lemon crystals. *Journal of Food Technology*, 12, 65–69.

Kris-Etherton, P.M., Taylor, D.S., Yu-Poth, S. (2000). Polyunsaturated fatty acids in the food chain in the United States. *American Journal of Clinical Nutrition,* 71, 179S–188S.

Lila, A.S.A., Ishida, T. (2017). Liposomal delivery systems: design optimization and current applications. *Biological and Pharmaceutical Bulletin*, 40, 1–10.

Lin, C.W., Ho, P.S., Chao, T.C. (1995). Microencapsulation of squid oil with hydrophilic macromolecules for oxidative and thermal stabilization. *Food Science*, 22, 141–148.

McClements, D.J. (2014). *Nanoparticle- and Microparticle-Based Delivery Systems: Encapsulation, Protection and Release of Active Compounds*. CRC Press: Boca Raton, FL. ISBN 9781482233162.

Mehran, M., Masoum, S., Memarzadeh, M. (2020). Improvement of thermal stability and antioxidant activity of anthocyanins of echium amoenum petal using maltodextrin/modified starch combination as wall material. *International Journal of Biological Macromolecules*, 148, 768–776.

Mihalcea, L., Turturic, A.M., Ghinea, I.O., Barbu, V., Ionita, E., Cotârlet, M. (2017). Encapsulation of carotenoids from sea buckthorn extracted by CO2 supercritical fluids method within whey proteins isolates matrices. *Innovative Food Science and Emerging Technologies*, 42, 120–129.

Mozafari, M.R., Khosravi-Darani, K., Borazan, G.G., Cui, J., Pardakhty, A., Yurdugul, S. (2008). Encapsulation of food ingredients using nanoliposome technology. *International Journal of Food Properties,* 11, 833–844.

Nami, Y., Lornezhad, G., Kiani, A., Abdullah, N., Haghshenas, B. (2020). Alginate-persian gum-prebiotics microencapsulation impacts on the survival rate of Lactococcus Lactis ABRIINW-N19 in orange juice. *LWT-Food Science & Technology*, 124, 109190.

Naranjo-Durán, A.M., Quintero-Quiroz, J., Rojas-Camargo, J., Ciro-Gomez, G.L. (2021). Modified-release of encapsulated bioactive compounds from annatto seeds produced by optimized ionic gelation techniques. *Scientific Reports,* 11, 13–17.

Nedovic, V.A., Obradovic, B., Leskosek-Cukalovic, I., Vunjak-Novakovic, G. (2001). Immobilized yeast bioreactor systems for brewing-recent achievements. In: Hofman, M. and Thonart, P. (Eds.), *Engineering and Manufacturing for Biotechnology*. Springer, Dordrecht, 277–292.

Okuro, P.K., de Matos Junior, F.E., Favaro-Trindade, C.S. (2013). Technological challenges for spray chilling encapsulation of functional food ingredients. *Food Technology and Biotechnology*, 51, 171–182.

Olusanya, T.O., Haj Ahmad, R.R., Ibegbu, D.M., Smith, J.R., Elkordy, A.A. (2018). Liposomal drug delivery systems and anticancer drugs. *Molecules*, 23, 907.

Otálora, M.C., Carriazo, J.G., Iturriaga, L., Nazareno, M.A., Osorio, C. (2015). Microencapsulation of betalains obtained from cactus fruit (opuntia ficus-indica) by spray drying using cactus cladode mucilage and maltodextrin as encapsulating agents. *Food Chemistry*, 187, 174–181.

Ozkan, G., Franco, P., de Marco, I., Xiao, J., Capanoglu, E. (2019). A review of microencapsulation methods for food antioxidants: principles, advantages, drawbacks and applications. *Food Chemistry*, 272, 494–506.

Pagington, J.S. (1986). β-cyclodextrin and its uses in the flavour industry. In: Birch, G.G. and Lindley, M.G. (Eds.), *Developments in Food Flavours*. Elsevier Applied Science, London.

Park, H.-J., Lee, Y., Eun, J.-B. (2016). Physicochemical characteristics of Kimchi powder manufactured by hot air drying and freeze drying. *Biocatalysts and Agricultural Biotechnology,* 5, 193–198.

Petraityte, S., Šipailiene, A. (2019). Enhancing encapsulation efficiency of alginate capsules containing lactic acid bacteria by using different divalent cross-linkers sources. *LWT - Food Science and Technology*, 110, 307–315.

Petrović, T., Nedović, V., Dimitrijević-Branković, S., Bugarski, B., Lacroix, C. (2007). Protection of probiotic microorganisms by microencapsulation. *Chemical Industry and Chemical Engineering,* 13, 169–174.

Pieczykolan, E., Kurek, M.A. (2019). Use of guar gum, gum arabic, pectin, beta-glucan and inulin for microencapsulation of anthocyanins from chokeberry. *International Journal of Biological Macromolecules*, 129, 665–671.

Pitigraisorn, P., Srichaisupakit, K., Wongpadungkiat, N., Wongsasulak, S. (2017). Encapsulation of lactobacillus acidophilus in moist-heat-resistant multi layered microcapsules. *Journal of Food Engineering*, 192, 11–18.

Poshadri, A., Kuna, A. (2010). Microencapsulation technology: A review. *Journal of Research ANGRAU*, 38, 86–102.

Quintanilla-Carvajal, M.X., Camacho-Díaz, B., Meraz-Torres, L., Chanona-Pérez, J., Alamilla-Beltrán, L., Jiménez-Aparicio, A., Gutiérrez-López, G. (2010). Nanoencapsulation: A new trend in food engineering processing. *Food Engineering Reviews,* 2, 39–50.

Ranjit, K., Baquee, A.A. (2013). Nanoparticle: An overview of preparation, characterization and application. *International Research Journal of Pharmacy,* 4, 47–57.

Risch, S.J. (1995). Encapsulation: Overview of uses and techniques. In: Risch, S.J. and Reineccius, G.A. (Eds.), *Encapsulation and Controlled Release of Food Ingredient*, ACS Symposium Series 590. American Chemical Society, Washington, DC, 1–7.

Schlameus, W. (1995). Centrifugal extrusion encapsulation. In: Risch, S.J. and Reineccius, G.A. (Eds.), *Encapsulation and Controlled Release of Food Ingredients*. ACS Symposium Series, vol. 590. ACS Publications, Washington, DC, 96–103.

Shade, C.W. (2016). Liposomes as advanced delivery systems for nutraceuticals. *Integrative Medicine: A Clinician's Journal*, 15, 33–36.

Shukla, S., Haldorai, Y., Hwang, S.K., Bajpai, V.K., Huh, Y.S., Han, Y. (2017). Current demands for food-approved liposome nanoparticles in food and safety sector. *Frontiers in Microbiology*, 17, 2398.

Souza, A.L., Hidalgo-Chávez, D.W., Pontes, S.M., Gomes, F.S., Cabral, L.M., Tonon, R.V. (2018). Microencapsulation by spray drying of a lycopene-rich tomato concentrate: Characterization and stability. *LWT - Food Science & Technology*, 91, 286–292.

Sparks, R.E., Jacobs, I.C., Mason, N.S. (1995). Centrifugal suspension-separation for coating food ingredients. In: Risch, S.J. and Reineccius, G.A. (Eds.), *Encapsulation and Controlled Release of Food Ingredients*. American Chemical Society, Washington, DC, Chapter 8, 87–95.

Subasi, B.G., Vahapoglu, B., Capanoglu, E. (2021). *Microencapsulation Methods for Food Antioxidants in Plant Antioxidants and Health*. Springer Nature, Switzerland, 1–37.

Subramani, T., Ganapathyswamy, H. (2020). An overview of liposomal nano-encapsulation techniques and its applications in food and nutraceutical. *Journal of Food Science & Technology*, 57, 3545–3555.

Timilsena, Y.P., Akanbi, T.O., Khalid, N., Adhikari, B., Barrow, C.J. (2019). Complex coacervation: Principles, mechanisms and applications in microencapsulation. *International Journal of Biological Macromolecules*, 121, 1276–1286.

Timilsena, Y.K., Haque, M.A., Adhikari, B. (2020). Encapsulation in the food industry: A brief historical overview to recent developments. *Food and Nutritional Sciences*, 11, 481–508.

Trucillo, P., Campardelli, R., Reverchon, E. (2018). A versatile supercritical assisted process for the one-shot production of liposomes. *Journal of Supercritical Fluids*, 146, 136–143.

Varavinit, S., Chaokasem, N., Shobsngob, S. (2001). Studies of flavour encapsulation by agents produced from modified sago and tapioca starches. *Starch/Starke*, 53, 281–287.

Venugopalan, V.K., Gopakumar, L.R., Kumaran, A.K., Chatterjee, N.S., Soman, V., Peeralil, S. (2021). Encapsulation and protection of Omega-3-rich fish oils using food-grade delivery systems. *Foods,* 10, 1566.

Whelehan, M., Marison, I.W. (2011). Microencapsulation using vibrating technology. *Journal of Microencapsulation*, 28, 669–688.

Yonekura, L., Sun, H., Soukoulis, C., Fisk, I. (2014). Microencapsulation of Lactobacillus acidophilus NCIMB 701748 in matrices containing soluble fibre by spray drying: Technological characterization, storage stability and survival after in vitro digestion. *Journal of Functional Foods*, 6, 205–214.

Yuliani, S., Bhandari, B., Rutgers, R., D'Arcy, B. (2004). Application of microencapsulated flavour to extrusion product. *Food Review International*, 20, 163–185.

Zaghari, L., Basiri, A., & Rahimi, S. (2020). Preparation and characterization of double-coated probiotic bacteria via a fluid-bed process: a case study on Lactobacillus reuteri. *International Journal of Food Engineering*, *16*(9).

CHAPTER 3

Characterization of Manifold Novel Polymers Used in Encapsulation

Nargis Yousf, Mudasir Bashir Mir, Furheen Amin, and Reshu Rajput

CONTENTS

DOI: 10.1201/9781003259183-3

3.1 INTRODUCTION

Biopolymers are ideal for microencapsulation of foods as they are edible, biodegradable, and inexpensive. Carbohydrates, proteins, and other polysaccharides act as a viable medium for the transport and protection of bioactive ingredients. Each biopolymer possesses both advantages and disadvantages. Among carbohydrates (monosaccharides, disaccharides, and polysaccharides), the polysaccharides are most commonly used for encapsulation. They are low-cost biopolymers (Mohan et al., 2015) that actively entrap bioactive components due to their massive molecular structure (Fatha et al., 2014). They have receptive groups that allow polysaccharide interaction with bioactive compounds. These reactive groups also improve functional properties of biopolymers including self-assertion structure for the encapsulate (Palao-Suay et al., 2016). These functional groups also possess some disadvantages like encapsulating some of the bioactive components and creating toxic products after reacting with them. Polysaccharides are versatile carbohydrates due to their chemical structure, reactive groups, molecular weight, and presence of hydrophilic and hydrophobic groups (Santiago and Castro, 2016). Other carbohydrates like gums and pectin are heterogeneous polysaccharides.

Polysaccharides are dietary fibers highly resistant to enzymes in the gastrointestinal tract but suitable for microbial fermentation in the colon. For encapsulation, plant-derived biopolymers, such as cellulosic material, alginic acid, carbohydrate, jelly, and guar gum, and compounds that are derived from animals, such as chitosan and dextran, are used. Food proteins are efficiently used for microencapsulation matrices. They exert a protective effect on ingredients that are bioactive. They are fully loaded with nutrients (Mehyar et al., 2014). Some proteins act as vehicles for storage, transport, and delivery of essential nutrients and other molecules throughout the organism (Considine et al., 2009). They also bind with hydrophilic and hydrophobic small molecules (Livney, 2010). They also possess functional properties like self-assembling, good gelation, and surface activity (Karaca et al., 2001). Generally, proteins that are extracted from animals like dairy product proteins, egg proteins, and gelatin are used for encapsulation (Tavares et al., 2014). The major benefits of dairy product proteins over plant-based proteins are high solubility, low molecular weight, and greater flexibility; however, plant proteins are less expensive and have more consumer appeal (Tavares et al., 2014). Dairy proteins are vehicles for encapsulation (Delcour and Hoseney, 2010). The most relevant alternatives to plant proteins are soy proteins followed by proteins from pulses (peas, lentil, or chick pea) and cereal grains (zein and barley; Nishinari et al., 2014).

3.2 PLANT- AND ANIMAL-BASED MATERIAL USED FOR ENCAPSULATION

3.2.1 Plant-Based Materials

3.2.1.1 Soyabean Proteins

Since soy proteins have positive effects on human health, are nutritionally valuable, and have outstanding functional qualities, they are crucial dietary proteins used in human food. They are made up of albumins and globulins, which make up between 50% and 90% of the total protein in seeds. According to their sedimentation value, they are classified into four different fractions: 2S, 7S, 11S, and 15S. Albumins are found to be present in the 2S fraction, while the 7S, 11S, and 15S fractions correspond to globulins. A crucial component of soy protein, soy protein isolate (SPI), contains 85%–90% protein on a dry basis (Tang, 2017; Nesterenko et al., 2010). Soy proteins possess good water solubility index, excellent fat and water absorption capacities, and other functional properties like foaming capacity, gel formation, emulsifying capacity, and film formation. When combined with proteins or polysaccharides, they are utilized as wall materials for bioactive substances (Delcour and Hoseney, 2010).

3.2.1.2 Cereal Proteins

Cereal grains are the main source of three major biomolecules that are divided into three different areas as starch, protein, and nonstarch polysaccharides, and they are grown worldwide (Gil, 2012). Cereal microtechnology and nanotechnology focus on the formation of extremely practical microstructures and nanostructures that develop from biopolymers of cereal. Proteins of cereal (such as protein from wheat and protein from barley) are essential sources of protein for diet. Cereal proteins have great functional qualities, including affordability, accessibility, and adaptability, which are essential for microencapsulation and nanoencapsulation technologies. They are divided into four groups based on how easily they dissolve:

a. Albumins; water-soluble proteins.
b. Globulins; proteins that can dissolve in the weak salt solution.
c. Prolamins; proteins that can dissolve in aqueous alcohol.
d. Glutelin proteins; proteins that can be dissolved in weak acids or alkalis (Joye, 2019; Shafaei et al., 2017).

3.2.1.3 Maize Protein

Zein protein is regarded as storage prolamin protein and constitutes about 35%–60% of total protein. Pure zein protein is used in different industrial applications as it is clear, tasteless, odorless, and edible in nature (Lorenzo et al., 2018). The main amino acids present in zein are leucine, glutamic acid, alanine, and proline, which are connected with bonds formed by disulfide. It has four primary protein fractions, namely, α, β, γ, and δ, which together make up around 80% of the total protein. δ fraction is the smallest fraction. Commercial alpha-zein (α-zein) protein is offered for sale. Due to the inclusion of nonpolar amino acids, it is insoluble in water but is (50%–95%) soluble in aqueous alcohol. Due to the absence of the crucial amino acids, such as tryptophan and lysine, it has poor solubility in water; hence, zein protein has been employed as a nanocarrier to encapsulate core components to improve their dispersion in water (Uthayakumaran and Wrigley, 2017; Flambeau et al., 2017). As these methods do not require post-production hardening, which helps to stabilize the integrity of the prepared material, and also because the systems are built, protein does not necessitate dehydration; prolamin protein is an useful substance, which is used to prepare microscale/nanoscale structures that are soluble in nature.

3.2.1.4 Wheat Proteins

About 80% of the total wheat seed protein comes from wheat, which is produced when starch polysaccharide is separated from wheat flour as a by-product. Both humans and animals eat it as a source of nourishment. Gliadin and glutenin are the two wheat proteins. The gliadin is a single-chain polypeptide (the average molecular weight ranges from 25 to 100 KDa) linked by intramolecular disulfide bonds and is soluble in aqueous alcohol (70% ethanol), while glutenin is a soluble fraction and is similar to gliadin composition, but they are linked via intermolecular disulfide bonds with molecular weight higher than 105 KDa (Sun et al., 2009). Wheat proteins have poor water solubility and viscoelasticity, which provide them with physical characteristics including gel- and film-forming abilities.

3.2.1.5 Barley Protein

Barley crops and their by-products are abundant sources of protein, making up 8%–13% and 20%–30% of total protein, respectively (Finnie and Svensson, 2009). Hordein (35%–55%) and glutelin (33%–40%) are the two protein fractions that make up barley proteins. Based on their amino acid compositions and electrophoretic mobility, the hordein fraction (alcohol removed) is split into five groups: A, B, C, D, and hordein. However, as glutelin is an alkali-soluble protein, it cannot be separated into a pure glutelin fraction that is uncontaminated by the hordein fraction. Barley proteins can be made into microcapsules and nanocapsules using the emulsifying-stabilization technique because they are hydrophobic and have excellent foaming, emulsifying, and film-forming capabilities (Zhao et al., 2015; Roy et al., 2010).

3.2.1.6 Pulse Proteins

Every year, during the winter season, pulse crops are grown as agricultural products. They are raised in the Middle East, Asia, and North America, particularly Canada (particularly India and Egypt). They are a good source of fiber, vital vitamins, minerals, and carbs as well as nutritional protein. They therefore provide health advantages such as preventing heart disease, type 2 diabetes, and lowering High density lipoprotein (HDL) cholesterol. Because they have a low risk of allergenicity and are not genetically engineered, pulse proteins and legumes including lentil, pea, and chickpea are utilized as alternatives to soy proteins (Ducel et al., 2004; Luo and Hu, 2017). According to their solubility, pulse proteins are divided into three categories: globulins and albumins out of which the former make up 70% of the total pulse protein and are soluble in water-salt solutions and the latter account for 10% of the total pulse protein. Apart from these glutelin which is soluble in dilute acid and alkali comprise remaining 10%–20% of the total pulse protein (Hall et al., 2017).

3.2.1.7 Pea Protein

Pea protein constitutes about 18%–30% of its seeds. It is composed of globulins (65%–80%), which includes convicilin and vicilin (Van der Spiegel et al., 2013). It also contains albumins and glutelin as two minor proteins. Cheap and nutrient dense, pea protein is a rich food. The 11S globulin protein has a molecular weight that is between 350 and 400 KDa, whereas convicilin and vicilin (d7S globulin) have a molecular weight of about 150 KDa (Puvvada et al., 2015). Pea proteins have emulsification and jelly properties; as a result, they can be used alone as well as in mixture with proteins or polysaccharides to create stable emulsions that increase the effectiveness of a microencapsulation/nanoencapsulation technique because they provide good distribution of particle size and encapsulation happens naturally due to the surfactant, foaming, and solubility properties.

3.2.1.8 Chickpea Protein

Chickpea protein is less expensive, biodegradable, biocompatible, and nontoxic, so it is used for encapsulation. As it has little to no negative effects or toxicity, it is safe for use in the food sector (Tomadoni et al., 2020; Rachwa-Rosiak et al., 2015). It contains gluten, albumin, globulin, and protamine. The main compositions of these proteins are 3.12%–6.89%, 8.39%–12.31%, 53.44%–60.29%, 19.38%–24.40%, and represents about 28.6% of the total chickpea crop (Livney, 2010; Tavares et al., 2014). It has strong emulsifying abilities, a capacity to be encapsulated, high nutritional content, and positive health impacts. In addition, it can create substantial viscoelastic coatings around oil droplets, improving their processing-related stability. It is used in stews, soups, salads, and other dishes and has long-lasting nutritional advantages (Ghalandari et al., 2015; Botrel et al., 2014; Katouzian and Jafari, 2016; Oluk and Karaca, 2016).

3.2.1.9 Lentil Protein

Crops of lentils are grown all over the world. Compared to both pea and chick pea crops, they have greater soluble dietary fiber. They have a high amount of protein which makes up 20.6%–31.4% of the entire lentil plant. The percentages of globulins (MW 320–380 KDa), albumin (MW 20 KDa), glutelins (MW 17–46 KDa), and prolamins (MW 16–64 KDa) are, respectively, 70%, 16%, 11%, and 3% (Boye et al., 2010). It is frequently utilized as a wall material to create stable capsules with a high encapsulation efficiency (EE) percent because of its strong solubility, drying, and emulsifying qualities (Sabaa et al., 2019).

3.2.2 Animal-Based Material

Due to the presence of both hydrophilic and hydrophobic amino acid residues, these substances are amphiphilic. They can be utilized alone or in combination with other kinds of animal proteins including milk proteins like whey proteins and casein, gelatine, and bovine serum albumin, as well as biopolymers, such as proteins and polysaccharides, as wall materials for both microencapsulation and nanoencapsulation processes (Li and Yao, 2009; Mainardes and Khalil, 2019). They are more soluble, have low molecular weight, and are highly flexible than plant proteins.

3.2.2.1 Milk Proteins

Milk proteins are casein and whey. They can bind with different substances through hydrophilic and hydrophobic groups as per their affinities (Wang et al., 2017). Due to their various physicochemical characteristics, they are used as wall materials for microencapsulation and nanoencapsulation of bioactive compounds. Their adaptability aids in the encapsulation of bioactive, hydrophilic, and hydrophobic substances, and their bioactive peptides have physiological effects. They furthermore possess useful qualities like pH responsiveness, self-assembly, and gel swelling capability. They are utilized as bioactive delivery methods as a result of these functional qualities (Mou et al., 2018; Jakki et al., 2016).

Casein or cheese production results in the production of whey proteins. The water-soluble proteins lactoglobulin, lactalbumin, and serum albumin are used to make them (Hoseyni et al., 2020). Because they include all nine essential amino acids and have a low lactose content, they are regarded as complete proteins. Whey protein concentrate (WPC) contains little fat and sugar. Ninety percent of WPC is made up of protein. Whey protein isolate (WPI) has a high protein level of about 90% and zero fat or lactose. A predigested type of whey protein is called whey protein hydrolysate (WPH), which has undergone partial hydrolysis. Because of their excellent encapsulation effectiveness and storage stability, whey proteins are utilized as wall coating materials in microencapsulation and nanoencapsulation techniques for the controlled release of bioactive ingredients such as oils, lipids, vitamins, and volatile chemicals (Thomas et al., 2017).

Casein is an amphiphilic milk protein (which constitutes 80% of total milk protein). It has numerous physicochemical properties like it is nontoxic, stable, biocompatible, biodegradable, has great binding with ions and molecules, and the protein is less expensive, an excellent emulsifier, and has self-assembly that increases its loading efficiency and encapsulation capacity (Ma et al., 2019; Dias et al., 2015). They are classified as complete proteins since they include all nine essential amino acids and have a low lactose level. Sugar and fat content in whey protein concentrate (WPC) are minimal. 90% of the protein in WPC is protein. Whey protein isolate (WPI) has no lactose or fat and a high protein content of roughly 90%. Whey protein hydrolysate (WPH), a predigested form of whey protein, has undergone partial hydrolysis. Whey proteins are used as wall coating materials in micro and nanoencapsulation techniques for the controlled release of bioactive ingredients such as oils, lipids, vitamins, and volatile chemicals because of their excellent encapsulation effectiveness and storage stability (Menin and Edwards-Lévy, 2011).

3.2.2.2 Bovine Serum Albumin

There is a globular protein called BSA (bovine serum albumin). It is extracted from either milk lactation cells or bovine serum, which then spreads it across the two (Gomez-Mascaraque et al., 2016a). It has a single chain of 583 amino acids with a molecular weight of 62.2 kDa, three domains (I, II, and III), two helical subdomains (A and B), and the ability to bind lipid, nucleotides, and metal ions (Gomez-Mascaraque et al., 2016b; Matsuno and Adachi, 1993). Because of its unusual ligand-binding capacity, biocompatibility, and biodegradability, as well as its nontoxicity, lack of immunogenicity, good stability, low cost, abundance, and ease of purification, its microcapsule and nanocapsule carriers have become more and more popular in recent years. It is negatively charged at pH 7.4, is a protein plasma, and is used for drug and antigen delivery as well as in the food industry (Jafari et al., 2008; Moomand and Lim, 2014).

3.2.2.3 Gelatin Protein

Since gelatine protein does not exist in nature, it is isolated from partially hydrolyzed collagen. The most prevalent protein in fish and cattle bones and skin is collagen. It is a denatured protein, as well. When collagen is extracted via acid hydrolysis, the resulting gelatine is referred to as type-A gelatine and has a positively charged pI (isoelectric point) that ranges from 7 to 9.4. Type-B gelatine is what is produced when it is extracted using alkaline hydrolysis, and its pI varies from 4.8 to 5.5. It possesses a negative charge, and the temperature at which it denatures is roughly 40°C. Gelatin protein is also utilized as a coating material in combination with various polysaccharides, such as pectin, alginate, and chitosan, to create hard and soft capsules in the food and pharmaceutical industries because of its amphoteric nature, which makes it a good wall material (Abd El-Salam and El-Shibiny, 2015).

3.3 SPECIFIC APPLICATION OF BIOPOLYMERS

The efficiency of encapsulation, ingredient protection, and release behavior are all impacted by the choice of wall material (Tesch et al., 2002). The selection of encapsulation matrix is most important for the development of microencapsulation structures. Natural polyphenols are bioactive molecules. They act as antioxidants and possess health benefits. Some are poorly bioavailable, unstable, and have a bad flavor (Gharsallaoui et al., 2007). Numerous scientific studies deal with the microencapsulation of polyphenols utilizing different materials and processing methods in order to get beyond these restrictions.

Pomegranate polyphenols had a higher level of encapsulation effectiveness than soy isolate. The wall material was maltodextrin. This maltodextrin during storage at a high temperature of 60°C provides greater protection (Kolanowski et al., 2006). Encapsulation efficiency and protection to bio-ingredients, for example, epigallocatechin gallate (polyphenol present in green tea), increase when encapsulated in gelatine than in chitosan by spray drying and electroplating (Al-Ismail et al., 2015). Similar efficiency for chitosan results when epigallocatechin gallate is encapsulated with other carbohydrate matrices.

The protein matrices used for the encapsulation of polyphenol are due to the complex formation with proteins (Gharsallaoui et al., 2007). When using gelatine as a delivery vehicle rather than chitosan, epigallocatechin gallate (EGCG) is more bioaccessible after digestion.

3.3.1 Oil and Flavor Encapsulation

To quantify the capacity of wall material to create a protective network during spray drying, a method based on emulsion drying rate determination was developed. Proteins are employed as matrices for the encapsulation of polyunsaturated fatty acids because the drying curve does not distinguish between different drying curves or provide information about the chemical properties of the wall material (Qian et al., 2012).

Excellent emulsifying capabilities must be present in the wall material used for oil encapsulation (Kaufmann et al., 2016), so proteins act as good emulsifiers due to their phallic nature. They have good encapsulation efficiency and good oxidation potential, so proteins are used as matrices for polyunsaturated fatty acid encapsulation (Perez-Masia et al., 2015).

Carbohydrates have poor retention and low surface activity (Kaufmann et al., 2016), so some polysaccharides are modified to overcome this limitation (Zhang et al., 2016). Gum arabic has small proteins that possess emulsifying properties (Yonekura and Nagao, 2009). The oxidation stability of some modified cellulose used to encapsulate fish oil generated by spray drying is not improved (Rich et al., 2007). Food systems rarely use formulations based on polysaccharides. Microencapsulation of linseed with gum arabic does not increase the thermal stability of linseed by spray drying but causes loss of alfalinolenic acids (Gbassi and Vandamme, 2012) when whey proteins are used. It improves the thermal stability of encapsulated oils.

In comparison to a maltodextrin and gum arabic combination, milk proteins, such as sodium caseinate and whey, have a higher encapsulation efficiency and better oxidative stability to protect seed oil. Gum arabic has good encapsulation efficiency and retention of cardamom oil than whey proteins during storage (El-Salam and El-Shibiny, 2015). Due to their strong emulsifying capabilities and ability to produce a protective layer at the O/W emulsion interface, proteins generally outperform polysaccharides as an encapsulating material for oils.

3.3.2 Encapsulation of Pigments

Pigments called carotenoids have positive health effects. Provitamin A activity is present in several of them (Rodrigues et al., 2011). The most abundant phytochemicals, along with polyphenols, are carotenoids (Lopez-Rubio et al., 2012). Carotenoids are dissolved in edible oils to improve their absorption in vitro and in vivo because they have low water solubility, chemical solubility, and consequently low bioaccessibility (Corona-Hernandez et al., 2013). So, in order to boost their bioaccessibility and disperse in an aqueous biopolymer solution or dispersion, O/W emulsion is generated prior to capsule preparation. Compared to dextran or chitosan, protein matrix has better encapsulation effectiveness for lycopene (Corona-Hernandez et al., 2013).

Encapsulation efficiency of carotenoids found in oleoresin from Rosa mosqueta increases when encapsulated in gelatine capsule rather than starch (Kolanowski et al., 2006). Gelatine also provides more protection for carotenoids, particularly trans-carotenoids, than starch. Alginate is used as a desirable encapsulation matrix, as it gives excellent protection to carotenoids against acidic conditions during digestion and protects beta-carotene from degradation during digestion (De Prisco and Mauriello, 2016). The encapsulation with alginate causes reduction in bioaccessibility due to insufficient release, so it should be incorporated with mixed cell polysaccharide (Muzaffar and Kumar, 2016).

3.3.3 Probiotics Encapsulation

Probiotics are bioactive components that are live organisms and provide health benefits to the host when consumed in adequate quantities, so they must be alive and active during consumption (DavidovPardo et al., 2015) and need microencapsulation.

The encapsulation matrix should be stable at low pH and get deranged at a pH of 6 (DavidovPardo et al., 2015).

Alginate is the wall material used for encapsulation of probiotic bacteria (Ifeduba and Akoh, 2016). Carrageen, which is an anionic polysaccharide with gelling capacity used at melting temperature, is problematic (Tessier and Niquet, 2007). So, milk protein is used as an alternative to alginate for encapsulation of probiotics (Ma et al., 2014). Whey protein with alginate is the adequate material for encapsulation of bioactive ingredients like probiotics (Al-Otaibi, 2009). WPC provides more protection to Bifidobacterium animalis subspecies Lactis (Bb12) under relatively high humidity storage conditions as compared to pullulan, which is a polysaccharide developed by electrospraying (Raoufinia et al., 2016). Electrosprayed whey protein decreases viability loss in vitro digestion by three logs (cfu/g). It is also used as a shield for Lactobacillus plantarum (Gullapalli and Mazzuchelli, 2017; Yao et al., 2017). Alginate microcapsules can be successfully incorporated into foods like ice cream, but *Lactobacillus* does not show good results as it required different protocols (Ladjal-Ettoum et al., 2016). Probiotic microencapsulation has been reviewed in a number of articles lately (Karaca et al., 2015).

3.3.4 Carbohydrate and Protein Mixture

Maltodextrins incorporated with proteins are called thermal defenders (Gharsallaoui et al., 2007). These products are produced by spray drying and are used for microencapsulation (El-Salam and El-Shibiny, 2015). Proteins, when integrated with dextrin, produced a suitable encapsulant for fish oil by enhancing its oxidative stability (Gallardo et al., 2013). Blending of pectin and pea protein increases oxidative stability as compared to pea protein alone (Gaysinsky et al., 2008). Protein polysaccharides do not always perform better as compared to individual components. Cardamom oil encapsulated with SPI than using protein and guar gum or carrageen delivers better results. Mixture of proteins and polysaccharides are used for encapsulation of bioactive components (Gbassi and Vandamme, 2012). This is the most exploited encapsulation technique in the food industry.

Proteins and carbohydrates can be adjoined by dry heating and by Maillard reaction, and these encapsulates are also used for encapsulation of food ingredients. Caseinate dextran–coated zein nanoparticles are used to encapsulate resveratrol to increase its bioaccessibility (Gomez-Mascaraque et al., 2015). Nanoparticles that are coated with Maillard complexes deliver stability to aggregation under gastrointestinal tract conditions compared to caseinate alone. Crosslinking gelatine and gum arabic by Maillard reaction increases the oxidative stability of stearidonic acid soybean oil (Ifeduba and Akoh, 2016). Maillard reaction products have antioxidant activity but possess undesirable color and odor (Amarowicz, 2009).

3.4 CASE STUDY

Aceclofenac, a nonsteroidal anti-inflammatory drug (NSAID), having a shelf life of 4.3 hours, was integrated with a Biopharmaceutics Classification System (BCS) class 2 drug into polymeric microcapsule (Arslan and Tirnaksiz, 2010). Aceclofenac microcapsule was put together by the ionic gelation technique using sodium alginate as the wall material in combination with polysaccharide derivatives, e.g., sodium carboxymethyl cellulose (SCMC), hydroxypropyl methylcellulose (HPMC), and methylcellulose (Kumar et al., 2011). The microcapsules were spherical with a microencapsulation efficiency of 83.25%–99.94%. They possessed good mucoadhesive attributes to enhance bioavailability for 15 hours and displayed a sustained release of aceclofenac for 15 hours.

Polysaccharides used for the microencapsulation of bicalutamide microspheres are developed using a polysaccharide guar gum via oil-in-water emulsion by a method called solvent diffusion (Dhas and Deshmukh, 2021). The composition formula of 2.5% of guar gum, 0.25% span, 80% as dispersing agent showed the optimum drug release of 94.22% within 24 hours, without drug interaction. It increases the bioavailability and sustained release of the drug. These properties are due to polymeric character of the drug and can be enhanced by nanoencapsulation. This will provide a channel for targeting of the nonsteroidal antiandrogen to a greater extent than microencapsulation.

Molecular Envelope Technology (MET) nanoparticles obtained from complex polysaccharide chains of N-palmitoyl-N monomethyl-N, N-dimethyl-N, N, N-trimethyl-6-O-glycol chitosan are used for the delivery of drugs to target sites of the blood brain. MET nanoparticles containing lomustin drug is used for the treatment of Glioblastoma multiforme (Fisusi et al., 2016). The MET uses polysaccharide for nanoencapsulation, optimized biodistribution, and pharmacodynamics, resulting in the reduction of the toxic effects of lomustine and providing better results for patients suffering from brain cancer. As the lomustine is protected from degradation, MET ensures sustained release of the drug at target sites due to PEGylation of the polysaccharide to increase circulation time within the body (Fisusi et al., 2016).

Repaglinide was developed without being encapsulated in ethyl cellulose for the treatment of type 2 diabetes and characterized by the solvent evaporation method (Lokhande et al., 2013). It exhibited high encapsulation efficiency. Repaglinide, which is nonencapsulated, is biodegradable, biocompatible, and is able to improve its pharmacological activity via modification of surface function and charge to promote cell entry.

Microneedles have a length of 50–900 μm, and they penetrate into the stratum corneum (Menon et al., 2021). They act as a barrier in microchannels for delivery of encapsulated bioactive components without causing any damage to the nerves and blood vessels. Etanercept is a human dimeric fusion protein, which is fully soluble and is a tumor necrosis factor (TNF) inhibitor. It binds to the TNF preventing the activation of the inflammatory cascade. It is a fusion protein of recombinant human TNF-receptor p75 fused with the Fc domain of human immunoglobulin G1 (IgG1). When etanercept, a substance made of acrylate-modified hyaluronic acid, is applied to the skin, it causes a release that travels through the bloodstream to the arthritic tissue where it has an anti-inflammatory impact by binding with TNF. The etanercept-loaded microneedles were produced by the micro molding method (Menon et al., 2021). The microneedles were detached from the mold and crosslinked by ultraviolet (UV) light to increase mechanical strength. Drug loading and in vitro bioactivity were evaluated. Microneedle dissolution, skin penetration, and skin recovery therapeutic effects were determined with mice. The microneedles showed high mechanical strength, quick recovery of skin after 120 minutes, complete dissolution of microneedles in the skin after 90 minutes, good biocompatibility, little interference with bioactivity of etanercept, and high anti-inflammatory efficacy. There was evidence of decrease in TNF-α and IL-6 in serum, protection of the joint from erosion, and the microneedle system showed good bioequivalence to the classical subcutaneous route.

3.5 CHARACTERIZATION OF BIOPOLYMERS FOR ENCAPSULATION

The stability of encapsulation systems is the essential tool to access the efficiency and optimize the developed structures, the composition, and the chemical structure of the "matrix" (encapsulating material) encapsulated bioactive compound. The bioactive and matrix materials' interactions and morphology of the encapsulating system (size distribution, shape, porosity, etc.) have a huge impact on the stability of the encapsulating systems.

3.5.1 Microscopy Techniques

Techniques related to imaging characterize the structure of encapsulating system and are based on microscopic instruments (Neu et al., 2010). These are qualitative techniques and give information on the shape, size, and aggregation state. Optical, laser, and electron microscopy methods are chosen. It depends on the composition of the encapsulated material and the size range of the encapsulate. Optical microscopy is easily reachable and a less expensive imaging technique. It has low resolution and low magnification. Electron microscopy is used as an alternative to optical microscopy due to higher magnifications; particles with size ranges from tenths of nanometers to tenths of micrometers are characterized by electron microscopy. Transmission electron microscopy (TEM) and scanning electron microscopy (SEM) are the electron microscopy techniques, which are most extensively used.

TEM is a technique based on the transmission of a beam of electrons through an ultrathin sample. Exhausting sample preparation is a major limitation of this technique, as the sample should be thin, electron transparent, and dry, and then it is subjected to the high vacuum present inside the instrument (Neu et al., 2010). To enhance the contrast between the sample and the background, it requires negative staining. A small amount of the sample is placed on the TEM grid and stained with uranyl acetate solution, resulting in high contrast.

3.5.2 Freeze-Fracture Technique

This technique provides information regarding the internal structure of the particle. The sample is placed on the TEM grid, which is frozen by rapid freezing and then fractured under constant cooling and vacuum. It is used for biological materials and moisture-sensitive foods.

3.5.3 Scanning Electron Microscopy

This technique provides more information regarding the surface structure of the particle than the freeze-fracture technique, but it has low resolution than the freeze-fracture technique. The sample preparation in this technique is not tedious. A beam of electrons is scanned when passed through the sample surface and gives information about composition and surface topography. SEM requires sample conductivity and high vacuum. The information given by the SEM is subjected to structural modifications by the drying and coating methods used for the sample.

3.5.4 Atomic Force Microscopy

Atomic force microscopy is a more recent technique. It comprises of a pointed probe that scans the sample's surface once it has been contained on the surface. It gives information about the surface topography and is determined by the reaction of the probe to the force imposed by the sample on it when scanned. It has high resolution than SEM and does not need sample conductivity. This technique is not suitable for soft and sticky samples. These samples require drying before analysis.

3.6 SEPARATION AND ANALYTICAL TECHNIQUES

3.6.1 Chromatography Technology

This technique is used for separation of matrix and encapsulation compound and quantification of active compound in the encapsulation system. It consists of a stationary phase and a mobile phase. Separation is done on the basis of different speeds at which different compounds travel through the column (Coskun, 2016). The quantification of active compound is done by detector-like fluorescence detector ultraviolet-visible (UV-vis) light absorbance detector and diffractometer. The main advantage of chromatography is stationary phase, mobile phase, and separation type, and the detector can be adjusted depending upon the properties and composition of the encapsulation system.

Gas-chromatography (CG) and liquid chromatography (LC) techniques can be differentiated depending on the physical state of mobile phase.

High-performance liquid chromatography (HPLC). This technique is used in food analysis. The sample is dissolved in a liquid, which is then pressurized and injected into a column containing a stationary phase (Coskun, 2016). The separation occurs on the basis of the molecular mass of the different compounds, e.g., size exclusion chromatography (SEC), gel permeation chromatography (GPC), or the charge (cation or anion exchange chromatograph). GC coupled with mass spectrometry (GC-MS) is a technique to separate and quantify compounds that can be vaporized without affecting their molecular structure (Coskun, 2016). In this case, after the compounds in the sample travel through the chromatographic column and are released (with different retention times), the coupled mass spectrometer breaks down the molecular structure of each component into ionized fragments, separates the ions of differing masses, and determines their relative abundance by measuring the intensity of the ion flux.

3.6.1.1 Application of Chromatography

It is used to determine the encapsulation efficiency of gallic acid/cyclodextrin complex incorporated with phytic acid nanofibers. GC-MS is used to evaluate the thyme essential oil in β-cyclodextrin capsules.

Chromatographic techniques are useful to determine the encapsulation efficiency. They do not provide information about the interactions between the matrix and the encapsulated compound or the release mechanism under specific conditions. There is no information on the structure of the system.

3.6.2 Physicochemical Methods

Spectroscopy and scattering techniques are used to study the structural changes taking place at the molecular, microstructural, and nanostructural levels. Small-angle scattering techniques when combined with traditional methods are used for the characterization of food systems.

3.6.2.1 Spectroscopy Techniques

This technique is used to study the interactions between a source of radiation (electromagnetic radiation) and the particles of sample. It measures the changes in the intensity or frequency of the radiation energy, nuclear magnetic resonance (NMR) spectroscopy, and Fourier transform infrared (FT-IR) spectroscopy, which are used to characterize the molecular interactions between the components in encapsulating systems and the molecular structure.

3.6.2.2 Nuclear Magnetic Resonance Technique

This technique measures the electromagnetic radiation, absorbed by an active nucleus, like ^{1}H or ^{13}C, at a certain frequency. The absorption of an atomic nucleus is conditioned by the surrounding atoms. It provides detailed information about the molecular structure and can be used as a sample, in the solid or solution state, which contains nuclei possessing spin. In the particular case of encapsulating systems, ^{1}H-NMR and ^{13}C-NMR are used to detect the complexes between β-cyclodextrins and essential oils or to determine the molecular state of curcuminoids encapsulated in poly(butyl) cyanoacrylate nanoparticles. This technique is used to determine the interaction between silica matrices and encapsulated ibuprofen molecules based on the ibuprofen molecular mobility deduced from ^{13}C and ^{1}H spectra to determine the interaction between water and polymer in whey protein/alginate beads containing riboflavin.

3.6.2.3 Fourier Transform Infrared Technique

This technique measures the amount of light absorbed by a sample at different wavelengths. It does not use a monochromatic beam of light; the sample is radiated by a beam composed of many wavelengths at once and data are subsequently processed to determine the absorption taking place at each measured wavelength. It confirms the encapsulation of the active compound into the matrix by identifying absorption bands (Kowalczuk and Pitucha, 2019). It can be used to determine the presence of vitamin C and vitamin B12 encapsulated into different biopolymers by spray drying. Resveratrol is included in chitosan-sodium tripolyphosphate microspheres along with tea tree oil and xylitol that were obtained through complicated coacervation (Kowalczuk and Pitucha, 2019).

3.6.2.4 Scattering Techniques

This technique analyzes the scattered radiation produced after interaction of light, X-rays, and neutrons, with the particles present in a sample. It provides information about shape and size orientation in ordered and disordered systems. The scattered radiation detected by detector light scattering techniques is used for the characterization of colloidal suspensions, such as food systems.

Static light scattering (SLC), also known as laser light scattering (LSC), measures particle size based on light scattering intensity, which changes with variation in scattering angle, size of particles, refractive indices of particles, and the medium. SLC measures the average intensity of scattered light arriving at the detector, although actually, the intensity shows fluctuations over short time scales. In the case of solutions, the Brownian motion of particles causes fluctuations in the intensity of the scattered light arriving at the detector.

Dynamic light scattering (DLS) is based on the measurements of these fluctuations as a function of time. By analyzing these fluctuations through correlation functions, DLS is commonly used to characterize the particle size of encapsulating systems based on microemulsions. The main drawback of SLC and DLS methods is that highly diluted samples are often required to avoid multiple scattering (i.e., the photons are scattered more than once between the light source and the detector). In addition to ensure a proper data interpretation, the particles should present a narrow size distribution and they must present the same refractive index (which is not true in the case of multicomponent systems).

Similar to small-angle X-ray scattering (SAXS), wide-angle X-ray scattering (WAXS) probes lower length scales due to the increased angle between the sample and detector. XRD is conducted in the reflection or in the transmission geometry. XRD is used in crystalline structure, and ultrasmall-angle scattering techniques (USAXS) and ultra small angle neutron scattering (USANS) can be utilized to characterize larger structural features. SAXS or SANS is used to characterize a particular encapsulating system. It depends upon the scattering length density (SLD), contrast between the sample and the surrounding element materials or between different sample components soybean lecithin/chitosan nanoparticles loaded with a lipophilic positively charged drug used in breast cancer therapy. By integrating SAXS (small-angle X-ray scattering), SANS (small-angle neutron scattering cryo-TEM, static light scattering, and other techniques, its structure was identified. Using the SAXS technique, the internal structure of poly (lactide-co-glucoside)-block-poly (ethylene glycol) (PLGA-PEG) nanoparticles was studied in liposomes that were loaded with antibiotics. This technique is also used to identify the release of anticancer drug carboplatin, which is hydrophilic when incorporated into the PLGA-PEG nanoparticles. Neutrons and X-rays were used to characterize microemulsions embedded in alginate hydrogels.

REFERENCES

Abd El-Salam, M. H. and El-Shibiny, S. (2015). Preparation and properties of milk proteins-based encapsulated probiotics: A review. *Dairy Science and Technology, 9*(5), 393–412.

Al-Ismail, K. M., Mehyar, G., Al-Khatib, H. S. and Al-Dabbas, M. (2015). Effect of microencapsulation of cardamom's essential oil in gum arabic and whey protein isolate using spray drying on its stability during storage. *Quality Assurance and Safety of Crops Foods, 7,* 613–620.

Al-Otaibi, M. (2009). Evaluation of some probiotic fermented milk products from Al-Ahsa markets Saudi Arabia. *American Journal of Food Science and Technology, 83*(19), 1–11.

Amarowicz, R. (2009). Antioxidant activity of Maillard reaction products. *European Journal of Lipid Science & Technology, 111*(2), 109–111.

Arslan, S. A. and Tirnaksiz, F. (2010). A nonsteroidal anti-inflammatory drug: Aceclofenac. *Fabad Journal of Pharmaceutical Sciences*, *35*(2), 105–118.

Botrel, D. A, de Barros Fernandes, R. V., Borges, S. V. and Yoshida, M. I. (2014). Influence of wall matrix systems on the properties of spray-dried microparticles containing fish oil. *Food Research International, 62*, 344–352.

Boye, J., Zare, F., & Pletch, A. (2010). Pulse proteins: Processing, characterization, functional properties and applications in food and feed. *Food Research International*, *43*(2), 414–431.

Considine, T., Flanagan, J., Boland, M., Singh, H. and Thompson, A. (2009). Interaction between milk proteins and micronutrients. In: Boland, M., Singh, H. and Thompson, A. (Eds.), *Milk Proteins: From Expression to Food.* Academic Press, Cambridge, MA, pp. 347–376.

Corona-Hernandez, R. I., Álvarez-Parrilla, E, Lizardi-Mendoza, J., Islas-Rubio, A. R. L. A., de la Rosa, R. L. and Wall-Medrano, F. A. (2013). Structural stability and viability of microencapsulated probiotic bacteria: A review. *Comprehensive Review in Food Science and Food Safety*, 12, 614–62.

Coskun, O. (2016). Separation tecniques: Chromatography. *Northern Clinics of Istanbul*, *3*(2), 156–160. doi: 10.14744/nci.2016.32757.

DavidovPardo, G., Pérez Cordia, M., Marín Arroyo, M. and McClements, A. D. (2015). Improving resveratrol bio accessibility using biopolymer nanoparticles and complexes: Impact of protein–carbohydrate Maillard conjugation. *Journal of Agriculture Food Chemistry, 63*, 3915–3923.

De Prisco, A. and Mauriello, G. (2016). Probiotication of foods: A focus on microencapsulation tool. *Trends in Food Science and Technology*, *48*, 27–39.

Delcour, J. A. and Hoseney, R. C. (2010). *Principles of Cereal Science and Technology*. AACC International, Eagan, MI.

Dhas, S. K. and Deshmukh, G. (2021). Formulation and evaluation of meloxicam microspheres for colon targeted drug delivery. *Asian Journal of Pharmaceutical and Clinical Research*, *7*, 45–51. doi: 10.22159/ajpcr.2021.v14i8.38482.

Dias, M. I., Ferreira, L. C. and Barreiro, I. C. (2015). Microencapsulation of bio actives for food applications. *Food and Function*, *6*(15), 1035–1052.

Ducel, V., Richard, D., Papineau, Y. and Boury, F. (2004). Adsorption kinetics and rheological interfacial properties of plant proteins at the oil and water interface. *Biomacromolecules*, *5*(6), 2088–2093.

El-Salam, M. and El-Shibiny, S. (2015). Preparation and properties of milk proteins-based encapsulated probiotics: A review. *Dairy Science and Technology*, 95(4), 393–412.

Fatha, M., Martin, A. and McClements, D. J. (2014). Nanoencapsulation of food ingredients using carbohydrate-based delivery systems. *Trends in Food Science & Technology*, *39*(14), 18–39.

Finnie, C. and Svensson, B. (2009). Barley seed proteomics from spots to structures. *Journal of Proteomics*, *72*(3), 315–324.

Fisusi, F. A., Siew, A., Chooi, K. W., Okubanjo, O., Garrett, N., Lalatsa, K., Serrano, D., Summers, I., Moger, J., Stapleton, P., Satchi-Fainaro, R., Schätzlein, A. G. and Uchegbu, I. F. (2016). Lomustine nanoparticles enable both bone marrow sparing and high brain drug levels: A strategy for brain cancer treatments. *Pharmaceutical Research*, *33*(5), 1289–1303. doi: 10.1007/s11095-016-1872-x.

Flambeau, M., Redl, A. and Respondek, F. (2017). Proteins from wheat: Sustainable production and new developments in nutrition-based and functional applications. In: Nadathur, S. R., Wanasundara, J. P. D. and Scanlin, L. (Eds.), *Sustainable Protein Sources*. Elsevier, Amsterdam, Netherlands, pp. 67–78.

Gallardo, G., Guida, L., Martinez, V., López, M. C., Bernhardt, D., Blasco, R., Pedroza Islas, R. and Hermida, L. G. (2013). Microencapsulation of linseed oil by spray drying for functional food application. *Food Research International*, *52*, 473–482.

Gaysinsky, S. P., Davidson, P. M., McClements, D. J. and Weiss, J. (2008). Formulation and characterization of Phyto phenol-carrying antimicrobial microemulsions. *Food Biophysics*, *3*(1), 54–65.

Gbassi, G. K. and Vandamme, T. (2012). Probiotic encapsulation technology: From microencapsulation to release into the gut. *Pharmaceutics*, *4*, 149–163.

Ghalandari, B., Divsalar, A., Eslami-Moghadam, M., Saboury, A. A., Haertlé, T., Amanlou, M., et al. (2015). Probing of thenteraction between Î2-lactoglobulin and the anticancer drug oxaliplatin. *Applied Biochemistry and Biotechnology*, *175*(2), 974–987.

Gharsallaoui, A., Roudaut, G., Chambin, O., Voilley, A. and Sauer, R. L. (2007). Applications of spray-drying in microencapsulation of food ingredients: An overview. *Food Research International*, *40*, 1107–1121.

Gil, A. M. (2012). Techniques for analysing wheat proteins. In: Cauvain, S. P. (Ed.), Breadmaking. Elsevier, Amsterdam, Netherlands, *99*(12), pp. 77–99.

Gomez-Mascaraque, L., GLagarón, J. M. and Lopez-Rubio, A. (2015). Electrosprayedgelatinsubmicroparticles as edible carriers for the encapsulation of polyphenols of interest in functional foods. *Food Hydrocolloids*, *49*, 42–52.

Gomez-Mascaraque, L. G., Sanchez, G. and Lopez-Rubio, G. (2016a). Impact of molecular weight on the formation of electro sprayed chitosan microcapsules as delivery vehicles for bioactive compounds. *Carbohydrate Polymers*, *150*(16), 121–130.

Gómez-Mascaraque, L. G., Soler, C. and Lopez-Rubio, A. (2016b). Stability and bio accessibility of EGCG within edible micro-hydrogels chitosan vs. gelatine, a comparative study. *Food Hydrocolloid, 61*(16), 128–138.

Gullapalli, R. P. and Mazzuchelli, C. L. (2017). Gelatin and non-gelatin capsule dosage forms. *Journal of Pharmaceutical Science, 106*(6), 1453–1465.

Hall, C., Hellen, C. and Garden-Robinson, J. (2017). Composition, nutritional value, and health benefits of pulses. *Cereal Chemistry, 94*(1), 11–31.

Hoseyni, S. Z., Jafari, S. M., Shahiri-Tabarestani, H., Ghorbani, M., Assadpou, E. and Sabaghi, M. (2020). Production and characterization of catechin-loaded electro spun nanofibers from Azivashgum, polyvinyl alcohol. *Carbohydrate Polymers, 235*(115979), 3471–3479.

Ifeduba, E. A. and Akoh, C. (2016). Microencapsulation of stearidonic acid soybean oil in Maillard reaction-modified complex coacervates. *Food Chemistry, 199*, 524–532.

Jafari, S. M., Assad Poor, E., He, Y. and Bhandari, Y. (2008). Encapsulation efficiency of food flavours and oils during spray drying. *Drying Technology, 26*(8), 816–835.

Jakki, S. L., Ramesh, Y. V., Gowthamarajan, K., Senthil, V., Jain, K. and Sood, S. (2016). Novel anionic polymer as a carrier for CNS delivery of anti-Alzheimer drug. *Drug Delivery, 23*(9), 3471–3479.

Joye, I. J. (2019). Cereal biopolymers for nano-and microtechnology: A myriad of opportunities for novel (functional) food applications. *Trends in Food Science and Technology, 83*, 1–11.

Karaca, A. C., Low, N. M. and Nickerson, M. (2001). Potential use of plant proteins in the microencapsulation of lipophilic materials in foods. *Trends in Food Science and Technology, 42*, 5–12.

Karaca, A. C., Low, N. H. and Nickerson, M. (2015). Potential use of plant proteins in themicroencapsulationoflipophilic materials in foods. *Trends in Food Science and Technology, 42*(1), 5–12.

Katouzian, I. and Jafari, S. M. (2016). Nano-encapsulation as a promising approach for targeted delivery and controlled release of vitamins. *Trends Food Science and Technology, 53*, 34–48.

Kaufmann, A., André, C. M., Schneider, Y. J., Hoffmann, L. and Bohn, T. (2016). Carotenoid and polyphenol bio accessibility and cellular uptake from plum and cabbage varieties. Food Chemistry, *197*, 325–332.

Kolanowski, W., Ziolkowski, M., Weisbord, J., Kunz, J. and Laufenberg, G. (2006). Microencapsulation of fish oil by spray drying-impact on oxidative stability. *European Food Research and Technology*, *23*, 336–342.

Kowalczuk, D. and Pitucha, M. (2019). Application of FTIR method for the assessment of immobilization of active substances in the matrix of biomedical materials. *Materials*, *12*(18). doi: 10.3390/ma12182972.

Kumar, A. A., Kumar, P. R., Kumar, A. A., Reddy, K. L., Murthy, T. E. G. K. and Rao, T. V. (2011). Formulation design of aceclofenac microcapsules by inotropic gelation technique, characterization studies and release kinetics. *Journal of Applied Pharmaceutical Science*, *1*(6), 127–132.

Ladjal-Ettoum, Y., Boudries, H., Chibane, M. and Romero, A. (2016). Pea, chickpea and lentil protein isolates: Physicochemical characterization and emulsifying properties. *Food Biophysics*, *11*(1), 43–51.

Li, J. and Yao, P. (2009). Self-assembly of ibuprofen and bovine serum albumin- dextran conjugates leading to effective loading of the drug. *Langmuir*, 25(11), 6385–6391.

Livney, Y. D. (2010). Milk proteins as vehicles for bioactives. *Current Opinionin Colloid & Interface Science, 15*(1), 73–83.

Lokhande, A. B., Mishra, S., Kulkarni, R. D. and Naik, J. B. (2013). Preparation and characterization of repaglinide loaded ethylcellulose nanoparticles by solvent diffusion technique using high pressure homogenizer. *Journal of Pharmacy Research, 7*(5), 421–426. doi: 10.1016/j.jopr.2013.04.049.

Lopez-Rubio, A., Sanchez, E., Wilkanowicz, S., Sanz, Y. and Lagaron. J. M. (2012). Electrospinning as a useful technique for the encapsulation of living Bifidobacteria in food hydrocolloids. *Food Hydrocolloids, 28*, 159–167.

Lorenzo, G., Sosa, M. and Califano, A. (2018.) Alternative proteins and pseudo cereals in the development of gluten-free pasta. In: Holban, A. M. and Grumezescu, A. M. (Eds.), *Alternative and Replacement Foods*. Elsevier, Amsterdam, Netherlands, po. 433–458.

Luo, Y. and Hu, Q. (2017). Food-derived biopolymers for nutrient delivery. In: Grumezescu, A. M. (Ed.), Nutrient Delivery. Elsevier, Amsterdam, Netherlands, pp. 251–291.

Ma, J. J., Mao, X. Y., Wang, Q., Yang, S., Zhang, D., Chen, S. W. and Li, Y. H. (2014). Effect of spray drying and freeze drying on the immunomodulatory activity, bitter taste and hygroscopicity of hydrolysate derived from whey protein concentrate. *LWT Food Science and Technology, 56(*14), 296–302.

Ma, F., Pang, X. and Tang, B. (2019). Alginate/chondroitin sulfate-based hybrid hydrogel with different molecular weight and its capacity to regulate chondrocytes activity. *Carbohydrate Polymer, 206*, 229–237.

Mainardes, R. M. and Khalil, N. M. (2019). Nanoparticles of bovine serum albumin for encapsulation of food ingredients. In: Jafari, S. M. (Ed.), *Biopolymer Nanostructures for Food Encapsulation Purposes*. Elsevier, Amsterdam, Netherlands, pp. 169–186.

Matsuno, R. and Adachi, S. (1993). Lipid encapsulation technology-techniques and applications to food. *Trends in Food Science and Technology, 4*(93), 256–261.

Mehyar, G. F., Al-Ismail, K. M., Al-Ghizzawi, H. M. and Holley R. A. (2014). Stability of cardamom (*Elettariacardamomum*) essential oil in microcapsules made of whey protein isolate, guar gum, and carrageenan. *Journal of Food Science and Technology 79*, C1939–C1949.

Menin, A. and Edwards-Lévy, F. (2011). Encapsulation of natural polyphenolic compounds: A review. *Pharmaceutics, 3*(11), 793–829.

Menon, I., Bagwe, P., Gomes, K. B., Bajaj, L., Gala, R., Uddin, M. N., D'souza, M. J. and Zughaier, S. M. (2021). Microneedles: A new generation vaccine delivery system. *Micromachines, 12*(4), 1–18. doi: 10.3390/mi12040435.

Mohan, A., Rajendran, S. R., He, Q. S., Bazinet, L. U. and Denigwe, C. C. (2015). Encapsulation of food protein hydrolysates and peptides: A review. *RSC Advances, 5*(15), 79270–79278.

Moomand, K. and Lim, L. (2014). Oxidative stability of encapsulated fish oil in electro spun zein fibres. *Food Research International, 62*, 523–532.

Mou, J., Wu, Y., Bi, M., Qi, X. and Yang, J. (2018). Polyanionic holothurian glycosamino glycans-doxorubicin nano complex as a delivery system for anticancer drugs. *Colloids and Surfaces Biointerfaces, 16*, 364–369.

Muzaffar, K. and Kumar, P. (2016). Effect of soya protein isolate as a complementary drying aid of maltodextrin on spray drying of tamarind pulp. Drying Technology, *34*, 142–148.

Nesterenko, A., Alric, I., Silvestre, F., and Durrieu, V. (2010). Vegetable proteins in microencapsulation: A review of recent interventions and their effectiveness. *Industrial Crops and Products, 42*(10), 469–479.

Neu, T. R., Manz, B., Volke, F., Dynes, J. J., Hitchcock, A. P. and Lawrence, J. R. (2010). Advanced imaging techniques for assessment of structure, composition and function in biofilm systems. *FEMS Microbiology Ecology, 72*(1), 1–21. doi: 10.1111/j.1574-6941.2010.00837.x.

Nishinari, K., Fang, Y., Guo, S. and Phillips, G. O. (2014). Soy proteins: A review on composition, aggregation and emulsification. *Food Hydrocolloids, 39*(14), 301–318.

Oluk, C. A. and Karaca, O. B. (2016). Functional food ingredients and nutraceuticals, milk proteins as nutraceuticals nanoscience and food industry. In: Grumezescu, A. M. (Ed.), NanoScience in Food Chemistry as Nutraceuticals. Elsevier, London, pp. 751–759.

Palao-Suay, R., Gómez-Mascaraque, L. G., Aguilar, M. R., Vázquez-Lasa, B. and Román, J. S. (2016). Self-assembling polymer systems for advanced treatment of cancer and inflammation. *Progress in Polymer Science, 53*(16), 207–248.

Perez-Masia, R., Lagaron, J. and Lopez-Rubio, A. (2015). Morphology and stability of edible lycopene-containing micro- and nano capsules produced through electro spraying and spray drying. *Food Bioprocess Technology, 8*, 459–470.

Puvvada, N., Rajput, S., Kumar, B. P., Sarkar S., Konar, S. and Brunt, K. R. (2015). Novel ZnO hollow-nanocarriers containing paclitaxel targeting folate-receptors in amalignant pH-microenvironment for effective monitoring and promoting breast tumour regression. *Scientific Reports,* 5, 1–15.

Qian, C., Decker, E., Xiao, H. and Mc Clements, J. (2012). Nano emulsion delivery systems: Influence of carrier oil on β-carotene bio accessibility. *Food Chemistry, 135*, 1440–1447.

Rachwa-Rosiak, D., Nebesny, E. and Budryn, G. (2015). Chickpeas-composition, nutritional value, health benefits, application to bread and snacks: A review. *Critical Review in Food Science and Nutrition, 55*(8), 1137–1145.

Raoufinia, R., Mota, A., Keyhanvar, N., Safari, F., Shamakhi, S. and AbdoAlizadeh, J. (2016). Overview of albumin and its purificationmethods. *Advanced Pharmaceutical Bulletin, 6*(4), 495–507.

Rich, G. T., Bailey, A. L., Faulks, R. M., Parker, M. L., Wickham, M. S. and Fellery-Travis, M. S. (2007). Solubilization of carotenoids from carrot juice and spinach in lipid phases: I: modelling the gastric lumen Pineiro, M. and Stanton, C., Probiotic bacteria: legislative framework requirements to evidence basis. *The Journal of Nutrition, 137*, 850S–853S.

Rodrigues, D., Rocha-Santos, T., Sousa, S., Gomes, A. M., Pintado, M. M., Xavier Malcata, F., Sousa Lobo, M., Silva, J. P., Costa, P. and Amaral, M. H. (2011). On the viability of five probiotic strains when immobilised on various polymers. International journal of Dairy Technology, *64*, 137–144.

Roy, F., Boye, J. A. and Simpson, B. K. (2010). Bioactive proteins and peptides in pulse crops: Pea, chickpea and lentil. *Food Research International*, *43*, 432–442.

Sabaa, M. W., Hanna, D. H., Elella, M. H. A. and Mohamed, R. (2019). Encapsulation of bovine serum albumin within novel xanthan gum based hydrogel for protein delivery. *Materials Science and Engineering*, *94*, 1044–1055.

Santiago, L. G. and Castro, G. R. (2016). Novel technologies for the encapsulation of bioactive food compounds. *Current Opinion in Food Science, 7*(16), 78–85.

Shafaei, Z., Ghalandari, B., Divsalar, A. and Saboury, A. A. (2017). Controlled release nutrition delivery based intelligent and targeted nanoparticle. In: Grumezescu, A. M. (Ed.), *Nutrient Delivery*. Elsevier, Amsterdam, Netherlands, 329–367.

Sun, S., Song, Y. and Zheng, Q. (2009). Rheological behavior of heat-induced wheat gliadin gel. *Food Hydrocolloids, 23*(3), 1054–1056.

Tang, C. (2017). Emulsifying properties of soy proteins: A critical review with emphasis on the role of conformational flexibility. *Critical Review in Food Science and Nutrition, 57*(12), 2636–2679.

Tavares, G. M., Croguennec, T., Carvalho, A. F. and Bouhallab, S. (2014). Milk proteins as 68 encapsulation devices and delivery vehicles: Applications and trends. *Trends in Food Science and Technology, 37*(14), 5–20.

Tesch, S., Gerhards, C. and Schubert, H. (2002). Stabilization of emulsions by OSA starches. *Journal of Food Engineering, 54*, 167–174.

Tessier, F. J. and Niquet, C. (2007). The metabolic, nutritional and toxicological consequences of ingested dietary Maillard reaction products. *Journal of Social and Biological Structures, 201*(7), 199–207.

Thomas, L. V., Rahul, V. and Nair, P. D. (2017). Effect of stiffness of chitosan-hyaluronic acid dialdehyde hydrogels on the viability and growth of encapsulated chondrocytes. International Journal of Biological Macromolecules, *104*, 1925–1935.

Tomadoni, B., Capello, C., Valencia, G. A. and Gutierrez, T. J. (2020). Self-assembled proteins for food applications: A review. *Trends in Food Science and Technology*, 101, 1–16.

Uthayakumaran, S. and Wrigley, C. (2107). Wheat: Grain-quality characteristics and management of quality requirements. In: Wrigley, C. (Ed.), *Cereal Grains*. Elsevier, Amsterdam, Netherlands, pp. 91–134.

Van der Spiegel, M., Noor dam, M. Y. and Van der Fels-Klerx, H. (2013). Safety of novel protein sources (insects, microalgae, seaweed, duckweed, and rapeseed) and legislative aspects for their application food and feed production. *Comprehensive Reviews in Food Science and Food Safety, 2*(6), 662–678.

Wang, H., Yang, B. and Sun, H. (2017). Pectin-Chitosan polyelectrolyte complex nanoparticles for encapsulation and controlled release of Nisin. *American Journal of Polymer Science and Technology, 3*(5), 82–88.

Yao, M., Wu, J., Li, B., Xiao, H., McClements, D. J. and Li, I. (2017). Microencapsulation of Lactobacillus salivarious Li01 for enhanced storage viability and targeted delivery to gut microbiota. *Food Hydrocollides*, *72*, 228–236.

Yonekura, L. and Nagao, A. (2009). Soluble fibres inhibit carotenoid micellization in vitro and uptake by Caco-2 cells. *Bioscience, Biotechnology, and Biochemistry, 73*, 196–199.

Zhang, Z., Zhang, R. and McClements, D. J. (2016). Encapsulation of β-carotene in alginate-based hydrogel beads: Impact on physicochemical stability and bioaccessibility. *Food Hydrocolloids*, *61*, 1–10.

Zhao, J., Tian, Z. and Chen, L. (2015). Effects of deamidation on structure and functional properties of barley hordein. *Journal of Agriculture and Food Chemistry, 58*(21), 11448–11455.

CHAPTER 4

Recent Development in Nanoencapsulation of β-Sitosterol and γ-Oryzanol and Food Fortification

Monika Choudhary, Amarjeet Kaur, and Prabhjot Kaur

CONTENTS

4.1 INTRODUCTION

Encapsulation is a novel procedure wherein active agent derived from one substance is enclosed into the wall material of another substance. The encapsulated substance is termed as active, fill, payload, or interior phase (Wandrey et al., 2010; Fang and Bhandari, 2010 whereas the encapsulating substance is named as capsule, coating, carrier material, shell, membrane, matrix, or exterior phase (Nedovic et al., 2011). At times, encapsulation is considered as a method wherein the bioactive constituents are entirely encased, enclosed, and guarded by a physical blockade without any protruding of bioactive elements.

In another way, encapsulation may also be defined as enclosing solids, fluids, or volatile compounds inside minute capsules that eventually discharge active ingredients in a controlled manner in a certain environment (Desai and Park, 2005). The dimensions of the particles produced range from a few nanometers (nm) to a few millimeters (mm). The exterior phase surrounding the cells enables for quick and effective segregation of the maker cells and the metabolites, which was first established in the field of biotechnology to make production processes more efficient.

DOI: 10.1201/9781003259183-4

These technologies which were invented over 60 years ago have gained attention of the pharmaceutical sector (particularly medication and vaccine delivery), but they have also performed relevant functions in the food sector. In the current era, the food business has demanded that useful ingredients be added to products. Encapsulation has emerged as an efficient safety procedure for these chemicals which are in general extremely prone to industrial, ecological, and/or gastrointestinal (GIT) conditions. Moreover, organoleptic properties, such as color, flavor, texture, and preservation attributes, are all managed by functional chemicals. Encapsulated agents have a wide range of possible benefits in the processing sector. One of these is to conserve the stability of bioactive constituents during processing without any interaction with food structure and sustaining the activity during storage. This approach is sustained till the final delivery or consumption of the product by end users (Lesmes and McClements, 2009). As a result, the bioactive component will remain fully functional. Encapsulation may also be used to alter the physical features of the novel substance so as to (a) formulate it for easy usage, (b) to segregate the compounds of the amalgam, which may otherwise react with each other, and (c) to manufacture enough concentration and allocation of an active ingredient in an identical manner. So, encapsulation has diverse use in the food and health sector. This technique is an effective medium through which various bioactive components such as antioxidants (gamma-oryzanol (GO), lutein, lycopene), micronutrients, fatty acids, phytosterols (β-sitosterol), and living cells such as probiotics can be incorporated into the desired foods (de Vos et al., 2010).

β-Sitosterol having molecular composition comparable to cholesterol is found in plant cell membranes. This phytosterol is abundantly present in lipid-dense plant foods such as olive oil, legumes, nuts, and oilseeds. β-sitosterol, campesterol, and stigmasterol are three primary phytosterols that make up around 65%, 30%, and 3% of human nutrition obtained from plant sources, respectively. β-sitosterol is a powerful compound found in higher plants and absorbed by mammals via their food (Babu and Jayaraman, 2020). The European Food Safety Authority (EFSA) has indicated that an intake of 1.5–2.4 g phytosterol and/or stanols per day decreases blood cholesterol levels, which ultimately lowers the risk of heart disease. This fact has been authorized by the Food and Drug Administration. β-sitosterol is widely regarded as a safe and effective nutritional supplement having a lengthy history of use as a medicine with no negative side effects (Paniagua-Pérez et al., 2005). One exception is a genetic disorder known as "phytosterolaemia," which is caused by a mutation in the ABCG5/G8 proteins, which play a critical part in the protein pump that allows sterols to enter into intestinal cells and hepatocytes. β-sitosterol has various roles to play in the human body which have been well documented in scientific reports. These include activities such as antinociceptive (Santos et al., 1995), anxiolytic (López-Rubalcava et al., 2006), analgesic (Dighe et al., 2016), immunomodulatory (Fraile et al., 2012), antimicrobial (Ododo et al., 2016), anticancer (Sharmila and Sindhu, 2017), anti-inflammatory (Paniagua-Pérez et al., 2017), antioxidant (Ponnulakshmi et al., 2019), and antidiabetic (Babu et al., 2020), and protective effect against non-alcoholic fatty liver disease (NAFLD) (Feng et al., 2018), lipid-lowering effect (Yuan et al., 2019), hepatoprotective effect (Abdou et al., 2019), protective effect against respiratory diseases (Park et al., 2019), and wound-healing effect (Abbas et al., 2019). It also has anti-inflammatory and antidiabetic properties. In the present pace of research, scientific reports on β-sitosterol have been significantly reduced resulting in its present status as an "orphan phytosterol" (Bin Sayeed et al., 2016). GO is a constituent of crude rice bran oil's unsaponifiable matter obtained during the process of rice milling (Lakkakula et al., 2004). GO is a combination of triterpene alcohols and phytosterol esterified to ferulic acid. The primary components of GO are 24-methylene cycloartanyl ferulate, cycloartenyl ferulate, campesteryl, and sitosteryl ferulate (Xu and Godber, 1999). GO has been documented as an effective scavenger of free radicals such as diphenyl picrylhydrazyl (DPPH), hydroxyl (OH), and superoxide ($O_2^{-}\cdot$) ions. In an oxidative state, it can protect cells from lipid peroxidation. GO has also been shown to have anti-inflammatory properties. GO is extensively studied for its biological properties such as anti-inflammatory, antidiabetic, antiaging, anticholesterol, and menopausal symptom relief. Moreover, oryzanol is also a powerful free radical (iron-driven hydroxyl radical) inhibitor that has been linked to antioxidant action in lipid stabilization (Patel and Naik, 2004). Furthermore, GO's uses are limited due to its hydrophobic nature, less bioavailability, and low absorption. But, GO's solubility may be improved, which may further allow it to be used in more foods, cosmetics, and other items.

4.2 WALL AND COATING MATERIALS USED FOR ENCAPSULATION

A variety of coating materials are used for the nanoencapsulation of bioactive compounds. An ideal coating material must have few essential properties for their efficient use for the aforementioned purpose. These are good rheological characteristics at high concentrations, ability to spread or emulsify active ingredients and stabilize the resulting emulsion, the ability to seal and keep the active ingredient within its structure, the ability to release the solvent utilized during encapsulation, solubility in solvents that are permitted in the food sector, must be nonreactive and must have food-grade status at a low cost. Moreover, the types, sources, and qualities of coating material also influence the encapsulating materials used. Only a small number of shell materials have been licensed as GRAS (generally recognized as safe) for their potential exploitation in the food sector. Basically, encapsulating agents are biopolymeric in nature. Examples of such compounds are gums, proteins, lipids, and their derivatives.

In carbohydrates, the most popular components are starch, maltodextrin, carboxymethyl cellulose, guar gum, chia seed gum, gum arabic, and cyclodextrin. For instance, native rice starch and pre-gelatinized rice starch are employed for microencapsulation of GO via spray-drying, whereas microchannel emulsification with medium chain triglyceride oil is used to encapsulate GO via microchannel emulsification (Singh, 2016). These compounds are used to encapsulate diverse ingredients like polyunsaturated fatty acids (PUFA)-rich edible oil, enzymes, vitamins, bioactive peptides, and flavors, as these are tasteless, immensely accessible, and have exceptional core safety capacity (Jafari et al., 2008; King and Perry, 1976). To boost starch's emulsifying potential, side chains of lipophilic succinic acid are added to the starch. Modified starches have been proven to provide superior protection and emulsion stability than native and waxy starches. In the spray-drying process, gum acacia has been considered the most efficient coating material for encapsulating gaseous compounds. Gum arabic's use as a singular wall material for encapsulating active ingredients has been limited due to its higher cost and limited availability compared to others. To lower the cost, it is used in combination with other materials such as mesquite gum and maltodextrin. Other wall materials in this category (chitosan, cyclodextrins, and alginates) are relatively novel and are being considered for diverse applications in the health and food industry. One of the most significant benefits of employing gums as encapsulants is their capability to survive under GIT conditions, allowing them to serve as a prebiotic source for the gut microflora. Furthermore, they can be used to make hypoallergenic items for persons who have a protein allergy (Timilsena et al., 2020).

Protein is a great encapsulating material because of its superior physicochemical and functional qualities such as emulsifying capability, film formation, and gel-forming capability. Gelatin is a common shell matrix for making extremely stable soft gels of fish oil, linolenic fatty acid, and vitamin D. For several years, animal proteins (sodium caseinate and whey protein isolate) and plant proteins (soy and pea proteins) have been employed as wall components. Whey protein has been studied as an excellent wall material for encasing delicate flavors and essential fatty acids. And, it has also proven to be better in terms of encapsulation efficiency (up to 89.6%) than other proteins like soy protein (up to 75.9%) (Kim and Morr et al., 1996; Jimenez et al., 2004). Scientists also discovered that spray-dried microcapsules stay stable for 60 days with high water activity (aw = 0.74–0.90). One of the biggest drawbacks of utilizing protein as an encapsulant is that some people are allergic to it. Several people have claimed that soy proteins, wheat proteins (e.g. gluten), and peanut proteins are highly allergic. This not only restricts their use but also necessitates a manufacturer notification of their existence in the intended meals on the label. Furthermore, proteins are susceptible to physical (structure and shape) changes, and their usefulness as coatings is highly dependent on factors such as ionic strength, emulsion temperature, and pH (McClements, 2004; Damodaran, 2005). An amalgamation of protein-based coatings with other materials (carbohydrate-based corn syrup solids, lactose, and maltodextrin) has been shown as an efficient way to reduce the environmental shock on encapsulant performance (Sheu and Rosenberg, 1998; Keogh, 2001). For instance, GO was effectively encapsulated into zein nanoparticles in one research study. The parameters such as pH, zein concentration (0.3, 0.4, and 0.5% w/v) and percent GO loading (30, 40, and 50% w/w) were studied as manufacturing parameters. The GO stability during the storage period was assessed by measuring yield, encapsulation efficiency, particle size, and zeta potential. Scanning electron micrographs and transmission electron microscopes were used to examine the morphology of gamma-oryzanol-loaded zein nanoparticles (GOZNs). For each zein concentration with three loading levels, the particle sizes of the GOZNs were in the range of 93.24–350.93, 144.13–833.27, and 145.27–993.13 nm, respectively.

During storage (60 days) at 4°C and 18°C, nanoencapsulated GO retained a higher percentage of GO than nonencapsulated GO. In an in vitro investigation, GO was found to be released continuously in simulated stomach and intestinal fluids. This finding suggested that encapsulating insoluble GO with the hydrophobic zein protein could improve the stability of insoluble GO (Rodsuwan et al., 2021).

Next, lipids being hydrophobic in nature are frequently utilized to encase hydrophilic compounds. Glycerides, fatty acids, waxes, and phospholipids are among the lipids that have been studied for their capacity to encapsulate dietary actives (Wandrey et al., 2010). Despite the fact that fat-based encapsulation is still a comparatively new and budding industry, it is rapidly gaining popularity as a method of administering active ingredients. Solid lipid nanoparticles, nanostructure fat carriers, nanoliposomes, and nanoemulsions are the four main forms of lipid-based delivery methods (Fathi et al., 2012). For example, researchers encapsulated β-sitosterol using nanostructured lipid carriers (NLCs) and then added it to butter to improve its nutritional content and antioxidant property. β-Sitosterol-loaded NLCs were synthesized using a hot homogenization process, and the encapsulation efficiency and other properties of the optimized preparations were examined using laser light scattering and scanning electron microscopy procedures before being included into the butter. The particle size of the optimally prepared NLCs was 165 nm with a uniform size distribution (PDI = 0.21), a zeta potential of 13.5 mV, and an encapsulation efficiency (EE) of 99.96%. During a 3-month storage period, β-sitosterol-loaded NLCs in butter showed good stability. Furthermore, both the acid and peroxide levels in butter remained unchanged. Such studies paved the way for the development of value-added dairy products and the use of nanoencapsulation in the dairy sector (Bagherpour et al., 2017).

4.3 ENCAPSULATION TECHNIQUES

Top-down and bottom-up approaches are being used for nanoparticles formulation in nanoencapsulation. The employment of particular devices that permit a decrease in the size and structure formation for the required use of the nanomaterials being generated is part of the top-down strategy. Materials are created using a bottom-up technique, which involves the self-assembly and self-association of molecules that are influenced by a variety of factors such as pH, temperature, and ionic power. The nanonization in the first way is attained through energy utilization and the former utilizes physicochemical regulation for the accumulation of ions, molecules, and monomers so as to frame nanoencapsulation. Various techniques used for nanoencapsulation of β-sitosterol and γ-oryzanol have been discussed below.

Emulsification involves the use of an interface agent to mix two immiscible liquids (surfactants). The full lipid unites into the liquid phase in minute droplets that are disseminated into the continuous phase throughout this process. Components, production procedures, emulsion type, and a variety of other factors can all influence the size of the droplet (Solans and Solé, 2012). The nanoemulsion size is typically between 20 and 200 nm, indicating a limited size distribution. Many studies have shown that the majority of water-in-oil (W/O) or oil-in-water (O/W) nanoemulsions are formed using high-energy dispersion or emulsification. Also, few studies have investigated that low-energy emulsification and condensation are more productive in nanoemulsion formation (Helgeson, 2016). As this system is not at equilibrium state, nanoemulsions cannot be generated instantly. It is primarily developed using high-energy techniques and various mechanical equipment strategies, including ultrasonography, high-pressure homogenizers, and high-shear stirrers. The procedures described are not complicated and the high-energy input results in minor-size droplets. Nanoparticle creation necessitates a large quantity of energy making it inefficient. In contrast, in a basic emulsification process, only a modest amount of energy (about 0.1%) is required (Uluata et al., 2016). Low-energy emulsification procedures in contrast to high-energy emulsification produce lower particle sizes because they use the system's inherent chemical energy. It only requires a minimal amount of energy as well as some simple churning to generate minuscule particle sizes. Smaller particle sizes on the other hand can be achieved using both strategies which are dependent on the blend of variables and substantially on the system. It has also been proven that high-energy procedures allow for the formation of nanoemulsions with larger oil-to-surfactant ratios than low-energy techniques (Yang et al., 2012). Various researchers have attempted to encapsulate β-sitosterol and γ-oryzanol through this technique.

In one study, W/O emulsions containing various salt concentrations were generated using water-in-oil phase with a blend of β-sitosterol and γ-oryzanol. To make a solid organogel, the β-sitosterol and γ-oryzanol molecules get self-assembled into tubular microstructures in pure oil. The water molecules in the emulsion on the other hand adhered to β-sitosterol generating monohydrate crystals that prevented tubule development and resulted in an undesirable emulsion gel. While adding salt to water phase, a reduction was observed in water activity that prevented the development of sitosterol monohydrate crystals even up to one year of storage. The tubular microstructure was encouraged when the emulsions were made with less polar oils which considerably increased the stiffness of the emulsion gel. The study's key finding was that lowering the water activity and/or utilizing low-polarity oils could help enhance the creation of oryzanol and sitosterol tubular microstructure in the emulsion (Sawalha et al., 2012). In another study, straight-through microchannel emulsification was used to encapsulate β-sitosterol and γ-oryzanol at reasonably high concentrations in distinct food-grade O/W emulsions. The formation of monodisperse droplets with great encapsulation efficiency and stability of β-sitosterol and γ-oryzanol was a novel component of this investigation. In the continuous phase, milli-Q water was mixed with Tween 20 (1% w/w), while the dispersed phase included each of both compounds (oryzanol and sitosterol) in medium chain triglycerides at 0.5%–4% (w/w). Different concentrations of oryzanol and sitosterol were used to successfully generate droplets. Sauter mean diameters of β-sitosterol- and γ-oryzanol-loaded O/W emulsions (1% w/w) ranged from 26 to 28 m, with the span factor width below 0.21. These emulsions sustained the stability at 4°C and 25°C during storage. At this temperature, emulsions stabilized with Tween 20 had encapsulation efficiencies of more than 80% (Khalid et al., 2017). Similarly, one study was conducted to assess the effects of different emulsification variables on physicochemical parameters, release behavior, and encapsulation effectiveness of β-sitosterol. At first, 19 β-sitosterol emulsions were made with various investigational settings (shear 3000 and 6000 rpm; time 2.5 and 5 minutes; pressure 100, 200, and 300 bar; cycle 1–3). During investigations, the researchers observed that the smallest average droplet size (139 nm) among all samples was obtained in β-sitosterol emulsion prepared at 6,000 rpm, 5 minutes, 300 bar, and three cycles. The polydispersity index and turbidity were both affected by shear and time. The stability and efficiency of encapsulation procedure improved with an increase in homogenization pressure and cycle. Among all samples, the emulsion prepared at a pressure of 100 bar with one cycle showed the highest encapsulation effciency (81.7%) (Momeny et al., 2017).

Emulsification-solvent evaporation involves the emulsification of the polymer solution into an aqueous phase and evaporation of the polymer solvent generates polymer precipitation as nanospheres in this modified form of the solvent evaporation process (Kumari et al., 2010). Poly(lactide), poly(lactide-co-glicolide), cellulose acetate phthalate, cellulose, poly(hydroxybutyrate), and polycaprolactone are the most commonly employed polymers in this approach. To make nanoparticles, high-speed homogenization must be used. Furthermore, this approach proved to be the most efficient for obtaining particle sizes <250 nm (Chen et al., 2019). For instance, few researches highlighted the self-stabilizing characteristics of γ-oryzanol in intense colloidal systems like nanodispersions as well as a new method for encapsulating it in calorie-free preparations. During the experimentation, purified water (no emulsifier) or solutions of modified lecithin (ML), sorbitan monooleate (T80), sodium caseinate (SC), or polyoxyethylene (Abbas et al., 2019) as emulsifiers were used to generate concentrated gamma-oryzanol nanodispersions using the solvent displacement method. The nanodispersions contained γ-Oryzanol content (1.69–1.82 mg mL) and showed encapsulation efficiency (89.39%–98.78%). Also, similar polydispersity index (0.1), particle sizes (126–157 nm), and distribution (monomodal) were observed under optimal processing conditions. Furthermore, regardless of the formulation of the liquid phase injected, they possessed negative potentials showing that both γ-oryzanol and emulsifiers caused nanoparticle production. The dissociation of hydroxyl groups of caffeic acid and/or ferulic acid on γ-oryzanol structures contributes to produce 46.7 mV overall potential of emulsifier-free nanodispersions (Taarji et al., 2022). Similarly, another study compared various strategies for making γ-Oryzanol-containing polymeric nanoparticles. The strategies included nanoprecipitation, nanoemulsion template method, and emulsification-solvent evaporation technique. To make γ-Oryzanol nanoparticles, researchers employed two distinct ratios of γ-Oryzanol-polymer (ethyl cellulose) (1:2 and 1:4) and several surfactants as well as solvents. Solvent evaporation had been used successfully to manufacture γ-Oryzanol-loaded nanoparticles with desirable properties (Ghaderi et al., 2014).

The supercritical fluid technique utilizes carbon dioxide (CO_2) as a solvent alternative in several industrial and chemical processes due to its large availability and affordability. The properties of supercritical CO_2 lie between fluids and volatile compounds. This method similar to spray-drying encapsulates thermally sensitive substances. In supercritical fluid, bioactive compounds and polymers are dissolved. The fluid used in the spraying operation is evaporated in conjunction with the constituent precipitation (Pohlmann et al., 2010). Different encapsulated or coated wall materials can be used to make nanosuspensions (fluid or powder form) of active chemicals using appropriate drying techniques (spray- and freeze-drying) for encapsulation (Anandharamakrishnan, 2014a). Spray-drying is a quick, flexible, low-cost, and constant drying method for producing powder from desired solutions with the highest level of produce repeatability (Sosnik and Seremeta, 2015). Fluid solutions are sprayed into a hot air chamber through a set narrow size atomizer in these devices. On a lab scale, one of the biggest disadvantages of typical spray dryers is their low processing yield. To address this issue, one nanospray drier was invented in 2009 with an improved drying trait on a submicron scale (Lee et al., 2011). The quick effectiveness of this conventional spray drier allows it to transform aqueous solutions into fine particles. Even sensitive bioactive chemicals can benefit from the rapid drying process, which has no distracting effects. In various study areas, such as functional food preparation, cosmetics, and pharmaceuticals, nanospray-drying has been used for nanoencapsulation. For instance, one study was conducted to characterize encapsulated gamma-oryzanol powder (EOP) using maltodextrin and whey protein concentrate (WPC) as coating materials. Before spray-drying, rice bran oil (RBO) with 2% gamma-oryzanol was encapsulated using WPC (0.2%–1.0%) in combination with maltodextrin (10%). The EOP was further evaluated for its reconstitution and physicochemical properties. It was found that WPC used at various concentrations had an effect on EOP features. Low water activity and moisture were found in the EOP coated with 1% WPC, which had a high gamma-oryzanol content and encapsulation efficiency. The EOP emulsion contained a bigger particle size as well as lesser solubility index as compared to the fresh counterpart, but the reconstituted EOP showed high creaming stability. The study concluded that spray-dried encapsulated RBO (2% γ-oryzanol) coated with WPC (1%) and maltodextrin (10%) gave satisfactory results. So, EOP has an excellent potential as a food constituent in food as well as the pharmaceutical industry. This is the most significant advancement in spray-drying with the advantages of ultrafine powders having nanosize particle dispersion and high throughput (Wangdee et al., 2021). Similarly, freeze-drying (lyophilization) is also employed in various situations to conserve and stabilize thermal-sensitive active components and natural particles for a long duration (Ray et al., 2006). This distinctive drying process yields high-quality end products that may be reconstituted without difficulty. Freeze-drying has a number of drawbacks, including a lengthy operation time (≥20 hours), more energy consumption, and an open porous structure. In most cases, freeze-drying is used in conjunction with other production procedures to separate the nanoparticles. Another concern with freeze-drying is the formation of pores as a result of the ice sublimation process. However, this method is commonly used to remove moisture from nanocapsules without disturbing their structure. On the other hand, spray freeze-drying may serve as a suitable substitute to conventional ophilization for reduction in pore size and drying time (Ezhilarasi et al., 2013).

Other techniques used for the encapsulation procedure are coacervation, inclusion, complexation, and nanoprecipitation. Coacervation prevents the phase segregation of one or mixed polyeletrolyte present in solution, as well as the successive deposition of new coacervate phase surrounding the active constituent. Cross-linkers such as transglutaminase or glutaraldehyde may be utilized to boost the robustness of coacervate (Zuidam and Shimoni, 2010). Based on the type of polymer utilized, the method may be classified as simple (single polymer) or complex (various polymers). Complex formation is influenced by a number of elements like biopolymer type, concentration, pH, and ionic power along with the connection between the complex created and the type of polymer (McClements and Rao, 2011). Complex formation could have come from a considerable contribution of hydrogen bonding and hydrophobic interactions. Due to extremely high payloads (up to 99%) and the potential results of controlled release based on mechanical pressure and temperature, coacervation is a unique and promising encapsulation breakthrough. The main challenge with this technique is marketing of coacervated food components owing to crosslinking usage, which should be done in accordance with the country's laws. However, a great number of crosslinking enzymes are now being produced (Anandharamakrishnan, 2014b). This approach is mostly used to encapsulate volatile compounds (essential oils) in order to keep the noxious smell, taste, and odor at bay. With wall materials, this sort of encapsulation offers a high level of stability, efficiency, confinement, and preservation. However, this technique may allow encapsulation of certain components such as ß-lactogloglobulin as well as ß-cyclodextrin (Ezhilarasi et al., 2013). Nanoprecipitation is also known as solvent displacement, as it involves an impulsive emulsification of an organic interior phase having dissolved polymers, medicines, and organic solvents into liquid exterior phase (Stella et al., 2017). Polymer precipitation from an organic solution and organic solvent diffusion into liquid media are the mechanisms used in this process. It is claimed that solvent displacement can be used to create nanospheres and nanocapsules (Ezhilarasi et al., 2013).

Furthermore, EE is the most essential parameter for evaluating the efficiency of any encapsulation method. During the encapsulation procedure, EE assesses the proportion of bioactive ingredient entrapped into the inert core. In basic terms, the encapsulation process is most efficient when practically all of the agents are entirely entrapped into the matrix which increases the active's stability. To put it another way, the method is successful in protecting the bioactive core to a high degree. Encapsulation of practically all of the oil inside the matrix is preferable in PUFA-rich oil encapsulation, leaving a tiny amount of oil on the matrix's surface. Surface fat also known as free fat is a type of fat that is found on the skin's surface. As soon as the capsules are subjected to normal environment the surface oil begins to oxidize rapidly. As a result, elevated surface oil is associated with microcapsule off-flavor and worse customer approval of the resulting product. Although many emulsion products contain substantially more surface oil than is ideal for safe microcapsule preservation, the surface oil level must be <0.1%; (w/w) equation can be used to calculate the EE when oil is encapsulated.

$$\mathrm{EE} = W_t - W_s$$

EE stands for encapsulation efficiency, and W_t denotes the total amount of oil in the microcapsule, while W_s denotes the amount of oil on the microcapsule's surface. EE is commonly measured in practically all microencapsulation studies because it is one of the most essential metrics for determining the efficacy of the encapsulation process. As a greater EE is linked to better core stability and shelf life, it is preferable to raise the EE by selecting the finest feasible core to wall ratio, dehydration methods, homogenization process, along with other necessary variables.

4.4 FOOD FORTIFICATION

Nanoencapsulation is an excellent approach to administer more than one nutrient or micronutrients concurrently into the digestive system in a bioavailable form yet in stable and inert environment without causing any degradation. Nutrients are incorporated as premixes, concentrates, or encapsulated agents with altered physicochemical properties, conducive for incorporating into selected food vehicles with no reduction in their bioavailability (Sekhon, 2010). Nanoencapsulation can also improvise the sensory attributes of the fortified foods by removing the unwanted tastes and colors from the fortificants and by avoiding the interactions between food carrier and fortificants. This technique enables the fortificants to be administered in suitable forms that imitate the physical properties (color, size, shape, and appearance) of the particular food carriers. Nanoencapsulation allows for several nutrient fortifications of diverse staple foodstuffs with element sizes ranging from several hundred microns to millimeters. To avoid particle separation, which may end in probable low- or high-dosing, nutrients must be added in forms that either adheres to the food constituents or in premixes that resemble the particle size and density of the food. Effective food fortification through nanoencapsulation requires the equal distribution of nutrients that are not noticeable to the end users (Mohammadi et al., 2015; Ezhilarasi et al., 2013). Thus, the whole delivery system must resemble the food in sensory qualities, and must not modify its taste. For instance, vitamin sprays that spread nanodroplets are used to improve nutrient absorption. Through encapsulation techniques, probiotics and other items are delivered to the human system via iron and zinc nanostructured capsules (Chaudhry et al., 2008).

Nanoingredients and additives (micronutrients, antioxidants, antimicrobials, etc.) are now accessible in nutraceuticals and nutritional supplements (Han et al., 2015; Mohammadi et al., 2016; Mosquera et al., 2014) (Table 4.1). Commercial names for supplements are nanoceuticals and nutrition-be-nanotech. In this area, nanocochleates and nano-sized powders are also used to increase nutrient absorption in a protective environment. Nanocochleates (50 nm in size) such as phosphatidylserine carrier derived from soya bean are considered safe (GRAS) for nanomicelle-based carriers in nutraceuticals. Calcium ions are added to small phosphatidylserine vesicles to protect them. Several micronutrients and antioxidants are said to be protected by the nanocochleate system during manufacturing

Table 4.1 Application of Nanoencapsulation in Nutraceuticals

Encapsulating Agents	Application as Nutraceutical
Alginate and chitosan	Target delivery of vitamins A, D, and E3, co-enzyme Q10, lutein, lycopene, β-carotene, and omega-3-fatty acids (Dalzon et al., 2016)
Nanosize ingredients and additives	Improves taste, absorption, and bioavailability (Malhotra et al., 2015)
Incorporated carriers consist of lycopene, beta-carotenes, and phytosterols	To prevent cholesterol accumulation (Ardalani et al., 2017)
Nanoselenium	Green tea helps in the uptake of selenium (Ardalani et al., 2017)
Nanocochleates	Stabilizes micronutrients and improves nutritional quality (Reed et al., 2014)
HydraCel	Enhances water absorption and nutrients in the blood and reduces surface tension (Pathakoti et al., 2017)
α-lactalbumin (nanoencapsulated)	Nutrient carriers (He et al., 2019)
Casein micelles	Delivers sensitive products (Jain et al., 2013)
Dextrin's and curcumin	Bioactive products (Głab and Boratynski, 2017)
Gelated nanotubes and nanoparticles	Viscosifying agent (Gao and Dave, 2007)
Nanocapsule infusion of plant-based steroids	Replaces meat cholesterol (Kedar et al., 2010)
Nanosize powders	Increases nutrient absorption (Bharti et al., 2015)
Nanocochleates	Nutrients are administered without causing change in color and taste (Chatterjee, 2016)
Nanodroplets (vitamin sprays)	Improved absorption of active molecules (Pradhan et al., 2015)

and storage. Nanotubes developed from hydrolyzed lactalbumin are another application that could provide a novel carrier for nanoencapsulation of several nutrients and nutraceuticals (Graveland-Bikker and De Kruif, 2006). Moreover, the nanodelivery-mediated nutraceuticals, synergistic action of nanomaterials in food protection, and the application of nanosensors in smart food packaging for monitoring the quality of the stored foods are being employed for assessing the impact of nanomaterials in biological systems (Table 4.2). For instance, an approach using bionanoen capsulated quercetin to increase the shelf life of tomatoes has been documented, which may be implemented in other fruits and vegetables as well. In addition, some commonly marketed nanoencapsulation-based products such as Aquanova, Neosino capsules, Nanogreen tea, Nutralease, and Canola active oil are gaining popularity among health conscious consumers. Likewise, many nano-processed products are being commonly sold in China, Japan, Australia, and the United States like nanoteas, nanoceuticals, slim shakes, oat nutritional drinks, fortified fruit juices, and nanocapsules containing tuna fish oil in breads.

Table 4.2 Commercialized Fortified Food Products Using Nanoencapsulation

Product Type	Product Name	Encapsulation Material Used	Application
Supplemented functional drink	ASAP Health Max 30 and other silver products	Silver NPs	Antibacterial (Center for Food Safety, 2017)
	MaatShop Crystal Clear Nano Silver		
	Silvix3		Antibacterial and antifungal effects as a surface disinfectant (Center for Food Safety, 2017)
	Nano Silver Sol		Antibacterial activity and sterilization (Center for Food Safety, 2017)
	MesoSilver		Highest bioavailability (Center for Food Safety, 2017)
	Colloidal Silver Liquid		Maintains immunity and defense for natural healing (Center for Food Safety, 2017)
	Nano Colloidal Silver	Silver NPs	Quality control (Center for Food Safety, 2017)
	Sovereign Silver	Silver hydrosols	
Sustain beverage	Novasol	Nanomicelle	Delivers antioxidant into food and functional beverages (Center for Food Safety, 2017)
Health benefits for toddlers	Nano Silver Baby Milk Bottle	Nanosilver	Storage (Center for Food Safety, 2017)
Fortified Jambu Juice	Daily Vitamin Boost	Silver nanoparticle	Rich source of 22 essential vitamins and minerals (Center for Food Safety, 2017)
Beverage	Oat Chocolate and Oat Vanilla Nutritional Drink	300 nm of iron particles	Enhances bioavailability (Walia et al., 2019)
Nanocochleate	Bioral	Calcium ions in GRAS phosphatidylserine from soya bean	A defensive administering system for antioxidants against enzymatic degradation (Chaudhry et al., 2008)

REFERENCES

Abbas, M.M., Al-Rawi, N., Abbas, M.A. and Al-Khateeb, I. (2019). Naringenin potentiated β-sitosterol healing effect on the scratch wound assay. *Research in Pharmaceutical Sciences, 14*(6), 566.

Abdou, E.M., Fayed, M.A., Helal, D. and Ahmed, K.A. (2019). Assessment of the hepatoprotective effect of developed lipid-polymer hybrid nanoparticles (LPHNPs) encapsulating naturally extracted β-Sitosterol against CCl 4 induced hepatotoxicity in rats. *Scientific Reports, 9*(1), 1–14.

Anandharamakrishnan, C. (2014a). Drying techniques for nanoencapsulation. In: *Techniques for Nanoencapsulation of Food Ingredients*. Springer, New York, pp. 51–60.

Anandharamakrishnan, C. (2014b). Nanoencapsulation of food bioactive compounds. In: *Techniques for Nanoencapsulation of Food Ingredients*. Springer, New York, pp. 1–6.

Ardalani, H., Avan, A. and Ghayour-mobarhan, M. (2017). Podophyllotoxin: A novel potential natural anticancer agent. *Avicenna Journal of Phytomedicine, 7*, 285–294.

Babu, S. and Jayaraman, S. (2020). An update on β-sitosterol: A potential herbal nutraceutical for diabetic management. *Biomedicine and Pharmacotherapy, 131*, 110702.

Babu, S., Krishnan, M., Rajagopal, P., Periyasamy, V., Veeraraghavan, V., Govindan, R. and Jayaraman, S. (2020). Beta-sitosterol attenuates insulin resistance in adipose tissue via IRS-1/Akt mediated insulin signaling in high fat diet and sucrose induced type-2 diabetic rats. *European Journal of Pharmacology, 873*, 173004.

Bagherpour, S., Alizadeh, A., Ghanbarzadeh, S., Mohammadi, M. and Hamishehkar, H. (2017). Preparation and characterization of Betasitosterol-loaded nanostructured lipid carriers for butter enrichment. *Food Bioscience, 20*, 51–55.

Bharti, C., Nagaich, U., Pal, A.K. and Gulati, N. (2015). Mesoporous silica nanoparticles in target drug delivery system: A review. *International Journal Pharmaceutical Investigation, 5,* 124–133.

Bin Sayeed, M.S., Karim, S.M.R., Sharmin, T. and Morshed, M.M. (2016). Critical analysis on characterization, systemic effect, and therapeutic potential of beta-sitosterol: A plant-derived orphan phytosterol. *Medicines, 3*(4), 29.

Center for Food Safety, (2017). (www.cente rforf oodsa fety.org) Retrieved from http://salsa3.salsalabs.com/o/1881/p/salsa/web/common/public/content?content_item_KEY=14112%20#showJoin.

Chatterjee, B. (2016). Synthetic Lycopene: The future but unaware fact. *International Journal of Clinical and Biomedical Research, 2,* 14–18.

Chaudhry, Q., Scotter, M., Blackburn, J., Ross, B., Boxall, A., Castle, L., Aitken, R. and Watkins, R. (2008). Applications and implications of nanotechnologies for the food sector. *Food Additives and Contaminants, 25*(3), 241–258.

Chen, X.P., Li, Y., Zhang, Y. and Li, G.W. (2019). Formulation, characterization and evaluation of curcumin-loaded PLGA-TPGS nanoparticles for liver cancer treatment. *Drug Design, Development and Therapy, 13*, 3569.

Dalzon, B., Lebas, C., Jimenez, G., Gutjahr, A., Terrat, C., Exposito, J., Verrier, B. and Lethias, C. (2016). Poly (lactic acid) nanoparticles targeting α5β1 integrin as vaccine delivery vehicle: A prospective study. *PLoS One, 11,* 1–18.

Damodaran, S. (2005). Protein stabilization of emulsions and foams. *Journal of Food Science, 70*(3), R54–R66.

de Vos, P., Faas, M.M., Spasojevic, M. and Sikkema, J. (2010). Encapsulation for preservation of functionality and targeted delivery of bioactive food components. *International Dairy Journal, 20*(4), 292–302.

Desai, K.G.H. and Park, H.J. (2005). Recent developments in microencapsulation of food ingredients. *Drying Technology, 23*, 1361–1394.

Dighe, S., Kuchekar, B. and Wankhede, S. (2016). Analgesic and anti-inflammatory activity of β-sitosterol isolated from leaves of Oxalis corniculata. *International Journal of Clinical Pharmacology Research, 6*(3), 109–113.

Ezhilarasi, P.N., Karthik, P., Chhanwal, N. and Anandharamakrishnan, C. (2013). Nanoencapsulation techniques for food bioactive components: A review. *Food and Bioprocess Technology, 6*(3), 628–647.

Fang, Z. and Bhandari, B. (2010). Encapsulation of polyphenols: A review. *Trends in Food Science & Technology, 21*(10), 510–523.

Fathi, M., Mozafari, M.R. and Mohebbi, M. (2012). Nanoencapsulation of food ingredients using lipid based delivery systems. *Trends in Food Science & Technology, 23*(1), 13–27.

Feng, S., Gan, L., Yang, C.S., Liu, A.B., Lu, W., Shao, P., Dai, Z., Sun, P. and Luo, Z. (2018). Effects of stigmasterol and β-sitosterol on nonalcoholic fatty liver disease in a mouse model: A lipidomic analysis. *Journal of Agricultural and Food Chemistry, 66*(13), 3417–3425.

Fraile, L., Crisci, E., Córdoba, L., Navarro, M.A., Osada, J. and Montoya, M. (2012). Immunomodulatory properties of beta-sitosterol in pig immune responses. *International Immunopharmacology, 13*(3), 316–321.

Gao, X. and Dave, S.R. (2007). *Advances in Experimental Medicine and Biology.* Springer, New York.

Ghaderi, S., Ghanbarzadeh, S. and Hamishehkar, H. (2014). Evaluation of different methods to produce nanoparticle containing gammaoryzanol for potential use in food fortification. *Pharmaceutical Sciences, 20*(4), 130–134.

Głab, T.K. and Boratynski, J. (2017). Potential of casein as a carrier for biologically active. *Topics in Current Chemistry, 375,* 1–20.

Graveland-Bikker, J.F. and De Kruif, C.G. (2006). Unique milk protein based nanotubes: Food and nanotechnology meet. *Trends in Food Science & Technology, 17*(5), 196–203.

Han, H.J., Lee, J.S., Park, S.A., Ahn, J.B. and Lee, H.G. (2015). Extraction optimization and nanoencapsulation of jujube pulp and seed for enhancing antioxidant activity. *Colloids and Surfaces B: Biointerfaces, 130*, 93–100.

He, X., Deng, H. and Hwang, H. (2019). The current application of nanotechnology in food and agriculture. *Journal of Food and Drug Analysis, 27,* 1–21.

Helgeson, M.E. (2016). Colloidal behavior of nanoemulsions: Interactions, structure, and rheology. *Current Opinion in Colloid & Interface Science, 25*, 39–50.

Jafari, S.M., Assadpoor, E., He, Y. and Bhandari, B. (2008). Encapsulation efficiency of food flavours and oils during spray drying. *Drying Technology, 26*(7), 816–835.

Jain, N.K., Mishra, V., Mehra, N.K., Jain, N.K., Mishra, V. and Mehra, N.K. (2013). Targeted drug delivery to macrophages. *Expert Opinion on Drug Delivery, 10,* 353–367.

Jimenez, M., Garcia, H.S. and Beristain, C.I. (2004). Spray-drying microencapsulation and oxidative stability of conjugated linoleic acid. *European Food Research and Technology, 219*(6), 588–592.

Kedar, U., Phutane, P., Shidhaye, S. and Kadam, V. (2010). Advances in polymeric micelles for drug delivery and tumor targeting. *Nanomedicine Nanotechnology Biology Medicine, 6,* 714–729.

Keogh, M.K., O'kennedy, B.T., Kelly, J., Auty, M.A., Kelly, P.M., Fureby, A. and Haahr, A.M. (2001). Stability to oxidation of spray-dried fish oil powder microencapsulated using milk ingredients. *Journal of Food Science, 66*(2), 217–224.

Khalid, N., Kobayashi, I., Neves, M.A., Uemura, K., Nakajima, M. and Nabetani, H. (2017). Encapsulation of β-sitosterol plus γ-oryzanol in O/W emulsions: Formulation characteristics and stability evaluation with microchannel emulsification. *Food and Bioproducts Processing, 102*, 222–232.

Kim, Y.D. and Morr, C.V. (1996). Microencapsulation properties of gum arabic and several food proteins: Spray-dried orange oil emulsion particles. *Journal of Agricultural and Food Chemistry, 44*(5), 1314–1320.

King, W. and Perry, P. (1976). Modified starch encapsulating agents offer superior emulsification, film forming, and low surface oil [Incorporation into dry food systems]. *Food Product Development, 10*, 54–57.

Kumari, A., Yadav, S.K. and Yadav, S.C. (2010). Biodegradable polymeric nanoparticles based drug delivery systems. *Colloids and Surfaces B: Biointerfaces, 75*(1), 1–18.

Lakkakula, N.R., Lima, M. and Walker, T. (2004). Rice bran stabilization and rice bran oil extraction using ohmic heating. *Bioresource Technology, 92*(2), 157–161.

Lee, S.H., Heng, D., Ng, W.K., Chan, H.K. and Tan, R.B. (2011). Nano spray drying: A novel method for preparing protein nanoparticles for protein therapy. *International Journal of Pharmaceutics, 403*(1–2), 192–200.

Lesmes, U. and McClements, D.J. (2009). Structure–function relationships to guide rational design and fabrication of particulate food delivery systems. *Trends in Food Science & Technology, 20*(10), 448–457.

López-Rubalcava, C., Piña-Medina, B., Estrada-Reyes, R., Heinze, G. and Martínez-Vázquez, M. (2006). Anxiolytic-like actions of the hexane extract from leaves of Annona cherimolia in two anxiety paradigms: Possible involvement of the GABA/benzodiazepine receptor complex. *Life Sciences, 78*(7), 730–737.

Malhotra, B., Keshwani, A.N.U. and Kharkwal, H. (2015). Natural polymer based cling films for food packaging. *International Journal of Pharmacy and Pharmaceutical Sciences, 7*, 10–18.

McClements, D.J. (2004). Protein-stabilized emulsions. *Current Opinion in Colloid & Interface Science, 9*(5), 305–313.

McClements, D.J. and Rao, J. (2011). Food-grade nanoemulsions: Formulation, fabrication, properties, performance, biological fate, and potential toxicity. *Critical Reviews in Food Science and Nutrition, 51*(4), 285–330.

Mohammadi, A., Hashemi, M. and Hosseini, S.M. (2015). Nanoencapsulation of Zataria multiflora essential oil preparation and characterization with enhanced antifungal activity for controlling Botrytis cinerea, the causal agent of gray mould disease. *Innovative Food Science & Emerging Technologies, 28*, 73–80.

Mohammadi, A., Jafari, S.M., Assadpour, E. and Esfanjani, A.F. (2016). Nano-encapsulation of olive leaf phenolic compounds through WPC–pectin complexes and evaluating their release rate. *International Journal of Biological Macromolecules, 82*, 816–822.

Momeny, E., Mirhosseini, H. and Sarker, M.Z.I. (2017). Effect of medium-high energy emulsification condition on physicochemical properties of β-sitosterol multiple emulsion. *Food and Bioprocess Technology, 10*(9), 1642–1654.

Mosquera, M., Giménez, B., da Silva, I.M., Boelter, J.F., Montero, P., Gómez-Guillén, M.C. and Brandelli, A. (2014). Nanoencapsulation of an active peptidic fraction from sea bream scales collagen. *Food Chemistry, 156*, 144–150.

Nedovic, V., Kalusevic, A., Manojlovic, V., Levic, S. and Bugarski, B. (2011). An overview of encapsulation technologies for food applications. *Procedia Food Science, 1*, 1806–1815.

Ododo, M.M., Choudhury, M.K. and Dekebo, A.H. (2016). Structure elucidation of β-sitosterol with antibacterial activity from the root bark of Malva parviflora. *Springer Plus, 5*(1), 1–11.

Paniagua-Pérez, R., Madrigal-Bujaidar, E., Reyes-Cadena, S., Molina-Jasso, D., Gallaga, J.P., Silva-Miranda, A., Velazco, O., Hernández, N. and Chamorro, G. (2005). Genotoxic and cytotoxic studies of beta- sitosterol and pteropodine in mouse. *Journal of Biomedicine and Biotechnology, 2005*(3), 242–247.

Paniagua-Pérez, R., Flores-Mondragón, G., Reyes-Legorreta, C., Herrera-López, B., Cervantes-Hernández, I., Madrigal-Santillán, O., Morales-González, J.A., Álvarez-González, I. and Madrigal-Bujaidar, E. (2017). Evaluation of the anti-inflammatory capacity of beta-sitosterol in rodent assays. *African Journal of Traditional, Complementary and Alternative Medicines, 14*(1), 123–130.

Park, Y.J., Bang, I.J., Jeong, M.H., Kim, H.R., Lee, D.E., Kwak, J.H. and Chung, K.H. (2019). Effects of β-sitosterol from corn silk on TGF-β1-induced epithelial–mesenchymal transition in lung alveolar epithelial cells. *Journal of Agricultural and Food Chemistry, 67*(35), 9789–9795.

Patel, M. and Naik, S.N. (2004). Gamma-oryzanol from rice bran oil: A review. *Journal of Scientific & Industrial Research, 63*(7), 569–578.

Pathakoti, K., Manubolu, M. and Hwang, H.M. (2017). Nanostructures: current uses and future applications in food science. *Journal of Food and Drug Analysis, 25,* 245–253.

Pohlmann, A.R., Schaffazick, S.R., Creczynski-Pasa, T.B. and Guterres, S.S. (2010). Preparation of drug-loaded polymeric nanoparticles and evaluation of the antioxidant activity against lipid peroxidation. In: Armstrong, D. (Ed.), *Free Radicals and Antioxidant Protocols*. Humana Press, Totowa, NJ, pp. 109–121.

Ponnulakshmi, R., Shyamaladevi, B., Vijayalakshmi, P. and Selvaraj, J. (2019). In silico and in vivo analysis to identify the antidiabetic activity of beta sitosterol in adipose tissue of high fat diet and sucrose induced type-2 diabetic experimental rats. *Toxicology Mechanisms and Methods, 29*(4), 276–290.

Pradhan, G.C., Dash, S. and Swain S.K. (2015). Barrier properties of nano silicon carbide designed chitosan nanocomposites. *Carbohydrate Polymers, 134,* 60–65.

Ray, S., Quek, S.Y., Easteal, A. and Chen, X.D. (2006). The potential use of polymer-clay nanocomposites in food packaging. *International Journal of Food Engineering, 2*(4), 1–13.

Reed, R.B., Faust, J.J., Yang, Y., Doudrick, K., Capco, D.G., Hristovski, K. and Westerho, P. (2014). Characterization of nanomaterials in metal colloid-containing dietary supplement drinks and assessment of their potential interactions after ingestion. *ACS Sustainable Chemistry & Engineering, 2,* 1616–1624.

Rodsuwan, U., Pithanthanakul, U., Thisayakorn, K., Uttapap, D., Boonpisuttinant, K., Vatanyoopaisarn, S., Thumthanaruk, B. and Rungsardthong, V. (2021). Preparation and characterization of gamma oryzanol loaded zein nanoparticles and its improved stability. *Food Science & Nutrition, 9*(2), 616–624.

Santos, A.R., Niero, R., Valdir Filho, C., Yunes, R.A., Pizzolatti, M.G., Delle Monache, F. and Calixtro, J.B. (1995). Antinociceptive properties of steroids isolated from Phyllanthus corcovadensis in mice. *Planta Medica, 61*(4), 329–332.

Sawalha, H., den Adel, R., Venema, P., Bot, A., Flöter, E. and van der Linden, E. (2012). Organogel-emulsions with mixtures of β-sitosterol and γ-oryzanol: influence of water activity and type of oil phase on gelling capability. *Journal of Agricultural and Food Chemistry, 60*(13), 3462–3470.

Sekhon, B.S. (2010). Food nanotechnology: An overview. *Nanotechnology, Science and Applications, 3,* 1.

Sharmila, R. and Sindhu, G. (2017). Evaluate the Antigenotoxicity and anticancer role of β-Sitosterol by determining oxidative DNA damage and the expression of phosphorylated mitogen-activated protein Kinases', C-fos, C-Jun, and endothelial growth factor receptor. *Pharmacognosy Magazine, 13*(49), 95.

Sheu, T.Y. and Rosenberg, M. (1998). Microstructure of microcapsules consisting of whey proteins and carbohydrates. *Journal of Food Science, 63*(3), 491–494.

Singh, H. (2016). Nanotechnology applications in functional foods; opportunities and challenges. *Preventive Nutrition and Food Science, 21*(1), 1.

Solans, C. and Solé, I. (2012). Nano-emulsions: formation by low-energy methods. *Current Opinion in Colloid & Interface Science, 17*(5), 246–254.

Sosnik, A. and Seremeta, K.P. (2015). Advantages and challenges of the spray-drying technology for the production of pure drug particles and drug-loaded polymeric carriers. *Advances in Colloid and Interface Science, 223*, 40–54.

Stella, B., Marengo, A. and Arpicco, S. (2017). Nanoparticles: an overview of the preparation methods from preformed polymers. *Istituto Lombardo-Accademia di Scienze e Lettere-Incontri di Studio.*

Taarji, N., Bouhoute, M., Kobayashi, I., Tominaga, K., Isoda, H. and Nakajima, M. (2022). Preparation and characterization of concentrated γ-Oryzanol nanodispersions by solvent displacement method: Effect of processing conditions on nanoparticles formation. *Food Hydrocolloids, 123*, 107161.

Timilsena, Y.P., Haque, M.A. and Adhikari, B. (2020). Encapsulation in the food industry: A brief historical overview to recent developments. *Food and Nutrition Sciences, 11*(06), 481.

Uluata, S., McClements, D.J. and Decker, E.A. (2016). Riboflavin-induced oxidation in fish oil-in-water emulsions: Impact of particle size and optical transparency. *Food Chemistry, 213*, 457–461.

Walia, N., Dasgupta, N., Ranjan, S., Ramalingam, C. and Gandhi, M. (2019). Food—rade nanoencapsulation of vitamins. *Environmental Chemical Letters, 17,* 991–1002.

Wandrey, C., Bartkowiak, A. and Harding, S.E. (2010). Materials for encapsulation. In: Zuidam, N.J. and Nedovic, V. (Eds.), *Encapsulation Technologies for Active Food Ingredients and Food Processing*. Springer, New York, pp. 31–100.

Wangdee, K., Decker, E.A. and Onsaard, E. (2021). Characterization of encapsulated γ-oryzanol powder by spray drying using whey protein and maltodextrin as wall materials. *Journal of Food Science and Technology, 59*(1), 355–365.

Xu, Z. and Godber, J.S. (1999). Purification and identification of components of γ-oryzanol in rice bran oil. *Journal of Agricultural and Food Chemistry, 47*(7), 2724–2728.

Yang, Y., Marshall-Breton, C., Leser, M.E., Sher, A.A. and McClements, D.J. (2012). Fabrication of ultrafine edible emulsions: Comparison of high-energy and low-energy homogenization methods. *Food Hydrocolloids, 29*(2), 398–406.

Yuan, C., Zhang, X., Long, X., Jin, J. and Jin, R. (2019). Effect of β-sitosterol self-microemulsion and β-sitosterol ester with linoleic acid on lipid-lowering in hyperlipidemic mice. *Lipids in Health and Disease, 18*(1), 1–11.

Zuidam, N.J. and Shimoni, E. (2010). Overview of microencapsulates for use in food products or processes and methods to make them. In: Zuidam, N.J. and Nedovic, V. (Eds.), *Encapsulation Technologies for Active Food Ingredients and Food Processing*. Springer, New York, pp. 3–29.

CHAPTER 5

Characterization of Nanocomposites for Curcumin

Anindita Paul, Rohan Sarkar, Dinesh Kumar Yadav, Pushpendra Koli, Aditi Kundu, and Supradip Saha

CONTENTS

DOI: 10.1201/9781003259183-5

5.1 INTRODUCTION

Curcumin is the main bioactive component of turmeric, accounting for 2%–5% in general. It comprises different congeners that are grouped together as curcuminoids. Curcuminoids are composed of three major compounds namely curcumin (C, 70%–85%), demethoxycurcumin (DC, 15%–20%), and bisdemethoxycurcumin (BDC, 2%–5%) (Nelson et al., 2017). Curcuminoids are often used for the treatment of coughs, liver disorders, fevers, jaundice, wounds, parasitic skin diseases, eczema, anemia, inflammatory joints, urinary diseases, and bacterial and viral infections (Li et al., 2007; Wilken et al., 2011; Koli and Yadav, 2017). Curcumin research encompasses a broad array of disciplines, including medicinal chemistry, pharmacology, therapeutics, clinical trials, and molecular biology. Curcumin has become a "model molecule" in the domains of pharmacology and molecular biology studies due to its multidimensional functional properties (Liang, 2013).

Chemically, curcumin is composed of two phenolic rings linked by two α, β-unsaturated carbonyl system (Figure 5.1). Curcumin is a polyphenol that is hydrophobic. It is insoluble in water although it has high solubility in polar solvents. Curcumin is classified as a diarylheptanoid because it contains two aromatic *O*-methoxyphenolic groups. The turmeric powder was blended and with gentle agitation using the nonpolar solvent, oil was extracted after being filtered and concentrated under *vacuum*. A polar solvent was used to extract the defatted powder. Evaporation of solvent produces orange-yellow powder which is curcuminoid (Figure 5.2) (Yadav et al., 2017). Curcumin extraction from dried rhizomes was carried out by vacuous extraction techniques such as soxhlet, ultrasonic, microwave, and supercritical CO_2 extraction. After comparing percent yield (%) and duration, microwave was shown to be a more efficient extraction technique than Soxhlet (Wakte et al., 2011).

Curcumin has long been used as a spice and as a medicinal herb (Bhat et al., 2019). It has a wide range of bioactivities including anti-inflammatory, hepatoprotective, antiviral, antioxidant, antifungal, antibacterial, cardioprotective, and anticancer properties. (Amalraj et al., 2017) (Figure 5.3) Also it has a wide spectrum of therapeutic properties and is widely used in Ayurvedic medicine. It has recently been discovered to have antitumor properties in several breakthrough studies. Curcumin has been shown to have an antiproliferative impact in a variety of cancers and inhibiting nuclear factor-kb. Curcumin also has an effect on many growth

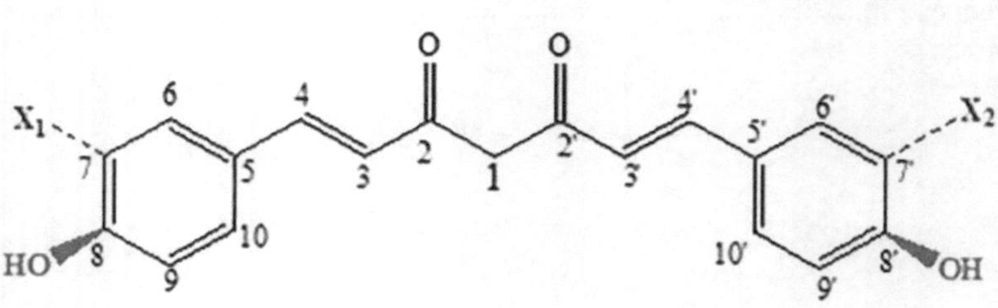

Figure 5.1 General framework of curcuminoids.

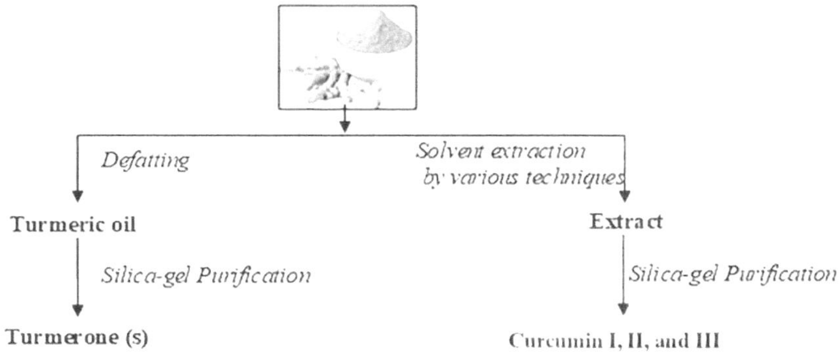

Figure 5.2 Extraction of curcuminoids and turmeric oil.

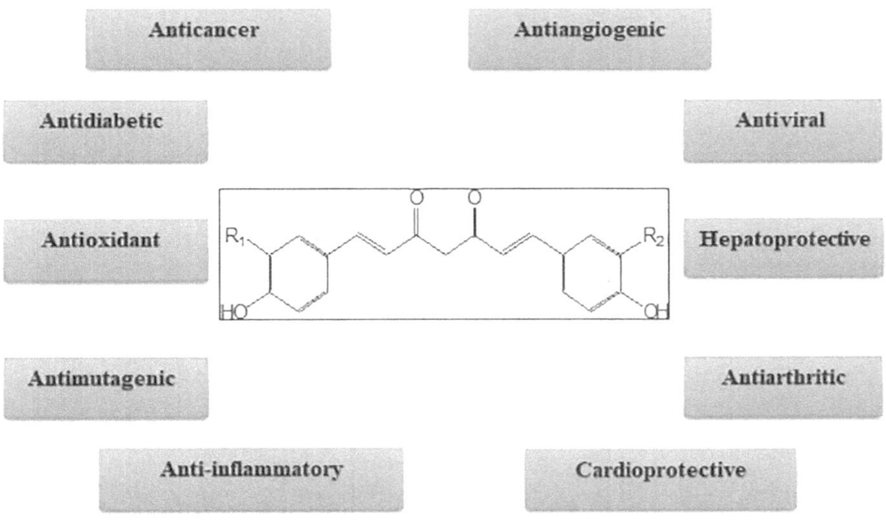

Figure 5.3 Biological activities of curcuminoids.

and transcription factors involved in tumor cell proliferation, mutagenesis process, apoptosis, metastatic cell cycle regulation, and angiogenesis. According to Wilken et al. (2011), curcumin has a chemotherapeutic effect in a variety of cancers. It was reported that curcumin inhibits the MDM2 protein, which is involved in human prostate cancer cell growth (Li et al., 2007). Curcumin has been shown to possess antioxidant and neuroprotective functions in rats with homocysteine neurotoxicity (Ataie et al., 2010). Furthermore, it was also found to be useful in anti-inflammatory treatments for diabetics. Analogues of curcumin were found to have better pharmacokinetic and anti-inflammatory properties than curcumin itself (Pan et al., 2012).

Safety and beneficial effects of curcumin consumption have been reported by several clinical studies. Curcumin has been the subject of extensive research for several decades due to its potential health benefits (Prasad et al., 2014). Several attempts have been made to incorporate the molecules as a nutraceutical agent into commercial food products (Ahmed et al., 2012). Though curcumin has immense health benefit effects, several challenges are associated with curcumin delivery. Poor aqueous solubility, chemical instability, short half-life, oral bioavailability, poor absorption in the gastrointestinal tract, substantial presystemic metabolism, quick systemic clearance, and low intrinsic activity are all key limitations of curcumin-based functional foodstuffs (Anand et al., 2007). Various studies have been carried out in which a variety of nanocarriers have been attempted for targeted delivery of curcumin. The major goal is to increase stability and bioaccessibility with improved efficacy. There are several ways to improve curcumin oral bioavailability including increasing aqueous solubility, increasing gastric retention time, increasing intestinal permeability, and improving the ability to cross lipid-rich membranes and also improving the lymphatic uptake (Figure 5.4) (Anand et al., 2007).

Several delivery strategies were explored for the development of curcumin nanoformulation, including nanoemulsions, nanosuspensions, solid lipid nanoparticles, liposomes, polymeric micelles, polymeric nanocomposites, inclusion

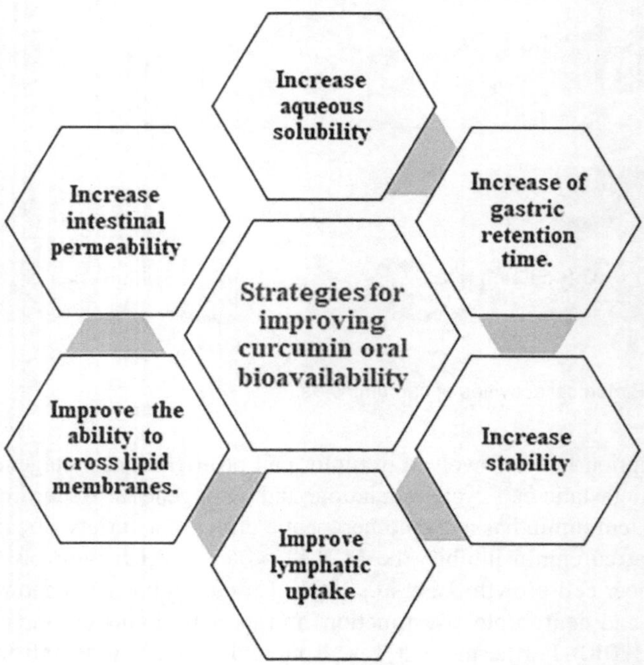

Figure 5.4 Approaches to enhance bioavailability of curcumin.

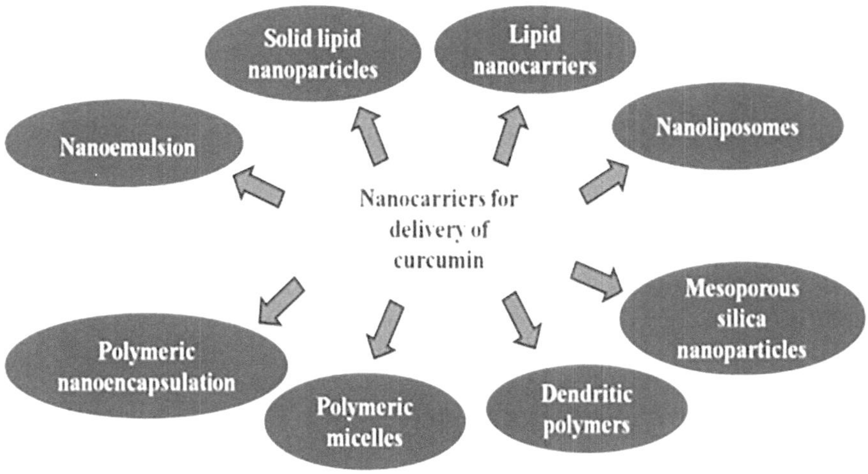

Figure 5.5 Nanoformulations developed for the improved delivery of curcumin.

complexes with cyclodextrins, dendrimers, and gold nanoparticles (Figure 5.5). A nanocomposite is a multicomponent solid substance with many phases at least one of which has a nanoscale dimension. Unique qualities of nanocomposites, such as their small size, higher surface-area-to-volume ratio, and quantum size effects have undoubtedly enhanced their usage. Theracurmin® (curcumin nanoparticles) was reported to enhance absorption in rats with a bioavailability equivalent to 40 times that of curcumin powder (Sasaki et al., 2011). Cheng et al. (2013) synthesized nano-sized curcumin and tested its bioavailability for improvement. It was observed that there was a greater concentration of curcumin in the plasma. Furthermore, mean residence time of the formulation was found to be longer than curcumin. Curcumin nanoformulations were found to enhance the bioavailability in animal and human models. It was reported in a study that liposome-encapsulated curcumin improved curcumin bioavailability by increasing gastrointestinal absorption (Takahashi et al., 2009). Curcumin's low oral bioavailability is a key issue limiting its use in nutraceuticals and functional foods. Curcumin bioavailability was seen to be nine times higher in novel organogel-based nanoemulsions than normal curcumin (Yu and Huang, 2012). Curcumin poly(lactic-co-glycolic acid) nanoparticles (PLGA-NPs) and cyclodextrin-encapsulated curcumin have been shown to increase the bioavailability of curcumin. These findings were explained by improved water solubility, increased absorption, longer residence time, longer half-life, and improved uptake in the intestinal cavity (Xie et al., 2011; Yadav et al., 2010). To overcome curcumin's poor aqueous solubility and bioavailability many nanoscale systems have been widely used for encapsulation and delivery (Hu and Luo, 2021).

5.2 NANOCARRIERS USED IN ENCAPSULATION

Though curcumin has immense health benefits, several challenges are associated with curcumin delivery. Photodegradation, very meager water solubility, lack of chemical stability, shorter lifespan, and faster metabolism are the reasons behind underperformance of the molecule (Rafiee et al., 2019). Different research works were conducted where a number of nanocarriers were tried to develop curcumin nanoformulation for targeted delivery. The main purpose is to improve the stability, bioaccessibility with improved efficacy. Attempts to prepare curcumin nanoformulations include several delivery systems like nanoemulsion, nanosuspension, solid lipid nanoparticles, liposomes, polymeric micelles, polymeric nanocomposites, inclusion complex with cyclodextrins, dendritic polymers, mesoporous nano gold and silver particles. Figure 5.5 shows a few prominent nanoformulations framed as carriers of curcumin.

Nanoemulsions are basically oil droplets dispersed in aqueous media. Due to smaller size with high surface area and kinetic stability, nanoemulsions are being utilized to encapsulate several bioactive components including curcumin (Saxena et al., 2018). On formulating curcumin into a nanoemulsion, its solubility was enhanced by 1,400 times (Li et al., 2016). Similarly, liposomes are able to hold lipophilic components like curcumin effectively in the bilayers made up of phospholipid and accelerate their intake at the cellular level (Mehanny et al., 2016) providing protection to the active compound from chemical and photodegradation. These are self-assembled vesicles (spherical in shape) with at least one phospholipid bilayer that separates the outer and inner aqueous media(Akhavan et al., 2018). Many research studies reported that liposomal formulation of curcumin improved bioavailability, solubility, and bioactivity by mediating sustained release and providing stability toward different external conditions (temperature, pH, salt) (Chen et al., 2012). But they are of limited use because of their short half-life circulation and susceptibility to hydrolysis and oxidation, which might result in the loss of their loading capacity and encapsulation efficiency of bioactive components. Solid lipid nanoparticles comprise lipid droplets containing bioactive compounds in a well-defined structure (Weiss et al., 2008). Main components of the solid lipid nanoparticle (SLN) are lipids, surfactants along with water, and cosurfactants. Several attempts were made to formulate curcumin with these solid lipid nanocarriers to improve its bioavailability and bioactivity (Caon et al., 2017). But some limitations, such as lower loading capacity and chances of explosion while storage, were associated with these nanocarriers. To overcome this problem, nano-sized lipid carriers were developed. The inner portion of solid lipid nanoparticles is prepared by solid lipid, but nanostructure lipid carriers contain both solid and liquid lipids (70:30 basis) (de Souza Simoes et al., 2017). Considerable solubility with stability and penetration of curcumin was achieved using this lipid nanocarrier-based delivery system (Chen et al., 2016). There are limited applications for these lipid-based nanocarriers for targeted delivery of curcumin oil. Some research works reported the development of nanoliposomes and solid lipid nanocarriers for loading of turmeric oil and their subsequent release (Yang et al., 2020). Applications of different delivery systems of curcumin have been discussed in brief and are mentioned in Table 5.1.

Table 5.1 Chemical Profile of Turmeric Curcuminoids

Curcuminoids	Chemical Formula	Mass (Da)	IUPAC Name	X_1	X_2
Curcumin-I (C)	$C_{21}H_{20}O_6$	368.1260	1,7-Bis (4-hydroxy-3-methoxyphenyl) -1,6- hepatadiene-3,5-dione	–OMe	–OMe
Curcumin-II (DC)	$C_{20}H_{18}O_5$	338.1154	1-(4-Hydroxy-3-methoxyphenyl)-7-(4-hydroxyphenyl) hepta-1,6-diene-3,5-dione	-H	-OMe
Curcumin-III (BDC)	$C_{19}H_{16}O_4$	308.1049	1,7-Bis(4-hydroxyphenyl)-1, 6-heptadiene-3,5-dione	–H	–H

The polymeric encapsulations as formulations of nanocarriers are prepared from biodegradable polymers like proteins (milk proteins, soy protein, gelatin), polysaccharides (sodium alginate, starch, chitosan, cyclodextrin), and synthetic polymers [poly(lactic acid), poly(D, L-lactide), poly(lactide-co-glycolide), poly(cyanoacrylate)]. In general, these polymers are useful to develop nanoencapsulations for nutraceutical product and drug delivery due to their variable nature including biocompatibility, nontoxicity, lower biodegradability (Esfanjani and Jafari, 2016) along with immunogenicity (Assadpour et al., 2017). Chereddy et al. (2013) loaded curcumin into the poly(lactide-co-glycolide) nanoparticles with efficiency of 88.2% of encapsulation. The nanoparticles displayed several benefits like enhanced water solubility, protection from light degradation, and controlled release over a longer period for the encapsulated formulation. Moreover, poly(lactide-co-glycolide)-curcumin nanocomposite showed two times higher activity of wound healing than curcumin. A similar result was observed in another study (Yoon et al., 2015) where improved efficiency for formulated curcumin was observed. In recent times, various research papers have been published regarding the application of different biopolymers like polysaccharide and protein to encapsulate curcumin. For instance, nanoparticles of zein were applied to encapsulate curcumin that showed photostability toward irradiation by ultraviolet and imparted stability at various pH levels (Patel et al., 2010). Similarly bovine serum albumin nanoparticles were prepared for delivery of curcumin with enhanced encapsulation efficiency and sustained release (Jithan et al., 2011). Curcumin can also be entrapped into complexes of nano-biopolymer developed by interaction between two biopolymers, i.e., polysaccharide and protein molecules. Mirpoor et al. (2017) formulated nanoparticles loaded with curcumin by interacting sodium alginate and β-lactoglobulin with protein-polysaccharide at a ratio of 4:3. Nanoencapsulation with 98% efficiency provided considerable safeguarding of curcumin toward different environmental actions. Certain amphiphilic polymers form micelles at aqueous media that entrap the targeted molecule that facilitates their release at the target site. Some polymer-based delivery systems for curcumin are mentioned in Table 5.2.

Apart from these formulations, there are other media like mesoporous silica nanoparticles, dendritic polymers that are compatible for curcumin delivery. Mesoporous silica nanoparticles possess honeycomb-like porous structure with high surface area and pore volume that provides unique qualities for delivery of a drug (Lu et al., 2007). Dendritic polymers also confer a complex three-dimensional network that exhibits different benefits for targeted delivery including the isolated space,

Table 5.2 Carrier Systems Used to Deliver Curcumin for Improved Bioefficacy (Self-Creation)

Types of Nanocarrier	Specifications	Functionality	References
Nanoemulsions	Emulsification with high-pressure homogenization using β-Lactoglobulin and triacylglycerols	improvement in availability of curcumin was correlated with chain length of lactoglobulin and triacylglycerols	Ahmed et al. (2012)
	Curcumin-loaded nanoemulsion with tocopherol acetate by ultrasonication and high-pressure homogenization	Controlled release for 24 hours with minimum cellular toxicity and hemolysis	Shukla et al. (2014)
	Soybean oil–based emulsion of curcumin with tween 20, 60, and 80 as emulsifiers by simple mechanical stirring	Expanding capacity to retain the stability of curcumin for 60 days	Borrin et al. (2016)
	Curcumin-loaded nanoemulsion using plant oil and whey protein isolate by using the microfluidization technique	Shelf life of curcumin was improved	Song and McClements (2021)
Nano-sized lipid carriers	Lipid nanocarriers prepared by polyethylene glycol monostearate, stearic acid, soy lecithin, and glyceryl monostearate by emulsion-evaporation solidification	Improved penetration amount on cumulative basis	Chen et al. (2016)
	Lipid nanocarriers with glyceryl monostearate, soybean phospholipids, and polyethylene	Water solubility of curcumin got improved significantly	Chanburee and Tiyaboonchai (2018)
	Nanocarriers prepared with glycerol behenate, oleic acid with span 80 and tween 80 as stabilizing agents	Sustainable release of curcumin was observed from the carrier material i.e., 73% of cumulative release within 24 hours	Rubab et al. (2021)
Liposomal nanoformulations	Curcumin nanoliposomes with phospholipid, cholesterol, and tween 80 prepared by dynamic high-pressure microfluidization following thin-film dispersion technique	Improvement in stability at different metal ion concentrations and pH values with enhanced uptake at cellular level together with antioxidant effect	Chen et al. (2015)
	Liposomes with phosphatidylcholine and α-phosphatidic acid with extrusion following reversed phase evaporation method	Increment in overall bioavailability as well as stability of loaded curcumin	Cuomo et al. (2018)
	Liposomal formulations with the technology of solvent-assisted active loading that is composed of phosphatidylglycerol, phosphatidylcholine, phosphoethanolamine, cholesterol, and polyethylene glycol	Increase in cellular uptake of curcumin with improved shelf life that showed promising effect over different tumor- induced cell lines	Karimi et al. (2020)

(Continued)

Table 5.2 (*Continued*) Carrier Systems Used to Deliver Curcumin for Improved Bioefficacy (Self-Creation)

Types of Nanocarrier	Specifications	Functionality	References
	Nanoliposome with soybean phospholipids using thin-film hydration-sonication method	Promising anticarcinogenic activity with no cytotoxicity observed	Sheikhpour et al. (2020)
Solid lipid nanoparticles	Curcumin nanoparticles with molten stearic acid by high-speed homogenization	Increased effect of cytotoxicity over IMR32 neuroblastoma cell line	Rahman et al. (2014)
	Curcumin solid lipid nanoparticles having modified surface with N-trimethyl chitosan	Higher stability-controlled release and higher oral bioavailability	Ramalingam and Ko (2015)
	Solid lipid nanoparticles of curcumin with lecithin and stearic acid by emulsion/ evaporation technique	Significant reduction in Hodgkin's lymphoma (HL) xenograft growth for nanoparticles compared to pure curcumin	Guorgui et al. (2018)
	Solid lipid nanoparticles of curcumin using tristearin and polyethylene-glycolylated emulsifiers	More than 12-fold increase in bioaccessibility with good mucous permeation	Ban et al. (2020)
Polymer-/ biopolymer- based nanoencapsulation	Polyester nanocarrier with poly(lactic-co-glycolic acid) prepared by the solvent evaporation method	Improvement in the bioavailability of curcumin, used to treat severe malignant breast cancer	Verderio et al. (2013)
	Curcumin nanoplex with chitosan by the precipitation method at ambient temperature	Better chemical stability with increase in bioavailability	Nguyen et al. (2015)
	Curcumin conjugated with zein-epigallocatechin gallate by the nanoprecipitation technique	Improving chemical stability, water dispersibility, and bioaccessibility	Liu et al. (2018)
	Curcumin complexation with ethylene diamine (EDA)-modified β-cyclodextrin using the solvent evaporation method	Better corneal penetration with improved biocompatibility	Liu et al. (2020)
	Encapsulation in sodium caseinate-laponite® nanocomposite	Controlled release under simulated gastric condition with increase in antioxidant activity	Qu et al. (2021)
Polymeric micelles	Polymeric nanomicelle with poly(lactide)-blockpoly(ethylene glycol) copolymer by the nanoprecipitation process	Sustained release of curcumin enhancing angiogenesis and promoting fibroblast proliferation during the healing process of a fully thickened wound	Alibolandi et al. (2017)

(*Continued*)

Table 5.2 (*Continued*) Carrier Systems Used to Deliver Curcumin for Improved Bioefficacy (Self-Creation)

Types of Nanocarrier	Specifications	Functionality	References
	Binding of curcumin with sodium caseinate followed by reassembling to micellar nanostructures	Increase in bioavailability with improved bioprotection and encapsulation efficiency	Ghayour et al. (2019)
	Functionalization of curcumin with micelles comprising dextran and poly(lactic-co-glycolic acid)	Good penetration observed inside bacterial biofilm that enhanced antibacterial activity	Barros et al. (2021)
Mesoporous silica nanoparticles	Glutathione and pH-responsive curcumin-silica nanoparticles coated with Fe (III)-tannic acid complex	pH-responsive release with increased cytotoxic effect over selected cancer cell lines	Kim et al. (2015)
	Novel bifunctionalized magnetic nanostars and silica nanostars loaded with curcumin	Significant increase in cytotoxic effect over selected cancer cell lines with better intracellular uptake	Huang et al. (2016)
Dendritic polymers	Curcumin-dendrimer conjugate	Improvement in water solubility with cytotoxicity over two selected breast cancer cell lines	Debnath et al. (2013)
	Curcumin conjugated with poly(amidoamine) Dendrimers	Solubility increased by 150–415 times with prolonged and sustainable release	Falconieri et al. (2017)

well-defined structure, nano-scaffolding, higher loading capacity, dense spheroidal shape, and the ability to carry different drugs due to their functionalities (Svenson, 2009). Some applications of these types of delivery systems are mentioned in Table 5.2.

5.3 TECHNIQUES TO CHARACTERIZE NANOCARRIERS USED FOR CURCUMIN DELIVERY

The diagnostic as well as therapeutic efficacy of nanomaterials is largely dependent upon their physiological action. In this regard, it's necessary to understand how different physicochemical parameters of nanomaterials affect their behavior and distribution within the targeted system (Lin et al., 2014). For that reason some robust techniques are required to study various physicochemical properties of nanomaterials and nanocarriers. A practical as well as rigorous approach to characterize the nanomaterials is a prerequisite for developing safe and rational nanocarriers and also for quality assurance of the products. Different ways used for the characterization of nanomaterials have been described comprehensively.

5.3.1 Dynamic Light Scattering (DLS)

Different physicochemical characteristics, *viz.*, hydrodynamic diameter, structure, shape, and aggregation state of nanomaterials can be assessed using DLS. It facilitates measuring the size distribution of smaller molecules/particles within micron to nano range using laser as the source for monochromatic light. The principle lies in monitoring the temporal fluctuations of the elastic scattering of light using photon or intensity auto-correlation function (ACF). This scattering, i.e., Rayleigh scattering is induced from Brownian motion of the molecules of a size much smaller than the wavelength of incident light (Brar and Verma, 2011). The fluctuation in intensity comprises a mixture of constructive as well as destructive interferences of scattered light by which particle size can be obtained from analysis of the ACF using the Stokes–Einstein equation (Brar and Verma, 2011). This technique is commonly practiced to characterize nanomaterials. Kazemi et al. (2021) developed chitosan-magnetite reduced graphene oxide–based curcumin nanocarrier that showed particle size distribution between 860 and 910 nm analyzed through DLS. Polydispersity index (PDI) and zeta potential value of 0.4 and of –34 to –37.3 mV were also obtained using DLS for the developed nanocarrier, respectively. The stability of nanomaterials can be ascertained from the critical values of PDI and zeta potential. In another study, human serum albumin nanoparticles were prepared using sodium deoxycholate and dithiothreitol for smart delivery of curcumin, where DLS was performed to determine nanoparticle size (Das et al., 2019). Curcumin nanoparticles were in the size range of 25–250 nm, out of which particle size of around 125 nm exhibited highest uptake and maximum cytotoxic effect. A poly(aspartic acid)-g-amylose-based nanostructure was prepared by Sattari et al. (2018) using helical inclusion complex for the effective delivery of curcumin, where a similar technique was used to determine the particle size and zeta potential which is indicative of the stability of developed nanomaterials. The results suggested that the particle size ranging between 160 and 185 nm with –30 to –40 mV zeta potential had positive influence on the stability of curcumin.

5.3.2 Transmission Electron Microscopy (TEM)

Unlike optical microscopy that uses light sources along with glass lenses in order to illuminate the specimens to develop magnified images, electron microscopy uses the beams of accelerated electrons with electromagnetic or electrostatic lenses to develop images having much higher resolution. The shorter wavelength of electrons than visible light facilitates the same. This is one of the most frequently used techniques to characterize nanomaterials in electron microscopy. It gives direct images along with chemical information of the nanomaterials at a resolution to the extent of atomic dimensions. There are a series of electromagnetic lenses that focus on the unscattered or scattered electrons followed by projected on a screen in order to generate a phase-contrast image, an amplitude-contrast image, or a shadow image with varied darkness, depending upon the density of unscattered electrons. Overall, electron microscopy (transmission and scanning electron microscopy) can display the shape, size, and heterogeneity of nanomaterials along with the degrees of dispersion and aggregation. However, transmission microscope has advantages over scanning; one for giving better resolution and capacity for some extra analytical measurements. Karimi et al. (2020) prepared and characterized nanoliposomal curcumin formulations with higher stability and loading capacity. The respective TEM image of the formulations (PG-LipCUR and LipCUR) showed that both were in almost spherical shape with a homogeneous size of around 100 nm maintaining their uniformity. Danafar et al. (2017) developed the polyethylene-glycolylated iron oxide-gold nanoparticles delivery system which was characterized by TEM to determine the morphology and size of particles. The average size measured was smaller than 30 nm. On the other hand, mesoporous silica nanoparticles were developed with tannic acid-iron coating for pH-mediated delivery of curcumin (Kim et al., 2015). The evidence for such coating over the surface of silica nanoparticles was provided by TEM images. The zoomed image of nanomaterial clearly showed hexagonal network of nanoparticles coated with tannic acid-iron complex with average thickness of 16.1 nm.

5.3.3 Scanning Electron Microscopy (SEM)

SEM is a method of surface imaging in which the incident beam of electrons scans the surface of sample and interacts with the molecules to generate signals representing the composition of atoms and topographic detail about the surface (Ratner et al., 2004). Detection of secondary electrons is the most common method in SEM that can achieve smaller than 1 nm of resolution. SEM technique can directly facilitate the determination of surface morphology of the developed nanomaterial.

The SEM images of silica nanoparticles and curcumin-loaded β-cyclodextrin modified mesoporous silica nanoparticles were illustrated by Abdous et al. (2017). It was observed that the synthesized conjugated nanoparticles had a mesoporous framework like that for silica nanoparticles with uniform pores and high morphological order along with cubical shaped honeycomb-like network. SEM images confirm that surface morphology and undesirable distortion of the nanoparticles might have happened during the conjugation process. Agrawal et al. (2021) developed nanolipid carrier loaded with curcumin to facilitate delivery. The morphological features of developed nanolipid carrier were studied using this SEM technique. The respective SEM images showed the surface topography of dispersed nanocarrier particles that were almost spherical, small with uniform-sized particles. It was also demonstrated that the outer surface of the nanolipid carrier was nearly spherical with imperfect porous or uneven surface which facilitated nanomaterials for better entrapment of active ingredient as well as its improved delivery. In another study, a drug delivery system for curcumin was developed by Anirudhan et al. (2021), where nanocellulose was oxidized and subsequently conjugated with folic acid to develop the delivery system. SEM image analysis of the oxidized nanocellulose displayed a polished surface with sheet-like structure. However, on conjugation with folic acid, the surface became smooth with a fluffy texture. The porous nature of the material made it feasible for easy diffusion of curcumin during loading and release.

5.3.4 Fourier Transform Infrared Spectroscopy (FTIR)

For nanomaterial characterization, FTIR spectroscopy is employed to observe the characteristic spectral bands that reveal the conjugation of nanomaterials and the biomolecule. It is also used to study the conformational position of the bound biomolecule. FTIR has been expanded to study the functional molecules that are grafted onto carbon nanotubes or nanospheres (Baudot et al., 2010). Curcumin-graphene quantum dot coated over Indium-Tin-Oxide electrode was developed and its FTIR spectrum showed the presence of characteristic bands that correspond to graphene quantum dot such as 2,980 and 1,393 cm^{-1}, representing C–H bond and $-CO_2$ group, respectively (Mars et al., 2018). Recently, chitosan-magnetite reduced graphene oxide–based curcumin nanocarrier was characterized by the FTIR technique that confirmed the conjugation of curcumin with the nanocomposite (Kazemi et al., 2021). The shifting of peaks corresponding to nanocomposite toward lower wavelength and also the appearance of peak at 1,261 and 1,087 cm^{-1}, representing the $C-O-OCH_3$ group indicated conjugation of curcumin with the nanocarrier. Moreover, appearance of a signal at 1,641 cm^{-1} related to –NH deformation and a change in peak at 1,087 cm^{-1} represented the ketone functional group of curcumin. Moussa et al. (2016) encapsulated curcumin in cyclodextrin-metal organic frameworks to enhance its stability. The FT-IR spectra displayed a large shift in the transmittance of the curcumin-loaded cyclodextrin-metal organic frameworks compared to the free one showing that single signal transmitting at 3,884 cm^{-1} correspond to –OH stretching of cyclodextrin moiety. On the contrary, the organic frameworks showed transmittance at 3,422 cm^{-1}, indicating a probable interaction between the framework and curcumin *via* –OH group.

5.3.5 X-Ray Diffraction (XRD)

XRD is one of the techniques of X-ray-based spectroscopic modalities which has been considered as the primary tool to resolve the structures of crystalline materials at the atomic level. The XRD can be described as the reflection of X-ray beams upon incidence over the crystalline planes of a specimen that follows Bragg's law (Cantor and Schimmel, 1980). Basically, it is used to characterize crystalline shape, size, and distortion of lattice based on elastic scattering of X-rays. It is a well-defined technique employed for the determination of structure of materials at an atomic scale.

Nosrati et al. (2018) developed magnetic iron oxide nanoparticles coated with bovine serum albumin as compatible carriers for curcumin. The developed nanocarrier was characterized using XRD. The position as well as the relative intensity of characteristic peaks (2θ of 18.28°, 35.62°, 53.12°, 62.88°) of the nanomaterials matched closely with the standard iron oxide nanoparticles that proved the successful synthesis of nanomaterials. In recent past, curcumin was encapsulated with poly-(lactic acid) nanoparticles using the emulsification-solvent evaporation technique (Rachmawati et al., 2016). The XRD patterns of curcumin and curcumin-loaded poly(lactic acid) nanoparticles were further investigated. The diffractogram corresponding to curcumin showed intense and sharp peaks thereby confirming its crystallinity. On the other hand, XRD graph for curcumin-loaded poly(lactic acid) nanoparticles displayed reduction in the number and intensity of peaks, indicating a decrease in crystallinity or the transformation into an amorphous phase. The decrease in peaks (compared to that for curcumin) confirmed the loading on the nanoparticles. Abdous et al. (2017) developed β-cyclodextrin-modified mesoporous silica nanoparticles as a delivery tool for curcumin and the nanoparticle was characterized in XRD. Peaks of 2θ at 0.78°, 0.95°, and 1.65° corresponding to mesoporous silica nanoparticles remained fixed after modification which suggested that the mesoporous nature remained intact in developed nanocarrier.

5.4 CHANGES IN BIOAVAILABILITY AND CLINICAL EFFICACY DUE TO ENCAPSULATION

Safety and beneficial effects of curcumin consumption have been reported by several clinical studies but lower bioavailability and stability issues are adhered with this molecule that limit its therapeutic efficacy at the full potential. During metabolism curcumin gets conjugated with several biomolecules and thats why its concentration in free form become very less in plasma and tissues (Anand et al., 2007). Various studies in humans and animals showed that curcumin imparts several problems regarding bioavailability like restricted distribution to tissues, lower levels in serum, rapid metabolism in intestine and liver along with a short half-life period (Schneider et al., 2015). In fact, these bioavailability and stability problems restrict the concentration of curcumin in gastrointestinal tract as well as activities including anticarcinogenic, antitumor, antimicrobial effect (Burgos-Morón et al., 2010). Various attempts are being made to solve the problem.

Application of nanotechnology for targeted delivery of bioactive components or beneficial drugs became one of the prominent solutions for many problems related to bioavailability and bioefficacy of therapeutics (Gou et al., 2011). The nanoformulations of curcumin possess a multi-targeting approach that facilitates to carry out a wide array of actions by these nanoformulations. These nanoformulations of curcumin have shown improved and sustained delivery at the targeted site (Yallapu et al., 2012). Nanocurcumin, therefore, proves to be an important therapeutic agent that helps to resolve and manage several illnesses in the human body. Reports from clinical trials displayed that nanoformulations can better the bioaccessibility of curcumin in a safe manner. Besides this, testing of nanoformulations as therapeutic agents is highly useful and is also important for upcoming clinical-based trials and also for use in the human system. Moreover, the application of nanoformulations in combinations is desirable to reduce the dose of the main therapeutic agent that can result in enhanced efficacy while reducing the systemic toxicity associated with them. Different nanocarriers for efficient delivery of curcumin have been discussed in Section 5.2. Various lipid-based carriers and polymeric encapsulations provide a promising medium to curcumin for its sustainable delivery and to confer bioactivity at its highest potential. In Table 5.3, the effects of various nanoformulations on bioavailability, and ultimately the clinical efficiency of curcumin, are summarized.

Table 5.3 Effects of Various Nanoformulations on Bioefficiency of Curcumin

Curcumin Nanoformulation	Role	References
Curcumin conjugated with lipid nanoparticles	Curcumin nanoparticles improved physicochemical properties of curcumin, including formation of amorphous state with reduction in particle size that improved the release of compound. *In vitro* experiments showed that curcumin nanoparticles enhanced the antihepatoma and antioxidant activities significantly	Yen et al. (2010)
Solid lipid nanoparticle of curcumin	The topical formulation showed good stability and spreadability when subjected to mechanical stress, no phase separation was observed with change in pH. Due to slow release, the nanoformulation was able to avoid penetration or permeation in the epidermis of pig ear during 18 hours	Zamarioli et al. (2015)
Liposomes loaded with curcumin	Solubility of curcumin improved with extended release that provided improved patient compliance. It also showed suppression on inflammation of cell line induced by lipopolysaccharide, providing prominent anti-inflammatory effect	Ng et al. (2018)
Quantum dots of curcumin-graphene conjugate	Sensing of APOe4 DNA, associated with Alzheimer's disease with ultrasensitivity and good reproducibility, repeatability, long storage stability, and selectivity	Mars et al. (2018)
Conjugation of curcumin with silica nanoparticles	Improvement in aqueous solubility, hence more bioavailable with increased cytotoxicity against normal as well as HeLa fibroblast cell lines	Gangwar et al. (2013)
Copper nanoparticles capped with curcumin	Investigated as inhibitors of pro-angiogenic cytokines, *in vivo* angiogenesis involved in tumor angiogenesis with inhibition of amplification of cell as well as migration of cell line for breast cancer, MDA-MB-231	Kamble et al. (2016)
Solid lipid nanoparticles	*In vitro* release of curcumin showed initial burst release, i.e., around 16.5% within 2 hours followed by release in controlled manner over 96 hours	Kumar and Punniamurthy (2017)
Nanostructured lipid carriers with taurocholic acid as a ligand	High encapsulation (>90%), loading capacity (around 3%) with sustained release of curcumin. Studies for *in situ* intestinal perfusion showed improved permeability coefficient and absorption rate of curcumin taurocholic acid nanolipid carrier	Tian et al. (2017)
Curcumin-loaded nanostructured lipid carriers	High encapsulation and loading capacity with sustained release of curcumin. The optimized nanolipid carriers showed stability of curcumin at 4°C and displayed a significant increase in the permeation of curcumin (around 2.5-fold) in the rabbit cornea as compared to the control	Lakhani et al. (2018)

(*Continued*)

Table 5.3 (*Continued*) Effects of Various Nanoformulations on Bioefficiency of Curcumin

Curcumin Nanoformulation	Role	References
Cationic liposome loaded with curcumin and modified with glycyrrhetinic acid	Compared to free curcumin, cationic liposome loaded with curcumin and modified with glycyrrhetinic acid showed better effects in improving different blood parameters while inhibiting growth of tumor, reducing tumor microvascular density, upregulating the expression of caspase-3 protein and mRNA in H22 tumor tissues together with downregulating the expression of mRNA and vascular endothelial growth factor (VEGF)-protein	Chang et al. (2020)
Peptide-modified curcumin-loaded liposome	*In vivo* imiquimod-induced psoriasis mouse model showed improved uptake of peptide-modified curcumin-loaded liposome at the intracellular level with enhanced inhibitory effect on HaCaT cells along with heightened transdermal and antipsoriatic ability of the nanocomposite	Yu and Huang (2012)
Curcumin encapsulated with polymethacrylate polymer and poly(lactide-co-glycolide) acid	Significantly enhanced transfuse of curcumin within Caco-2 cell monolayers in suspension. The nanoparticles reduced tumor necrosis factor-α secretion substantially by lipid polysaccharide-activated macrophages with significant reduction in neutrophil infiltration while keeping the colonic structure intact similar to the control group under a murine dextran sodium sulfate–induced colitis model	Beloqui et al. (2014)
Encapsulation of curcumin within cyclodextrin-metal organic network	Stability in the framework was enhanced by three times of magnitude as compared to sole curcumin molecule	Moussa et al. (2016)
Hydrophilic block forming micelles of curcumin molecules conjugated with poly(lactic acid) by tris (hydroxymethyl) aminomethane linker	Higher loading capacity with improved efficacy against cancer-inducing HepG2 cells	Yang et al. (2012)
Nanocapsules of curcumin with poly(lactic-co-glycolic acid)-based polymeric oil	Nanocapsules internalized more effectively in CT26 cells compared to free curcumin that exhibited good therapeutic action *in vitro*, leading to apoptosis followed by cell cycle blocking while tested in mice. Besides, the formulated nanocapsules imparted extended blood circulation compared to the nonpolymerized nanocapsules and accumulated in CT26-tumors	Klippstein et al. (2015)
Curcumin entrapped in poly(ethylene glycol)-poly (lactic-co-glycolic acid) nanoparticles conjugated to folic acid	Enhanced significantly the efficacy of curcumin by downregulating several survival signals induced by cervical cancer cells and improved the potency in hindering the growth of tumor in cervical cancer xenografts	Thulasidasan et al. (2017)

(*Continued*)

Table 5.3 (*Continued*) Effects of Various Nanoformulations on Bioefficiency of Curcumin

Curcumin Nanoformulation	Role	References
Nanoemulsion of curcumin-loaded glucomannan octenyl succinate	*In vitro* as well as *in vivo* study revealed that the nanoemulsion provided considerable protection for curcumin and furnished targeted delivery. The nanoformulation also showed good storage stability and thermal processing	Meng et al. (2020)
Curcumin packed within κ-carrageenan	Good stability with high encapsulation and controlled release of curcumin was achieved. *In vitro* studies at the cellular level showed that the composites exhibited higher cytotoxicity over selected cancer cell lines and were involved in inducing cellular apoptosis efficiently in A549 cancer cells. There was also a decrease in the potential of mitochondrial membrane that facilitated cellular apoptosis while being treated with this formulation	Sathuvan et al. (2017)
Entrapment of curcumin diethyl disuccinate within alginate/chitosan nanoparticles	Exhibited good stability upon exposure to ultraviolet light and simulated digestive fluids with a sustained-release profile that facilitated improved bioavailability, bioaccessibility, cellular uptake, and cytotoxicity against HepG2 cells	Sorasitthiyanukarn et al. (2021)

5.5 APPLICATION IN THE FOOD INDUSTRY

Various studies have demonstrated different pharmaceutical efficacies of curcumin on different diseases, such as oxidative damage, inflammatory diseases, coronary disorders, cardiovascular diseases, arthritis, cancer, depression, and other chronic diseases related to age (Figure 5.6) (Mehanny et al., 2016). Despite its

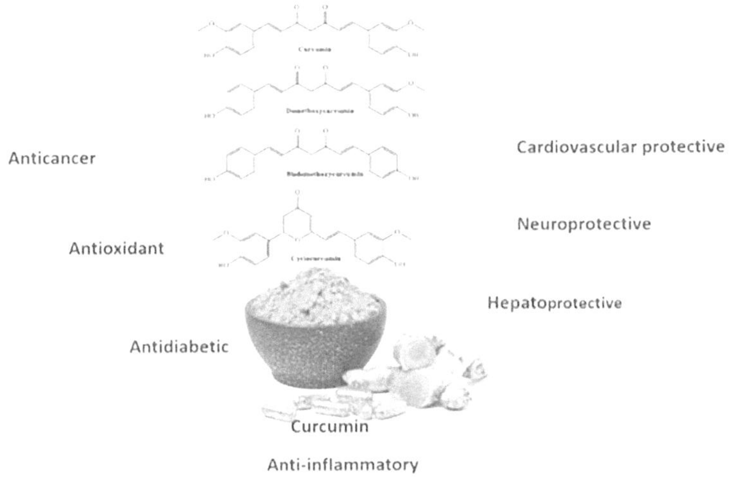

Figure 5.6 A wide array of versatile curcumin.

tremendous bioactivity and health benefits, the application of curcumin in nutraceutical, pharmaceutical, and food industries is limited and not up to its potential (Hu et al., 2020). Considering the epidemiological events caused by the appearance of the COVID-19 virus, in the last 2 years, there has been an intensification of studies on the methods to obtain curcumin (Pawar et al., 2018) and its bioactive properties (Ausili et al., 2021). The rhizome extract of the plant, curcumin powder, has over time been used in Southeast Asia in food preservation (Ibanez and Blázquez, 2021). Recently, the recognition of curcumin as belonging to the category of antifungal and antibacterial agents of natural origin has led to the diversification. Apart from its application in the area of pharmaceutical/food industries (Chuacharoen and Sabliov, 2019), newer applications have evolved in the sectors, for example, as a sensor/pH indicator (Dalvandi et al., 2020) in food or in packaging industries (Balbinot-Alfaro et al., 2019) (Figure 5.7).

Curcumin can be used as a food additive, leading to the extension of the shelf life of food (Olszewska et al., 2020). For the purpose of preservation and extending the shelf life of meat and fish products, packaging films are used based on gelatin (Balbinot-Alfaro et al., 2019), carboxymethylcellulose (Dalvandi et al., 2020), or vinyl acetate

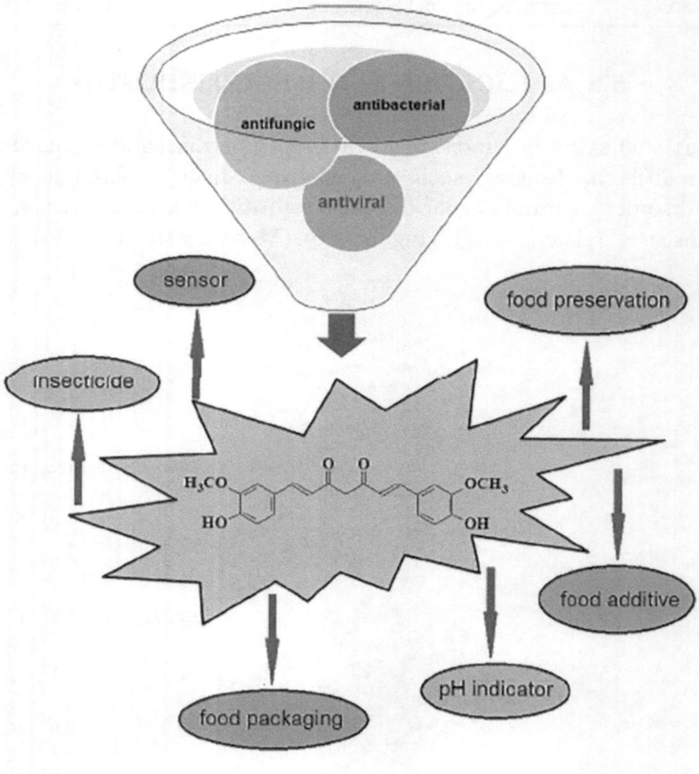

Figure 5.7 Applications of curcumin in food due to its antiviral, antibacterial, and antifungal properties.

(Chen et al., 2020), with curcumin as an antimicrobial agent. Curcumin embedded in polyvinyl acetate films has been used to obtain thin coatings with antimicrobial photodynamic activity under white light irradiation. Embedding curcumin in polymer matrices prevents the isomerization processes, increasing the stability and fluorescence lifetime of curcumin (Chen et al., 2020). From the angle of the natural additives used in the food and nutraceutical industry, curcumin is the ideal example due to its antioxidant, antifungal, antiviral, and antibacterial properties. However, its disadvantage, as with most phytochemicals, is its sensitivity to photodegradation and its low solubility in aqueous media. To improve these shortcomings, research is continuing to find curcumin delivery systems that are nontoxic and compatible with the use environments.

5.6 DELIVERY OF ACTIVE INGREDIENTS

5.6.1 Different Delivery System of Different Products

Curcumin is known in traditional medicine, especially in India, for use in treating fever, skin infections, and to facilitate digestion due to its impressive anti-inflammatory, antiviral, and antibacterial properties. In the last 2 years, because of the COVID-19 pandemic, the congestion of hospitals has led to an increased level of nosocomial infections (Zhou et al., 2020; Abbas et al., 2021). This situation has triggered new approaches regarding the dispersion of biofilm and in finding new ways to prevent the initial formation by modifying the surfaces (Jamal et al., 2018) (Figure 5.8).

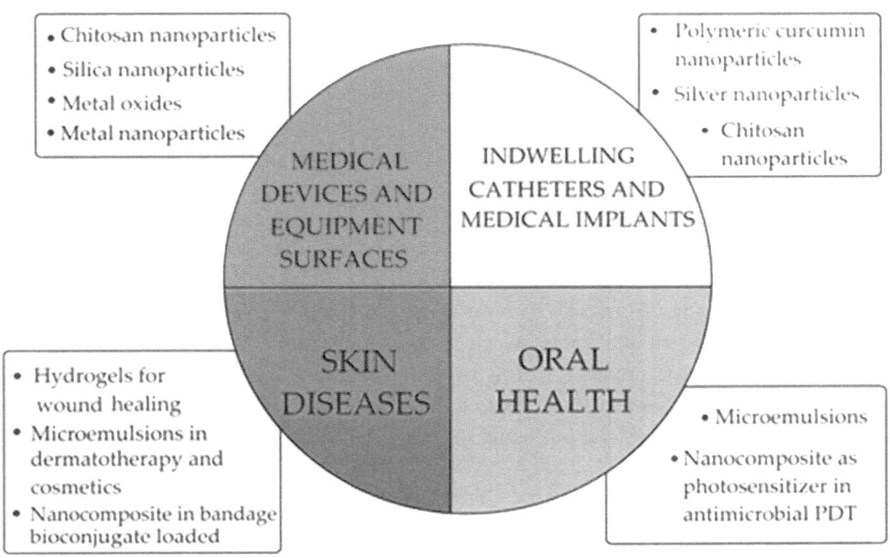

Figure 5.8 The main fields of use of curcumin in the medical field and its different delivery systems.

Table 5.4 Types of Composite Materials Used as a Support Loaded with Curcumin Used in Bandages or Dressings (Self-created)

No.	Polymer Support	Target Microorganisms/Potential Application
1.	Chitosan/pluronic membranes	*S. aureus*, *P. aeruginosa*/healing applications
2.	Chitosan/polycaprolactone	*S. aureus*/healing properties
3.	Hyaluronic acid modified pullulan polymers	*E. coli*, *S. aureus*/accelerating skin wound healing
4.	Lactide-co-glycolide/chitosan/β-cyclodextrin/ poly(vinyl alcohol)	Blend films
5.	Chitosan-collagen, gelatin, sodium alginate	*S. aureus*, *E. coli*
6.	Polyvinyl pyrrolidone(PVP)-cerium nitrate hexahydrate	*S. aureus*, *E. coli*/ dressing material-anti-scar property
7.	Cellulose hydroxypropyl-β-cyclodextrin-silver nanoparticle	*S. aureus*, *P. aeruginosa*, *C. auris*/healing properties
8.	Metal oxides-NPs-cotton	*S. aureus*, *E. coli*
9.	Cellulose-zinc oxide	*S. aureus*, *T. rubrum*/skin infections
10.	Graphene oxide	*S. aureus*, *E. coli*
11.	3-Methyl (hexadecyloxycarbonylmethyl) imidazolium Bromide	Hydrogel used for the wound healing
12.	Sodium alginate	Wound healing applications
13.	Thiocarbohydrazide gelatin nanofibers	*E. coli*/wound healing applications
14.	Bacterial nanocellulose	*S. aureus*, *E. coli*
15.	Oleic acid based polymeric bandage	Wound healing

Therefore, a nanocomposite based on metal oxides (Masoule et al., 2019) or chitosan nanoparticles containing curcumin have been tested.

The encouraging results of investigations on the action of curcumin in skin treatments has led to the development of another field, that of dressings and biomaterials carrying drugs for treating infected wounds (Table 5.4). Studies have continued to design different types of structures and platforms which allow them to be loaded with drugs or adjuvants to speed up the healing process. Another type of film obtained from poly nanofibers (lactide-co-glycolide) loaded with curcumin and heparin have been successfully tested for wound healing in diabetic mice (Liao et al., 2021). The polyvinyl alcohol-chitosan-curcumin-cyclodextrin films also showed antioxidant activity. Based on the histological tests, of the three types of structures, chitosan-gelatin-curcumin released curcumin the fastest, resulting in the greatest healing effect on wounds and the highest antibacterial action. The nanocomposite with alginate had the most porous structure, thus presenting the best results on biodegradation, while the collagen composite was the most stable.

5.6.2 Future Developments of Curcumin-Based Materials

Due to the multitude of properties manifested by curcumin analogs, the diversity of applications of the natural compound will be further developed as many issues related to the extraction, solubility, and stability of the parent compound will be solved. Due to its exceptional anti-inflammatory and antioxidant properties, modern applications of curcumin in the medical field targeting the treatment of neurodegenerative diseases (Gao et al., 2020) and cancer will be further developed. Simultaneously, different studies have been conducted on the protective efficiency of curcumin based on its antioxidant properties against the neurotoxicity produced by methamphetamine, organophosphorus insecticides, and aluminum. *In vivo* results have shown that the mechanisms of action of curcumin include the prevention of lipid peroxidation and an increase in the antioxidant capacity of different enzymes, superoxide dismutase and glutathione peroxidase, that protect cells from damage caused by reactive oxygen species (ROS). In the case of neurological diseases, nanocurcumin treatments have been evaluated *in vitro/in vivo* and in clinical trials. For the development of targeted therapy in the treatment of neuronal diseases, curcumin was encapsulated in micelles, liposomes, or polymeric nanoparticles. Thus, the efficacy of neuronal membrane penetration and the mechanisms of action of "nano"-curcumin in the treatment of different diseases (Pandey et al., 2020).

5.7 CONCLUSIONS

Curcumin and its derivatives were reported to exhibit several biological activities due to the presence of highly reactive unsaturated carbonyl moiety in the skeleton. Bioavailability of curcuminoids has been a subject of interest for scientists all over the world and current research thrust is directed toward the increased bioavailability of curcumin. To overcome these issues, several efforts had been made to improve the bioavailability of curcumin by using curcumin nanoformulation included in several delivery systems, such as nanoemulsion, nanosuspension, solid lipid nanoparticles, liposomes, polymeric micelles, polymeric nanocomposites, inclusion complex with cyclodextrins, dendrimers, mesoporous silver, and gold nanoparticles. These nanocarriers will be game changers in broadening the applications of curcumin in the field of food and pharmaceutical industries. Also, several efforts were carried out for converting curcumin biomolecule into more bioavailable nanoformulations or nanocomposites.

REFERENCES

Abbas, M., Nunes, T. R., Martischang, R., Zingg, W., Iten, A., Pittet, D., & Harbarth, S. (2021). Nosocomial transmission and outbreaks of coronavirus disease 2019: The need to protect both patients and healthcare workers. *Antimicrobial Resistance & Infection Control*, *10*(1), 1–13.

Abdous, B., Sajjadi, S. M., & Mamani, L. (2017). β-Cyclodextrin modified mesoporous silica nanoparticles as a nano-carrier: Response surface methodology to investigate and optimize loading and release processes for curcumin delivery. *Journal of Applied Biomedicine*, *15*(3), 210–218.

Agrawal, M., Saraf, S., Pradhan, M., Patel, R. J., Singhvi, G., & Alexander, A. (2021). Design and optimization of curcumin loaded nano lipid carrier system using Box-Behnken design. *Biomedicine & Pharmacotherapy*, *141*, 111919.

Ahmed, K., Li, Y., McClements, D. J., & Xiao, H. (2012). Nanoemulsion-and emulsion-based delivery systems for curcumin: Encapsulation and release properties. *Food Chemistry*, *132*(2), 799–807.

Akhavan, S., Assadpour, E., Katouzian, I., & Jafari, S. M. (2018). Lipid nano scale cargos for the protection and delivery of food bioactive ingredients and nutraceuticals. *Trends in Food Science & Technology*, *74*, 132–146.

Alibolandi, M., Mohammadi, M., Taghdisi, S. M., Abnous, K., & Ramezani, M. (2017). Synthesis and preparation of biodegradable hybrid dextran hydrogel incorporated with biodegradable curcumin nanomicelles for full thickness wound healing. *International Journal of Pharmaceutics*, *532*(1), 466–477.

Amalraj, A., Pius, A., Gopi, S., & Gopi, S. (2017). Biological activities of curcuminoids, other biomolecules from turmeric and their derivatives: A review. *Journal of Traditional and Complementary Medicine*, *7*(2), 205–233.

Anand, P., Kunnumakkara, A. B., Newman, R. A., & Aggarwal, B. B. (2007). Bioavailability of curcumin: Problems and promises. *Molecular Pharmaceutics*, *4*(6), 807–818.

Anirudhan, T. S., Manjusha, V., & Sekhar, V. C. (2021). A new biodegradable nano cellulose-based drug delivery system for pH-controlled delivery of curcumin. *International Journal of Biological Macromolecules*, *183*, 2044–2054.

Assadpour, E., Jafari, S. M., & Esfanjani, A. F. (2017). Protection of phenolic compounds within nanocarriers. *CAB Reviews*, *12*(057), 1–8.

Ataie, A., Sabetkasaei, M., Haghparast, A., Moghaddam, A. H., & Kazeminejad, B. (2010). Neuroprotective effects of the polyphenolic antioxidant agent, Curcumin, against homocysteine-induced cognitive impairment and oxidative stress in the rat. *Pharmacology Biochemistry and Behavior*, *96*(4), 378–385.

Ausili, A., Gómez-Murcia, V., Candel, A. M., Beltrán, A., Torrecillas, A., He, L., Jiang, Y., Zhang, S., Teruel, J. A., & Gómez-Fernández, J. C. (2022). A comparison of the location in membranes of curcumin and curcumin-derived bivalent compounds with potential neuroprotective capacity for Alzheimer's disease. *Colloids and Surfaces B: Biointerfaces*, *199*, 111525.

Balbinot-Alfaro, E., Craveiro, D. V., Lima, K. O., Costa, H. L. G., Lopes, D. R., & Prentice, C. (2019). Intelligent packaging with pH indicator potential. *Food Engineering Reviews*, *11*(4), 235–244.

Ban, C., Jo, M., Park, Y. H., Kim, J. H., Han, J. Y., Lee, K. W., Kweon, D. H., & Choi, Y. J. (2020). Enhancing the oral bioavailability of curcumin using solid lipid nanoparticles. *Food Chemistry*, *302*, 125328.

Barros, C. H., Hiebner, D. W., Fulaz, S., Vitale, S., Quinn, L., & Casey, E. (2021). Synthesis and self-assembly of curcumin-modified amphiphilic polymeric micelles with antibacterial activity. *Journal of Nanobiotechnology*, *19*(1), 1–15.

Baudot, C., Tan, C. M., & Kong, J. C. (2010). FTIR spectroscopy as a tool for nano-material characterization. *Infrared Physics & Technology*, *53*(6), 434–438.

Beloqui, A., Coco, R., Memvanga, P. B., Ucakar, B., des Rieux, A., & Préat, V. (2014). pH-sensitive nanoparticles for colonic delivery of curcumin in inflammatory bowel disease. *International Journal of Pharmaceutics*, *473*(1–2), 203–212.

Bhat, A., Mahalakshmi, A. M., Ray, B., Tuladhar, S., Hediyal, T. A., Manthiannem, E., … & Sakharkar, M. K. (2019). Benefits of curcumin in brain disorders. *BioFactors*, *45*(5), 666–689.

Borrin, T. R., Georges, E. L., Moraes, I. C., & Pinho, S. C. (2016). Curcumin-loaded nanoemulsions produced by the emulsion inversion point (EIP) method: An evaluation of process parameters and physico-chemical stability. *Journal of Food Engineering*, *169*, 1–9.

Brar, S. K., & Verma, M. (2011). Measurement of nanoparticles by light-scattering techniques. *TrAC Trends in Analytical Chemistry*, *30*(1), 4–17.

Burgos-Morón, E., Calderón-Montaño, J. M., Salvador, J., Robles, A., & López-Lázaro, M. (2010). The dark side of curcumin. *International Journal of Cancer*, *126*(7), 1771–1775.

Cantor, C. R., & Schimmel, P. R. (1980). Other scattering and diffraction techniques: Light scattering. In: Cantor, C. R. (Ed.), *Biophysical Chemistry Part II: Techniques for the Study of Biological Structure and Function*. San Francisco, CA: W. H. Freeman, pp. 838–845.

Caon, T., Mazzarino, L., Simões, C. M. O., Senna, E. L., & Silva, M. A. S. (2017). Lipid-and polymer-based nanostructures for cutaneous delivery of curcumin. *AAPS PharmSciTech*, *18*(3), 920–925.

Chanburee, S., & Tiyaboonchai, W. (2018). Enhanced intestinal absorption of curcumin in Caco-2 cell monolayer using mucoadhesive nanostructured lipid carriers. *Journal of Biomedical Materials Research Part B: Applied Biomaterials*, *106*(2), 734–741.

Chang, M., Wu, M., & Li, H. (2020). Antitumor effects of curcumin and glycyrrhetinic acid-modified curcumin-loaded cationic liposome by intratumoral administration. *Evidence-Based Complementary and Alternative Medicine*, 2020, 4504936.

Chen, Y., Wu, Q., Zhang, Z., Yuan, L., Liu, X., & Zhou, L. (2012). Preparation of curcumin-loaded liposomes and evaluation of their skin permeation and pharmacodynamics. *Molecules*, *17*(5), 5972–5987.

Chen, X., Zou, L. Q., Niu, J., Liu, W., Peng, S. F., & Liu, C. M. (2015). The stability, sustained release and cellular antioxidant activity of curcumin nanoliposomes. *Molecules*, *20*(8), 14293–14311.

Chen, P., Zhang, H., Cheng, S., Zhai, G., & Shen, C. (2016). Development of curcumin loaded nanostructured lipid carrier based thermosensitive in situ gel for dermal delivery. *Colloids and Surfaces A: Physicochemical and Engineering Aspects*, *506*, 356–362.

Chen, L., Song, Z., Zhi, X., & Du, B. (2020). Photoinduced antimicrobial activity of curcumin-containing coatings: Molecular interaction, stability and potential application in food decontamination. *ACS Omega*, *5*(48), 31044–31054.

Cheng, K. K., Yeung, C. F., Ho, S. W., Chow, S. F., Chow, A. H., & Baum, L. (2013). Highly stabilized curcumin nanoparticles tested in an in vitro blood–brain barrier model and in Alzheimer's disease Tg2576 mice. *The American Association of Pharmaceutical Scientists*, *15*(2), 324–336.

Chereddy, K. K., Coco, R., Memvanga, P. B., Ucakar, B., des Rieux, A., Vandermeulen, G., & Préat, V. (2013). Combined effect of PLGA and curcumin on wound healing activity. *Journal of Controlled Release*, *171*(2), 208–215.

Chuacharoen, T., & Sabliov, C. M. (2019). Comparative effects of curcumin when delivered in a nanoemulsion or nanoparticle form for food applications: Study on stability and lipid oxidation inhibition. *LWT*, *113*, 108319.

Cuomo, F., Cofelice, M., Venditti, F., Ceglie, A., Miguel, M., Lindman, B., & Lopez, F. (2018). In-vitro digestion of curcumin loaded chitosan-coated liposomes. *Colloids and Surfaces B: Biointerfaces*, *168*, 29–34.

Dalvandi, F., Almasi, H., Ghanbarzadeh, B., Hosseini, H., & Karimian Khosroshahi, N. (2020). Effect of vacuum packaging and edible coating containing black pepper seeds and turmeric extracts on shelf-life extension of chicken breast fillets. *Journal of Food and Bioprocess Engineering*, *3*(1), 69–78.

Danafar, H., Sharafi, A., Askarlou, S., & Manjili, H. K. (2017). Preparation and characterization of PEGylated iron oxide-gold nanoparticles for delivery of sulforaphane and curcumin. *Drug Research*, *67*(12), 698–704.

Das, R. P., Gandhi, V. V., Singh, B. G., Kunwar, A., Kumar, N. N., & Priyadarsini, K. I. (2019). Preparation of albumin nanoparticles: Optimum size for cellular uptake of entrapped drug (Curcumin). *Colloids and Surfaces A: Physicochemical and Engineering Aspects*, *567*, 86–95.

de Souza Simoes, L., Madalena, D. A., Pinheiro, A. C., Teixeira, J. A., Vicente, A. A., & Ramos, Ó. L. (2017). Micro-and nano bio-based delivery systems for food applications: In vitro behavior. *Advances in Colloid and Interface Science*, *243*, 23–45.

Debnath, S., Saloum, D., Dolai, S., Sun, C., Averick, S., Raja, K., & Fata, J. (2013). Dendrimer-curcumin conjugate: A water soluble and effective cytotoxic agent against breast cancer cell lines. *Anti-Cancer Agents in Medicinal Chemistry*, *13*(10), 1531–1539.

Esfanjani, A. F., & Jafari, S. M. (2016). Biopolymer nano-particles and natural nano-carriers for nano-encapsulation of phenolic compounds. *Colloids and Surfaces B: Biointerfaces*, *146*, 532–543.

Falconieri, M. C., Adamo, M., Monasterolo, C., Bergonzi, M. C., Coronnello, M., & Bilia, A. R. (2017). New dendrimer-based nanoparticles enhance curcumin solubility. *Planta Medica*, *83*(5), 420–425.

Gangwar, R. K., Tomar, G. B., Dhumale, V. A., Zinjarde, S., Sharma, R. B., & Datar, S. (2013). Curcumin conjugated silica nanoparticles for improving bioavailability and its anticancer applications. *Journal of Agricultural and Food Chemistry*, *61*(40), 9632–9637.

Gao, C., Chu, X., Gong, W., Zheng, J., Xie, X., Wang, Y., Yang, M., Li, Z., Gao, C., & Yang, Y. (2020). Neuron tau-targeting biomimetic nanoparticles for curcumin delivery to delay progression of Alzheimer's disease. *Journal of Nanobiotechnology*, *18*(1), 1–23.

Ghayour, N., Hosseini, S. M. H., Eskandari, M. H., Esteghlal, S., Nekoei, A. R., Gahruie, H. H., Tatar, M., & Naghibalhossaini, F. (2019). Nanoencapsulation of quercetin and curcumin in casein-based delivery systems. *Food Hydrocolloids*, *87*, 394–403.

Gou, M., Men, K., Shi, H., Xiang, M., Zhang, J., Song, J., Long, J., Wan, Y., Luo, F., Zhao, X., & Qian, Z. (2011). Curcumin-loaded biodegradable polymeric micelles for colon cancer therapy in vitro and in vivo. *Nanoscale*, *3*(4), 1558–1567.

Guorgui, J., Wang, R., Mattheolabakis, G., & Mackenzie, G. G. (2018). Curcumin formulated in solid lipid nanoparticles has enhanced efficacy in Hodgkin's lymphoma in mice. *Archives of Biochemistry and Biophysics*, *648*, 12–19.

Hu, Q., & Luo, Y. (2021). Chitosan-based nanocarriers for encapsulation and delivery of curcumin: A review. *International Journal of Biological Macromolecules*, *179*, 125–135.

Hu, Q., Hu, S., Fleming, E., Lee, J. Y., & Luo, Y. (2020). Chitosan-caseinate-dextran ternary complex nanoparticles for potential oral delivery of astaxanthin with significantly improved bioactivity. *International Journal of Biological Macromolecules*, *151*, 747–756.

Huang, P., Zeng, B., Mai, Z., Deng, J., Fang, Y., Huang, W., Zhang, H., Yuan, J., Wei, Y., & Zhou, W. (2016). Novel drug delivery nanosystems based on out-inside bifunctionalized mesoporous silica yolk–shell magnetic nanostars used as nanocarriers for curcumin. *Journal of Materials Chemistry B*, *4*(1), 46–56.

Ibanez, M. D., & Blázquez, M. A. (2021). Curcuma longa l. Rhizome essential oil from extraction to its agri-food applications: A review. *Plants*, *10*(1), 44.

Jamal, M., Ahmad, W., Andleeb, S., Jalil, F., Imran, M., Nawaz, M. A., Hussain, T., Ali, M., Rafiq, M., & Kamil, M. A. (2018). Bacterial biofilm and associated infections. *Journal of the Chinese Medical Association*, *81*(1), 7–11.

Jithan, A. V., Madhavi, K., Madhavi, M., & Prabhakar, K. (2011). Preparation and characterization of albumin nanoparticles encapsulating curcumin intended for the treatment of breast cancer. *International Journal of Pharmaceutical Investigation*, *1*(2), 119.

Kamble, S., Utage, B., Mogle, P., Kamble, R., Hese, S., Dawane, B., & Gacche, R. (2016). Evaluation of curcumin capped copper nanoparticles as possible inhibitors of human breast cancer cells and angiogenesis: A comparative study with native curcumin. *AAPS PharmSciTech*, *17*(5), 1030–1041.

Karimi, M., Gheybi, F., Zamani, P., Mashreghi, M., Golmohammadzadeh, S., Darban, S. A., Badiee, A., & Jaafari, M. R. (2020). Preparation and characterization of stable nanoliposomal formulations of curcumin with high loading efficacy: In vitro and in vivo anti-tumor study. *International Journal of Pharmaceutics*, *580*, 119211.

Kazemi, S., Pourmadadi, M., Yazdian, F., & Ghadami, A. (2021). The synthesis and characterization of targeted delivery curcumin using chitosan-magnetite-reduced graphene oxide as nano-carrier. *International Journal of Biological Macromolecules*, *186*, 554–562.

Kim, S., Philippot, S., Fontanay, S., Duval, R. E., Lamouroux, E., Canilho, N., & Pasc, A. (2015). pH-and glutathione-responsive release of curcumin from mesoporous silica nanoparticles coated using tannic acid–Fe (iii) complex. *RSC Advances*, *5*(110), 90550–90558.

Klippstein, R., Wang, J. T. W., El-Gogary, R. I., Bai, J., Mustafa, F., Rubio, N., Bansal, S., Al-Jamal, W. T., & Al-Jamal, K. T. (2015). Passively targeted curcumin-loaded pegylated PLGA nanocapsules for colon cancer therapy in vivo. *Small*, *11*(36), 4704–4722.

Koli, P., & Yadav, D. K. (2017). Popular Kheti magazine. *Curcumin: A Miracle Molecule,* 5(1), 69–73.

Kumar, P. S., & Punniamurthy, N. (2017). Formulation development and characterization of curcumin loaded solid lipid nanoparticles for improved aqueous solubility and bioavailability. *The Pharma Innovation*, *Part A*, *6*(4), 7.

Lakhani, P., Patil, A., Taskar, P., Ashour, E., & Majumdar, S. (2018). Curcumin-loaded nanostructured lipid carriers for ocular drug delivery: Design optimization and characterization. *Journal of Drug Delivery Science and Technology*, *47*, 159–166.

Li, M., Zhang, Z., Hill, D. L., Wang, H., & Zhang, R. (2007). Curcumin, a dietary component, has anticancer, chemosensitization, and radiosensitization effects by down-regulating the MDM2 oncogene through the PI3K/mTOR/ETS2 pathway. *Cancer Research*, *67*(5), 1988–1996.

Li, J., Hwang, I. C., Chen, X., & Park, H. J. (2016). Effects of chitosan coating on curcumin loaded nano-emulsion: Study on stability and *in vitro* digestibility. *Food Hydrocolloids*, *60*, 138–147.

Li, H., Zhang, X., Zhao, C., Zhang, H., Chi, Y., Wang, L., Zhang, H., Bai, S., & Zhang, X. (2022). Entrapment of curcumin in soy protein isolate using the pH-driven method: Nanoencapsulation and formation mechanism. *LWT*, *153*, 112480.

Liang, G. (2013). Editorial (hot topic: recent progress and novel insights in curcumin research "from chemistry to clinical use). *Current Pharmaceutical Design*, *19*(11), 1965–1965.

Liao, H. T., Lai, Y. T., Kuo, C. Y., & Chen, J. P. (2021). A bioactive multi-functional heparin-grafted aligned poly (lactide-co-glycolide)/curcumin nanofiber membrane to accelerate diabetic wound healing. *Materials Science and Engineering: C*, *120*, 111689.

Lin, P. C., Lin, S., Wang, P. C., & Sridhar, R. (2014). Techniques for physicochemical characterization of nanomaterials. *Biotechnology Advances*, *32*(4), 711–726.

Liu, F., Ma, D., Luo, X., Zhang, Z., He, L., Gao, Y., & McClements, D. J. (2018). Fabrication and characterization of protein-phenolic conjugate nanoparticles for co-delivery of curcumin and resveratrol. *Food Hydrocolloids*, *79*, 450–461.

Liu, C. H., Lee, G. W., Wu, W. C., & Wang, C. C. (2020). Encapsulating curcumin in ethylene diamine-β-cyclodextrin nanoparticle improves topical cornea delivery. *Colloids and Surfaces B: Biointerfaces*, *186*, 110726.

Lu, J., Liong, M., Zink, J. I., & Tamanoi, F. (2007). Mesoporous silica nanoparticles as a delivery system for hydrophobic anticancer drugs. *Small*, *3*(8), 1341–1346.

Mars, A., Hamami, M., Bechnak, L., Patra, D., & Raouafi, N. (2018). Curcumin-graphene quantum dots for dual mode sensing platform: Electrochemical and fluorescence detection of APOe4, responsible of Alzheimer's disease. *Analytica Chimica Acta*, *1036*, 141–146.

Masoule, S. F., Pourhajibagher, M., Safari, J., & Khoobi, M. (2019). Base-free green synthesis of copper (II) oxide nanoparticles using highly cross-linked poly (curcumin) nanospheres: Synergistically improved antimicrobial activity. *Research on Chemical Intermediates*, *45*(9), 4449–4462.

Mehanny, M., Hathout, R. M., Geneidi, A. S., & Mansour, S. (2016). Exploring the use of nanocarrier systems to deliver the magical molecule; curcumin and its derivatives. *Journal of Controlled Release*, *225*, 1–30.

Meng, F. B., Zhang, Q., Li, Y. C., Li, J. J., Liu, D. Y., & Peng, L. X. (2020). Konjac glucomannan octenyl succinate as a novel encapsulation wall material to improve curcumin stability and bioavailability. *Carbohydrate Polymers*, *238*, 116193.

Mirpoor, S. F., Hosseini, S. M. H., & Yousefi, G. H. (2017). Mixed biopolymer nanocomplexes conferred physicochemical stability and sustained release behavior to introduced curcumin. *Food Hydrocolloids*, *71*, 216–224.

Moussa, Z., Hmadeh, M., Abiad, M. G., Dib, O. H., & Patra, D. (2016). Encapsulation of curcumin in cyclodextrin-metal organic frameworks: Dissociation of loaded CD-MOFs enhances stability of curcumin. *Food Chemistry*, *212*, 485–494.

Nelson, K. M., Dahlin, J. L., Bisson, J., Graham, J., Pauli, G. F., & Walters, M. A. (2017). The essential medicinal chemistry of curcumin: Miniperspective. *Journal of Medicinal Chemistry*, *60*(5), 1620–1637.

Ng, Z. Y., Wong, J. Y., Panneerselvam, J., Madheswaran, T., Kumar, P., Pillay, V., Hsu, A., Hansbro, N., Bebawy, M., Wark, P., & Chellappan, D. K. (2018). Assessing the potential of liposomes loaded with curcumin as a therapeutic intervention in asthma. *Colloids and Surfaces B: Biointerfaces*, *172*, 51–59.

Nguyen, M. H., Yu, H., Kiew, T. Y., & Hadinoto, K. (2015). Cost-effective alternative to nano-encapsulation: amorphous curcumin–chitosan nanoparticle complex exhibiting high payload and supersaturation generation. *European Journal of Pharmaceutics and Biopharmaceutics*, *96*, 1–10.

Nosrati, H., Sefidi, N., Sharafi, A., Danafar, H., & Manjili, H. K. (2018). Bovine Serum Albumin (BSA) coated iron oxide magnetic nanoparticles as biocompatible carriers for curcumin-anticancer drug. *Bioorganic Chemistry*, *76*, 501–509.

Olszewska, M. A., Gędas, A., & Simões, M. (2020). Antimicrobial polyphenol-rich extracts: Applications and limitations in the food industry. *Food Research International*, *134*, 109214.

Pan, Y., Zhu, G., Wang, Y., Cai, L., Cai, Y., Hu, J., & Wei, T. (2012). Attenuation of high-glucose-induced inflammatory response by a novel curcumin derivative B06 contributes to its protection from diabetic pathogenic changes in rat kidney and heart. *The Journal of Nutritional Biochemistry*, *24*(1), 146–155.

Pandey, A., Chaturvedi, M., Mishra, S., Kumar, P., Somvanshi, P., & Chaturvedi, R. (2020). Reductive metabolites of curcumin and their therapeutic effects. *Heliyon*, *6*(11), e05469.

Patel, A., Hu, Y., Tiwari, J. K., & Velikov, K. P. (2010). Synthesis and characterisation of zein–curcumin colloidal particles. *Soft Matter*, *6*(24), 6192–6199.

Pawar, H. A., Gavasane, A. J., & Choudhary, P. D. (2018). A novel and simple approach for extraction and isolation of curcuminoids from turmeric rhizomes. *Natural Product Chemistry Research*, *6*(1), 2.

Prasad, S., Gupta, S. C., Tyagi, A. K., & Aggarwal, B. B. (2014). Curcumin, a component of golden spice: From bedside to bench and back. *Biotechnology Advances*, *32*(6), 1053–1064.

Qu, B., Xue, J., & Luo, Y. (2021). Self-assembled caseinate-laponite® nanocomposites for curcumin delivery. *Food Chemistry*, *363*, 130338.

Rachmawati, H., Yanda, Y. L., Rahma, A., & Mase, N. (2016). Curcumin-loaded PLA nanoparticles: Formulation and physical evaluation. *Scientia Pharmaceutica*, *84*(1), 191–202.

Rafiee, Z., Nejatian, M., Daeihamed, M., & Jafari, S. M. (2019). Application of curcumin-loaded nanocarriers for food, drug and cosmetic purposes. *Trends in Food Science and Technology*, *88*, 445–458.

Rahman, M. H., Ramanathan, M., & Sankar, V. (2014). Preparation, characterization and in vitro cytotoxicity assay of curcumin loaded solid lipid nanoparticle in IMR32 neuroblastoma cell line. *Pakistan Journal of Pharmaceutical Sciences*, *27*(5), 1281–1285.

Ramalingam, P., & Ko, Y. T. (2015). Enhanced oral delivery of curcumin from N-trimethyl chitosan surface-modified solid lipid nanoparticles: Pharmacokinetic and brain distribution evaluations. *Pharmaceutical Research*, *32*(2), 389–402.

Ratner, B. D., Hoffman, A. S., Schoen, F. J., & Lemons, J. E. (2004). *Biomaterials Science: An Introduction to Materials in Medicine*. San Diego, CA: Academic Press, pp. 162–174.

Rubab, S., Naeem, K., Rana, I., Khan, N., Afridi, M., Ullah, I., Shah, F. A., Sarwar, S., ud Din, F., Choi, H. I., & Zeb, A. (2021). Enhanced neuroprotective and antidepressant activity of curcumin-loaded nanostructured lipid carriers in lipopolysaccharide-induced depression and anxiety rat model. *International Journal of Pharmaceutics*, *603*, 120670.

Sasaki, H., Sunagawa, Y., Takahashi, K., Imaizumi, A., Fukuda, H., Hashimoto, T. T., Wada, H., Katanasaka, Y., Kakeya, H., Fujita, M., & Hasegawa, K. (2011). Innovative preparation of curcumin for improved oral bioavailability. *Biological and Pharmaceutical Bulletin*, *34*(5), 660–665.

Sathuvan, M., Thangam, R., Gajendiran, M., Vivek, R., Balasubramanian, S., Nagaraj, S., Gunasekaran, P., Madhan, B., & Rengasamy, R. (2017). κ-Carrageenan: An effective drug carrier to deliver curcumin in cancer cells and to induce apoptosis. *Carbohydrate Polymers*, *160*, 184–193.

Sattari, S., Dadkhah Tehrani, A., & Adeli, M. (2018). pH-responsive hybrid hydrogels as antibacterial and drug delivery systems. *Polymers*, *10*(6), 660.

Saxena, V., Hasan, A., Sharma, S., & Pandey, L. M. (2018). Edible oil nanoemulsion: An organic nanoantibiotic as a potential biomolecule delivery vehicle. *International Journal of Polymeric Materials and Polymeric Biomaterials*, *67*(7), 410–419.

Schneider, C., Gordon, O. N., Edwards, R. L., & Luis, P. B. (2015). Degradation of curcumin: From mechanism to biological implications. *Journal of Agricultural and Food Chemistry*, *63*(35), 7606–7614.

Sheikhpour, M., Sadeghizadeh, M., Yazdian, F., Mansoori, A., Asadi, H., Movafagh, A., & Shahraeini, S. S. (2020). Co-administration of curcumin and bromocriptine nano-liposomes for induction of apoptosis in lung cancer cells. *Iranian Biomedical Journal*, *24*(1), 24.

Shukla, P., Dwivedi, P., Gupta, P. K., & Mishra, P. R. (2014). Optimization of novel tocopheryl acetate nanoemulsions for parenteral delivery of curcumin for therapeutic intervention of sepsis. *Expert Opinion on Drug Delivery*, *11*(11), 1697–1712.

Song, H. Y., & McClements, D. J. (2021). Nano-enabled-fortification of salad dressings with curcumin: Impact of nanoemulsion-based delivery systems on physicochemical properties. *LWT*, *145*, 111299.

Sorasitthiyanukarn, F. N., Muangnoi, C., Rojsitthisak, P., & Rojsitthisak, P. (2021). Chitosan-alginate nanoparticles as effective oral carriers to improve the stability, bioavailability, and cytotoxicity of curcumin diethyl disuccinate. *Carbohydrate Polymers*, *256*, 117426.

Svenson, S. (2009). Dendrimers as versatile platform in drug delivery applications. *European Journal of Pharmaceutics and Biopharmaceutics*, *71*(3), 445–462.

Takahashi, M., Uechi, S., Takara, K., Asikin, Y., & Wada, K. (2009). Evaluation of an oral carrier system in rats: Bioavailability and antioxidant properties of liposome-encapsulated curcumin. *Journal of Agricultural and Food Chemistry*, *57*(19), 9141–9146.

Thulasidasan, A. K. T., Retnakumari, A. P., Shankar, M., Vijayakurup, V., Anwar, S., Thankachan, S., Pillai, K. S., Pillai, J. J., Nandan, C. D., Alex, V. V., & Anto, R. J. (2017). Folic acid conjugation improves the bioavailability and chemosensitizing efficacy of curcumin-encapsulated PLGA-PEG nanoparticles towards paclitaxel chemotherapy. *Oncotarget*, *8*(64), 107374.

Tian, C., Asghar, S., Wu, Y., Chen, Z., Jin, X., Yin, L., Huang, L., Ping, Q., & Xiao, Y. (2017). Improving intestinal absorption and oral bioavailability of curcumin via taurocholic acid-modified nanostructured lipid carriers. *International Journal of Nanomedicine*, *12*, 7897.

Verderio, P., Bonetti, P., Colombo, M., Pandolfi, L., & Prosperi, D. (2013). Intracellular drug release from curcumin-loaded PLGA nanoparticles induces G2/M block in breast cancer cells. *Biomacromolecules*, *14*(3), 672–682.

Wakte, P. S., Sachin, B. S., Patil, A. A., Mohato, D. M., Band, T. H., & Shinde, D. B. (2011). Optimization of microwave, ultra-sonic and supercritical carbon dioxide assisted extraction techniques for curcumin from Curcuma longa. *Separation and Purification Technology*, *79*(1), 50–55.

Weiss, J., Decker, E. A., McClements, D. J., Kristbergsson, K., Helgason, T., & Awad, T. (2008). Solid lipid nanoparticles as delivery systems for bioactive food components. *Food Biophysics*, *3*(2), 146–154.

Wilken, R., Veena, M. S., Wang, M. B., & Srivatsan, E. S. (2011). Curcumin: A review of anti-cancer properties and therapeutic activity in head and neck squamous cell carcinoma. *Molecular Cancer*, *10*(1), 1.

Xie, X., Tao, Q., Zou, Y., Zhang, F., Guo, M., Wang, Y., Wang, H., Zhou, Q., & Yu, S. (2011). PLGA nanoparticles improve the oral bioavailability of curcumin in rats: Characterizations and mechanisms. *Journal of Agricultural and Food Chemistry*, *59*(17), 9280–9289.

Yadav, D. K., Sharma, K., Dutta, A., Kundu, A., Awasthi, A., Goon, A., Banerjee, K., & Saha, S. (2017). Purity evaluation of curcuminoids in the turmeric extract obtained by accelerated solvent extraction. *Journal of AOAC International*, *100*(3), 586–591.

Yadav, V. R., Prasad, S., Kannappan, R., Ravindran, J., Chaturvedi, M. M., Vaahtera, L., Parkkinen, J., & Aggarwal, B. B. (2010). Cyclodextrin-complexed curcumin exhibits anti-inflammatory and antiproliferative activities superior to those of curcumin through higher cellular uptake. *Biochemical Pharmacology*, *80*(7), 1021–1032.

Yallapu, M. M., Jaggi, M., & Chauhan, S. C. (2012). Curcumin nanoformulations: A future nanomedicine for cancer. *Drug Discovery Today*, *17*(1–2), 71–80.

Yang, R., Zhang, S., Kong, D., Gao, X., Zhao, Y., & Wang, Z. (2012). Biodegradable polymer-curcumin conjugate micelles enhance the loading and delivery of low-potency curcumin. *Pharmaceutical Research*, *29*(12), 3512–3525.

Yang, B., Jiang, J., Jiang, L., Zheng, P., Wang, F., Zhou, Y., Chen, Z., Li, M., Lian, M., Tang, S., & Wang, Q. (2020). Chitosan mediated solid lipid nanoparticles for enhanced liver delivery of zedoary turmeric oil in vivo. *International Journal of Biological Macromolecules*, *149*, 108–115.

Yen, F. L., Wu, T. H., Tzeng, C. W., Lin, L. T., & Lin, C. C. (2010). Curcumin nanoparticles improve the physicochemical properties of curcumin and effectively enhance its antioxidant and antihepatoma activities. *Journal of Agricultural and Food Chemistry*, *58*(12), 7376–7382.

Yoon, I. S., Park, J. H., Kang, H. J., Choe, J. H., Goh, M. S., Kim, D. D., & Cho, H. J. (2015). Poly (D, L-lactic acid)-glycerol-based nanoparticles for curcumin delivery. *International Journal of Pharmaceutics*, *488*(1–2), 70–77.

Yu, H., & Huang, Q. (2012). Improving the oral bioavailability of curcumin using novel organogel-based nanoemulsions. *Journal of Agricultural and Food Chemistry*, *60*(21), 5373–5379.

Yu, F., Zhang, Y., Yang, C., Li, F., Qiu, B., & Ding, W. (2021). Enhanced transdermal efficiency of curcumin-loaded peptide-modified liposomes for highly effective antipsoriatic therapy. *Journal of Materials Chemistry B*. *9*(24), 4846–4856

Zamarioli, C. M., Martins, R. M., Carvalho, E. C., & Freitas, L. A. (2015). Nanoparticles containing curcuminoids (Curcuma longa): Development of topical delivery formulation. *Revista Brasileira de Farmacognosia*, *25*, 53–60.

Zhou, Q., Gao, Y., Wang, X., Liu, R., Du, P., Wang, X., Zhang, X., Lu, S., Wang, Z., Shi, Q., & Yang, K. (2020). Nosocomial infections among patients with COVID-19, SARS and MERS: A rapid review and meta-analysis. *Annals of Translational Medicine*, *8*(10), 629.

CHAPTER 6

Bioavailability Constraints of Nanoencapsulated Oils from Chia Seeds and Fish Oils

Rosy Bansal, Renu Sharma, and Kawaljeet Kaur

CONTENTS

6.1 INTRODUCTION

The process of enclosing particles in a protective layer covering or containment material to safeguard a sensitive ingredient or nucleus from undesirable responses is known as particle encapsulation. This entails encasing small particle cores in a protective wall with the goal of preserving the original products' organoleptic and

DOI:10.1201/9781003259183-6

physicochemical qualities while also improving the palatability of volatile odorous substances. In the food sector, encapsulation of flavors and smells is a fast-growing procedure. Before being used as flavoring agents, many fragrance compounds must be transformed into solid products. Nanoencapsulation and microencapsulation (ME) technology in the food sector is a very promising area that will have a significant impact on a wide range of products including functional foods, packaging, preservatives, antioxidants, flavors, and perfumes. Richard Feynman, a physicist at Caltech first proposed the concept of nanotechnology in 1959. He never said "nanotechnology," but he implied that precise manipulation of atoms and molecules will be achievable in the future. Nanotechnology was coined by Norio Taniguchi in 1974 to describe the ability to engineer materials precisely at the nanoscale. In actuality, "Engineer Materials" is now commonly understood to involve the design, characterization, manufacture, and use of materials, and the scope has been broadened to include devices and systems rather than just materials in recent years. Biotechnology, nanotechnology, and information technology have all been combined to create new possibilities. Food nanotechnology traces its roots back to Pasteur's introduction of the pasteurization method (1,000 nm) to kill spoilage germs which marked the beginning of a revolution in food processing and quality enhancement. Despite this, the "nano-world" has captured the attention of the food industry in the last two decades resulting in a new range of materials and processes aimed at improving not only the organoleptic characteristics of food but also other aspects such as nutritional characteristics, safety, and packaging among others. The food and beverage industry is a multitrillion-dollar global enterprise. All of the big food corporations are always seeking for ways to increase production efficiency, food safety, and food quality. Extensive Research and development (R&D) projects are currently underway with the ultimate goal of attaining a competitive advantage and market share.

UV-Vis spectroscopy (UV-Vis) is used for the quantification of scattered and absorbed light passing through a sample. In this technique, a material is positioned between a source of light and a photodetector, and the UV/visible light intensity is monitored both prior to and after the sample is exposed to the light beam. To calculate the wavelength-dependent extinction spectrum of the sample, these measurements are contrasted at each wavelength. In most cases, the data is represented graphically as an extinction function of wavelength. To ensure that the sample extinction spectrum does not contain spectral features from the buffer, each spectrum is corrected using a buffer blank (Titus et al., 2019).

Infrared (IR) spectroscopy is predicated on the measurement of electromagnetic radiation's absorption at wavelengths between 4,000 to 400 cm−1 (Mourdikoudis et al., 2018). There are various degrees of rotational and vibrational energy in molecules. Following the absorption of photons of various wavelengths, vibrational states change. When a molecule absorbs IR light, the dipole moment is altered in some way, making the molecule IR active. The emitted radiation from an infrared source is moved by an interferometer composed of a beam splitter, a fixed mirror, and a moving mirror. The wavelength of the light being emitted is determined by the interferometer using interference patterns. An IR spectrum is produced by exposing a material to IR radiation and measuring the intensity of the radiation that passes through at a specific wavenumber. Molecular structures and interactions can be inferred from a captured spectrum by looking at the positions of bands that are connected to the kind, strength, and nature of bonds and particular functional groups (Eid, 2022).

Changing the texture of food components, encapsulating food components or additives, producing new tastes and sensations, managing the release of flavors, and/or boosting the bioavailability of nutritional components are some of the primary developments of nanotechnology in food science. Nanofood also has applications in improving food color, extending shelf life and preservation, detecting microorganisms and antimicrobial properties, and developing intelligent packaging materials. Furthermore, nanofood encompasses not just processed foods but also entire fields ranging from farming through packaging. Encapsulation is a method of enclosing delicate chemicals in a coating or a wall material. The wall material shields the delicate chemicals from harmful reactions and regulates their release.

The encapsulation of small solid particles, liquid droplets, or gases in a coating (1–1,000 μm) is known as ME. Natural substances such as essential oils (EOs) or vegetal extracts containing polyphenols with well-known antibacterial characteristics can be successfully entrapped using ME. This is a crucial starting point for enterprises, as it allows them to experiment with novel, natural, and safe materials or packaging solutions capable of extending the shelf life of perishable fresh items (vegetables, fruits, meat, etc.) without compromising their quality or hygiene. ME can be used in food packaging to mask disagreeable smells and odors, as well as to provide barriers between sensitive bioactive ingredients and the environment (represented by food or oxygen). This approach can be used to protect bioactive substances such as vitamins, antioxidants, proteins, and carbohydrates in the development of functional meals with improved functioning and stability (Rein et al., 2013). The most significant advantages of nanotechnology for food ingredient ME are as follows:

a. An increase in surface area may improve taste and food ingredient bioavailability: this is especially relevant for flavoring ingredients with low solubility and/or low flavor and odor detection thresholds.
b. Improved solubility of water-insoluble substances such as omega solubilization with a micelle-based method.
c. Optically transparent nanoemulsions and microemulsions with oil droplet sizes less than 100 nm are optically transparent which is significant in beverage applications.
d. During spray drying, more ingredient retention (volatile organic carbon decrease) is achieved. For example, molecular inclusion complexes based on amylose and cyclodextrins are closer to real molecular solutions.
e. Homogeneity in system parameters such as density.
f. In nanoemulsion/microemulsion forms, higher activity levels of encapsulated constituent, e.g., antimicrobials.

6.2 BIOAVAILABILITY OF FOOD COMPONENTS

Bioavailability is the rate and extent to which a bioactive molecule or a medicine is absorbed and becomes available at the site of action from a pharmacological standpoint. Bioavailability is the fraction of a given food that the body can use and is thus a measure of nutritional efficacy. Bioavailability refers to a number of processes, including food matrix release, absorption, distribution, metabolism, and excretion (Pateiro et al., 2021).

6.2.1 Generalizations on Coatings and Carrier Materials

Nanoliposomes, nanoemulsions, nanoparticles, and nanofibers are examples of materials that can be utilized as building blocks to make nanostructures. Both inorganic and organic compounds are used as nanomaterials in culinary applications. Engineered nanomaterials (ENMs) are divided into three types:

I. **Inorganic**: Transition metals, such as silver and iron, alkaline earth metals, such as calcium and magnesium, and nonmetals, such as selenium and silicates, are used in food, food additives, food packaging, and food storage. Metal nanoparticles are most commonly used in food packaging.
II. **Surface-functionalized materials**: Surface-functionalized nanoparticles enhance the matrix's functioning in specific ways (e.g., antimicrobial activity). They are used to bind with the polymer matrix in food packaging materials in order to provide mechanical strength or a barrier against the movement of gases, volatile components (such as flavors), or moisture.
III. **Organic engineered nanomaterials**: Organic nanoparticles are employed in food products because they boost vitamin and antioxidant bioavailability in the body by increasing uptake and absorption.

Nanoencapsulation uses techniques such as nanoemulsification and nanostructuration to safeguard ingredients in micro and deliver a finished product. These chemicals are useful components that are rarely used in their natural state. To begin with, nanoencapsulation acts as a carrier for transporting the functional ingredient to the intended action site. Second, they safeguard the functional ingredient during production, storage, and use from chemical or biological deterioration. Finally, it must be able to regulate the release of the active substance. Also, the delivery method must be compatible with the rest of the system's components as well as the ultimate product's physicochemical and qualitative characteristics. The most essential elements impacting the efficacy of functional additives in food are the characteristics of the delivery system. Nanocarrier systems are often based on carbohydrates, proteins, or lipids. Due to the need for sophisticated chemical or heat treatments, carbohydrate- and protein-based nanocapsules cannot be fully scaled up. Lipid-based nanocarriers, on the other hand can be mass-produced industrially and have higher encapsulation efficiency and lower toxicity.

6.2.2 Bioavailability Characteristics of Different Nanocarriers (Detsi et al., 2020)

i. **Nanosystems based on lipids**: Antioxidant effectiveness is improved by these systems, which improve solubility and bioavailability, *in vitro* and *in vivo* stability, and prevent undesirable interactions with other food components.

Nanoliposomes, nanocochleates, and archaeosomes are the three main lipid nanoencapsulation technologies that can be employed to protect and transport meals and nutraceuticals.

ii. **Nanoliposomes**: Encapsulation and controlled release of food items, as well as improved bioavailability, stability, and shelf life of delicate chemicals, are all applications for nanoliposomes. Nanoliposomes are used to transport nutrients, nutraceuticals, enzymes, food additives, and antimicrobials in food.

iii. **Colloidosomes**: Colloidosomes are hollow capsules made of particles one-tenth the size of a human cell that self-assemble. Any substance's molecule can be placed inside this shell (fat blockers, medicine, and vitamins). Soy lecithin is the primary structural component of aqueous nanodispersions that transport large amounts of water-insoluble actives.

iv. **Nanocochleates**: Nanocochleates are nano-coiled particles that wrap around micronutrients, and they have the ability to maintain and protect a wide range of micronutrients, as well as the potential to boost the nutritional content of processed meals.

v. **Nanoemulsions**: When high-pressure valves, homogenizers, or microfluidizers are used, emulsions with droplet sizes of less than 100–500 nm are frequently produced; these emulsions are referred to as "nanoemulsions." Droplets, the interfacial area, or the continuous phase can all be used to include functional food components. Nanoemulsions have been found to be suitable candidates for delivering food nutrients that are poorly water soluble, such as fish oil and lipophilic vitamins. Food emulsions are commonly made with food-grade ingredients (such as proteins, polysaccharides, and phospholipids) and processing methods (such as homogenization and mixing) (Das et al., 2020).

6.3 NANOENCAPSULATION APPLICATIONS IN THE FOOD INDUSTRY

The ME method can be employed in the food sector for a variety of reasons, which Ferreira and Nunes (2019) summarized as follows (Table 6.1):

- The core material is protected from degradation by reducing its responses to the external environment.
- The main material's evaporation or transfer rate to the external environment is reduced.
- The physical characteristics of the original material are modified to allow easier handling.
- The core material's release is tailored to occur slowly over time or at a specific time.
- The core material's undesirable flavors or tastes are masked.

High-oleic palm oil (HOPO), fish oil, and chia seed oil were investigated to determine the best microfluidization, production, and storage conditions for nanoemulsions, promising new ways to protect the oil against lipid oxidation while also improving solubility and stability. Microfluidization, manufacturing, and storage conditions for nanoemulsions generated from HOPO should be optimized to retain the oils' beneficial characteristics, potentiate their use in the food business. In bioactive oil, chia mucilage can be employed as a wall material.

Table 6.1 Applications of Nanoencapsulated Oils in Food Industry (Self-creation)

S.No.	Nano-Component	Action	Application	References
1.	Fish oil	Yogurt, fruit juice, and other food enrichment; protection against oxidation and strong odors	Foods with superior nutritional value and with high bioavailability; product approach containing no saturated fatty acids but rich in unsaturated fatty acids; beverage industry or as a food seasoning, for example, in fish sauce	Ghorbanzade et al., (2017)
2.	High-oleic palm oil (HOPO)	Obtain the most favorable microfluidization, formation, and storage conditions for the nanoemulsions obtained from HOPO	Potentiate its use in the food industry to preserve the oils' favorable properties	Ricaurte et al. (2016)
3.	Chia seed oil	Protect the oil against lipid oxidation and improve the solubility and stability	Chia mucilage can be used as a wall material in bioactive oil nanoencapsulations for application in the food area to substitute synthetic polymers	De Campo et al. (2017)

6.4 CHIA SEED OIL

Chia (*Salvia hispanica L.*) is a Lamiaceae herbaceous plant native to southern Mexico that is used as whole seeds, flour, seed oil, and mucilage. Chia seed oil contains polyunsaturated fatty acids (PUFA), primarily linoleic and linolenic acids. Chia oil has a larger amount of linolenic acid (64.4%) than walnut oil (11.93%) and flaxseed oil (flaxseed oil) (53.5%). When exposed to light, air, moisture, and heat, unsaturated fatty acids become unstable, resulting in changes in quality and a reduction in shelf life due to the production of off-flavors and changes in oil sensory properties. Encapsulation of chia seed oil is a promising solution in this regard, as it protects the oil from lipid oxidation and improves its solubility (Firtin et al., 2020). Natural antioxidants in chia seed oil, such as tocopherols, phytosterols, carotenoids, and phenolic compounds, keep the oil stable despite its high PUFA concentration. Natural antioxidants that defend against reactive oxygen and nitrogen species damage have sparked a lot of curiosity.

De Campo et al. (2017) produced chia seed oil nanoparticles to test the protective potential of chia seed mucilage as a wall material. Physical stability and high thermal stability were observed in chia seed oil nanoparticles after storage under accelerated circumstances. The nanoparticles were successfully obtained, as evidenced by the high encapsulation efficiency and loading capacity reached, as well as the development of chia seed oil characteristic bands in chia seed oil nanoparticles, as determined by Fourier transform infrared (FT-IR). The findings showed that chia seed oil nanoencapsulation with chia seed mucilage as a wall material improved oil oxidative stability during storage at accelerated conditions for 28 days, implying that chia seed mucilage can be used as a wall material for bioactive oils' nanoencapsulation in the food industry.

6.5 FISH OIL

Fish oil is a key dietary source of PUFAs, particularly eicosapentaenoic acid (EPA) and docosahexaenoic acid (DHA), which are thought to have particularly powerful biological activities and health impacts. The total PUFA profile of fish oil varies depending on the fish species, sex, maturity, nutrition, and habitat, all of which can influence their health benefits. The use of health-promoting omega-3 PUFAs in supplements, pharmaceuticals, and functional foods is limited. Because these unsaturated fatty acids are prone to oxidative breakdown, there is a risk of unwanted odors ("rancidity") being generated, lowering consumer acceptance. Furthermore, some of the lipid oxidation reaction products are hazardous, which might lead to long-term health problems if taken on a frequent basis. Scientists examining the nutritional value of fish oil supplements found that the majority of them were severely oxidized, demonstrating the potential severity of the problem. The need of assessing the oxidation status of commercial fish oil supplements to ensure they are safe is the need of the hour.

Advanced encapsulation technologies, which entail turning omega-3 PUFAs into colloidal forms such as liposomes, lipid droplets, or biopolymer particles, which are subsequently put into foods or beverages, may typically overcome these problems. These colloid materials can be transformed to a powdered form in some applications to make them easier to handle, store, and use, as well as to boost their oxidation resistance. A range of processing techniques, including spray, freeze, and fluidized bed drying, can be used to accomplish this. The use of the right encapsulation technology can improve the chemical stability, water dispersibility, and bioavailability of omega-3 oils significantly. The limited solubility of fish oils in water is another barrier to the creation of omega-3-enriched foods and beverages. Finally, depending on the way in which omega-3 PUFAs are supplied, their bioavailability may be poor and vary. Bulk forms, for example, are adsorbed more slowly and to a lesser amount than emulsified forms.

Before fish oil may be successfully integrated into commercially functional meals and supplements, a number of obstacles must be addressed. Because of the hydrophobic effect, fish oil contains long-chain triacylglycerols, which have a very low solubility in water (Hosseini et al., 2021). As a result, they can't just be mixed with water; instead, they need to be transformed into colloidal form before being used in aqueous-based food and beverage items. Because omega-3 PUFAs have numerous double bonds (–C=C–), they are prone to fast lipid oxidation. Certain environmental variables, including light, heat, oxygen, transition metals, and some enzymes, enhance the rate and degree of lipid oxidation. The oxidation of fish oil results in reaction products with unpleasant odors (rancidity), making them unsuitable for consumption (Vieira et al., 2017). As a result, the rate at which fish oil deteriorates during storage has a significant impact on its shelf life. Furthermore, some of the lipid oxidation reaction products have been proven to be toxic, which could lead to negative health outcomes if oxidized fish oil is consumed on a regular basis. As a result, effective solutions for preventing the oxidation of fish oil in food and supplement items are required. To achieve this purpose, a variety of methods have been developed, including managing storage conditions, packing, and adding antioxidants and chelating agents, as well as sophisticated encapsulation technologies.

Another potential issue with some fish oil formulations is their extremely varied oral bioavailability, which is influenced by the molecular form of the omega-3 fatty acids as well as the composition of the food matrix. The limited solubility of omega-3 oils in aqueous gastrointestinal fluids inside the human gut, as well as their vulnerability to chemical breakdown during passage through the stomach, are the main causes of this impact. Fish oils are absorbed after passing via the mouth, esophagus, stomach, and small intestine after intake. They're exposed to a wide range of digestive enzymes, pH conditions, mechanical forces, flow profiles, mineral ions, bile salts, and other gastrointestinal contents as a result.

Nanoemulsions have recently been used in a variety of sectors, including the food and pharmaceutical industries, to provide an effective nutrition delivery mechanism (Lammari et al., 2021). Vitamin D (a lipophilic molecule) was encapsulated in fish oil for increased oral bioavailability in a study done by Walia et al. (2017). The ultrasonication process was used to create an oil-in-water nanoemulsion with a droplet size range of 300–450 nm and a shelf life of more than 90 days. A phase diagram was used to explore the effects of oil, water, and surfactant concentration. The encapsulation effectiveness of the prepared nanoemulsion ranged from 95.7% to 98.2%. Furthermore, when a nanoemulsion was passed through a simulated gastrointestinal tract, it demonstrated that the bioavailability of the vitamin was higher than when it was not encapsulated. As a result, the formulation can be employed to deliver diverse lipophilic chemicals as a drug delivery vehicle. No one has yet developed an effective nano-vehicle for the delivery of vitamin D nor has anybody examined the effectiveness of the delivery system. According to research, the bioavailability of omega-3 oils in emulsified form is better than in bulk form, which could be owing to the larger surface area accessible for lipases to adsorb in emulsions. Controlling the initial colloidal state of fish oils, such as the content, size, and surface properties of the particles in which they are encapsulated, can thus significantly affect their bioavailability profiles. Furthermore, the bioavailability of omega-3 oils can be enhanced by incorporating absorption enhancers in supplement formulations that boost epithelial cell permeability.

The bioavailability of bioactive chemical components is affected differently depending on the matrix and processes used to encapsulate oils. *In vitro* digestion models have been utilized to investigate the bioavailability of microencapsulated PUFAs in recent years. Oil bioavailability is typically measured using a percentage of free fatty acid (FFA) release after digestion. Reduced droplet size appears to have a good effect on FFA release, according to research (Venugopalan et al., 2021).

6.6 CHARACTERIZATION AND STABILITY OF NANOENCAPSULATED OILS FROM CHIA SEEDS AND FISH

Nanostructures developed via various methods are required to be characterized in terms of particle size, binding nature, and stability (Arpagaus et al., 2018). The characterization methods are mainly categorized into two approaches, namely, physical and chemical characterization. Particle size, zeta potential, thermal properties, and morphology of the nanocapsules are assessed through physical characterization methods while moisture content, fatty acids profile, and chemical stability of nanostructures are generally determined through chemical methods. Identification of these properties enables in better understanding of the interaction of nanoparticles with the other molecules and surrounding environment (Sharma et al., 2019).

A number of analytical tools are available for determining the efficiency of encapsulation as well as characterization of resultant nanostructures. Various instrumental analysis techniques including spectroscopic, microscopic and chromatographic methods are available for analyzing concentration, shape, and dimensions of nanocarriers (Habib et al., 2012). High-resolution electron microscopy methods, such as scanning electron microscopy (SEM), transmission electron microscopy (TEM), and atomic force microscopy (AFM), can provide information about the state of aggregation, structure, particle distribution, and sorption behavior of nanocarriers (Tarhan and Jafari, 2020; Tiede et al., 2008). Various spectroscopic tools including dynamic light scattering, static light scattering, and X-ray photoelectron spectroscopy are recognized for analyzing particle size, size distribution, and surface characteristics like charge and elemental composition of nanocapsules (Ganje et al., 2019; Jia et al., 2015; Montes et al., 2019; Sharma et al., 2019). Other spectroscopic techniques, such as UV-visible, infrared, and fluorescence spectroscopies, can also be employed for characterizing the nanocarriers containing bioactive components.

6.6.1 Physical Characterization Methods

Among the various physical properties of nanocapsules, particle size and its distribution are considered to be one of the most significant features as it directly affects the stability of the bioactive oils. In spite of the availability of various methods such as electro zone particle counters, image analysis, and sedimentation for particle size determination, a laser diffraction method is usually employed for determining particle size of the bioactive oil-loaded nanocapsules (Dodds, 2013). This method does not quantity particle size directly, but it determines diffraction pattern made after the passage of laser light through the dispersion. The diffraction of the light on passing through the particle is inversely proportional to the size of the particle. Thus, the diffraction pattern is measured using wide-angle and backscattering detectors. The signals thus obtained are used to calculate the particle size using Mie theory (Dodds, 2013). Light adsorption and refractive index of the media are also required for particle size determination.

Zeta potential, a physical characteristic corresponding to the nanoparticle's electrical potential, is usually applied to indicate suspension stability in colloidal dispersions. This parameter is significantly affected by the nanocapsule components, pH, concentration, and conductivity of the medium in which they are distributed. It represents the magnitude of electrostatic, repulsive, or attractive forces between particles. The particles in suspension with larger zeta potential values will tend to repel each other and thus lower the stability. Generally, the values of zeta potential $>30\,mV$ and $<-30\,mV$ prevent particles' aggregation and thus promote high stability (Mohanraj and Chen, 2006). In the case of nanoencapsulated oils, variability in the zeta potential is generally observed due to wall material characteristics. De Campo et al. (2017) found a zeta potential of $-11.58 \pm 1.87\,mV$ for encapsulated chia oil with chia seed mucilage as wall material. A negative zeta potential might be owing to the presence of carboxylate anion of carboxylic acidic functional groups (Herculano et al., 2015).

Thermogravimetric analysis (TGA) or differential scanning calorimetry (DSC) is commonly used to measure thermal parameters that reveal the physical stability of nanocapsules. In DSC analysis, nanocapsules involve energy variations due to heating or cooling. This analysis helps to analyze melting points, glass transition temperatures, and lipid oxidation points (Gordon et al., 2015). TGA study, on the other hand, records the weight loss rate, which indicates the deterioration of wall and core materials during heating (Gómez-Mascaraque and López-Rubio, 2016; Hădărugă et al., 2016). Light microscopy (Jafari et al., 2007; Joos et al., 2015), SEM (Moomand and Lim, 2014), TEM (Henna Lu et al., 2011), and AFM (Henna Lu et al., 2011) help to determine morphology of the nanocapsules. The loading capacity and entrapment competence of nanoencapsules are determined by the appropriate binding of bioactives. The stability of nanocapsules is aided by consistent environmental conditions. Controlled alterations in certain factors may result in the release of active compounds in a desired manner (Tontul et al., 2017).

6.6.2 Chemical Characterization Methods

The chemical interactions between bioactive oil and wall materials are determined using FT-IR (Esquerdo et al., 2015). Identification of probable interactions between the core material and encapsulant is a key factor for a successful encapsulation procedure (Abaee et al., 2017; Katouzian and Jafari, 2019). In this respect, FT-IR spectroscopy is an emerging tool that provides reliable information about the structure and conformation of encapsulated structures (Esfanjani et al., 2015; Jafari and Esfanjani, 2017; Picó, 2016). The various advantages of spectroscopic technique include determination of the number and amount of constituents from the number and area of the IR peaks, respectively, characterization of diverse components, and also assessing the consistency of a sample. Thus, the FT-IR spectroscopic technique is known to be a valuable method for characterizing the nanoencapsulation systems (Silva et al., 2012). Nuclear Magnetic Resonance (NMR) spectroscopy, another powerful analytical characterization technique, has also been employed in many nanoencapsulation research works for quantitative and qualitative analyses of enclosed components along with their enclosing materials. Moreover, this spectroscopic method helps in determining the nanoencapsulation efficiency, the most important characteristic that is required to be optimized during the production process of nutraceutical delivery systems.

The chemical stability of bioactive oils is often assessed by analyzing the fatty acid profile and oxidation products during shelf life or oxidation tests. Owing to the presence of a high proportion of fatty acids that are sensitive to environmental factors, the determination of fatty acids' profile is highly recommended in the case of bioactive oils. The gas chromatography technique is generally employed for its determination. Degree of oxidation is expressed in terms of peroxides, anisidine, and volatiles (hexanal, nonanal, etc.). During an oxidation test, the amount of these products in nanoencapsulated bioactive oils is often compared to the amount of bulk oils (Gómez-Mascaraque and López-Rubio, 2016; Moomand and Lim, 2014).

Nanoencapsulation techniques offer several opportunities in food applications. The various functional components incorporated into food products through nanoencapsulation will show an improved shelf life, sensory and textural characteristics, and health benefits. The different structural, morphological, and physicochemical properties of bioactive ingredients contained within various nanocarriers are very significant in the whole encapsulation process and the resultant nanostructures. The encapsulation efficiency, stability, and the releasing tendency of nanoencapsulated food ingredients can be assessed in terms of size, surface charge, surface composition, morphology, chemical composition, and thermal stability. For the determination of these parameters in food bioactive-loaded nanocapsules, a number of novel and advanced techniques are available that are already discussed briefly.

6.7 BIOAVAILABILITY CONSTRAINTS

Awareness about the associated risks of consuming fatty foods has amplified demand for low-calorie products, preferably those richer in dietary fiber. This reduces the amount of fat-soluble nutrients including vitamins A, D, E, and K, antioxidants, and polyunsaturated fatty acids (Considine et al., 2007). Omega-3 is a type of polyunsaturated fatty acid that has been shown to be beneficial to cardiovascular health. The most significant omega-3 fatty acids are linoleic acid, EPA, and DHA, which are found predominantly in seafood and plant products such as chia seeds, walnuts, rapeseeds, and flaxseeds (Lopez-Huertas, 2010). These components are very susceptible to oxidative degradation and generate an unpleasant odor and taste due to the presence of unsaturation in their structure.

Fish oil is highly prone to oxidation due to the high degree of unsaturation of omega-3 long-chain PUFAs. The presence of five to six multiple bonds can accelerate the oxidation reactions in the oil, if safety measures are not implemented during its manufacturing, storage, transportation, and also during processing of fish oil–fortified foods. The various factors, such as moisture, light, oxygen, heat, pH, metal ions, and reactive chemicals, are known for promoting oxidation reactions. These reactions led to the generation of harmful chemical substances, development of off-flavor, and also loss of vital nutrients (Hogan et al., 2003; Park et al., 2004). Although hydrogenation processes can be employed to lower the degree of unsaturation and thus to enhance the oxidative stability of fish oils, it is not suggested as the process involves the generation of nutritionally unfavorable transisomers and also the loss of valuable effects of omega-3 (Lee and Ying, 2008).

Another major problem associated with using fish oil is their unpleasant fishy odor and taste. The products obtained from oxidation of omega-3 fatty acids have been clearly recognized with fishy and metallic essence. These compounds are mainly two isomeric forms of 2,4,7-decatrienal (Lee and Ying, 2008). The various compounds including 4 (cis)-heptenal (fishy odor), 1,5-octadien-3-one, 1-octen-3-one, 2,6 (trans, cis)-nonadienal, 2,4 (trans, trans)-heptadienal (cucumber odor), and 1-penten-3-one (plastic, leather odor) are reported as the chief contributors to the unpleasant, fishy off-flavor of fish oil (Jacobsen et al., 2000). These off-flavors, which are commonly described as fishy, metallic, and rancid, are thought to be one of the most significant deterrents to increasing consumption of fish oil. The major problem for food manufacturers in formulating food items enriched with fish oil is to overcome rapid oxidative deterioration of oil and also to mask fishy aroma and flavor in the final product (Lee and Ying, 2008). Similarly, in chia seed oil, the PUFAs, primarily linoleic and linolenic acid, also develop off-flavors due to oxidative degradation in the presence of environmental factors such as light, oxygen, moisture, and heat (Tonon et al., 2011). This rancidity due to unsaturated fatty acids led to reduced quality, shelf life, and alterations in oil sensory properties (Martinez et al., 2015).

The extremely variable oral bioavailability is also another major challenge in formulating food products with fish oil, which is affected by the molecular form of the omega-3 fatty acids as well as the type of the food matrix (Ahmad et al., 2020). The main cause of this effect is the relatively low solubility of omega-3 oils in the aqueous gastrointestinal fluids inside the human gut, as well as their susceptibility to chemical degradation during passage through the gut. Research reports revealed that the bioavailability of omega-3 oils can be improved on ingesting omega-3 oils in an emulsified form rather than a bulk form (McClements, 2021), which can be due to the availability of larger surface area for lipases absorption in emulsions. Monitoring the initial colloidal state of fish oil, such as the size, composition, and surface features of the particles in which they are encapsulated, can therefore have a significant impact on their bioavailability. The issues of low water solubility, poor chemical stability, and low/variable bioavailability of omega-3 PUFAs can be resolved with the advanced encapsulation technologies (Venugopalan et al., 2021). Keeping in view the nutritional value of edible oils, there has been an increasing interest in edible oil modification technologies in recent years. Various technologies have been discovered by researchers to improve the properties of the oils and make them suited for specific purposes.

6.8 CURRENT STATUS OF NANOENCAPSULATED CHIA SEEDS AND FISH OIL APPLICATIONS

The field of food nanotechnology experienced its greatest growth around 2010 as nanotechnology research and development of food-grade additives suited for the nanoparticle process were conducted in the late nineties. Nanoencapsulation is considered to be a logical evolution of technology in the food industry (Mozafari et al., 2006; Chaudhry et al., 2008). Furthermore, nanoencapsulation is characterized by improving the bioavailability of the encapsulated active ingredients and shielding them from various natural and processing effects including chemical effects (Chau et al., 2007; De Vos et al., 2010), enzymatic effects and also preventing the physical instability perceived during the processing of any food products. Oil encapsulation is considered to be an efficient, viable method for improving physical stability and bioactivity, reducing toxicity, and also providing protection from environmental oxidation reactions that are mainly responsible for reducing sensory and nutritional qualities of any oil. Specifically, in the food industry, this technology enhances the qualities of processed food such as flavor retention, antioxidation, shelf life, color, and off-odor; extends food product storage time; and protects ingredients from the environment, reducing flavor loss during preservation, and controlling the release of bioactive substances (Ferreira and Nunes, 2019; Choi et al., 2009).

Fish oil is the most commonly employed oil in both microencapsulation and nanoencapsulation. It contains unsaturated fatty acids and PUFAs in abundance. Majority of the fatty acids are produced by humans. However, the human body is unable to synthesize omega-6 (n-6) and omega-3 (n-3) fatty acids which are essentially required in human nutrition. The vegetable oils including PUFAs play a significant role in lowering the risk of cardiovascular, neurological disorders and cancer diseases (Rodríguez et al., 2016; Frankel et al., 2013). In spite of having several health benefits, hasty decomposition and strong fishy odor limits its usage in the food industry. For these reasons, Vahidmoghadam et al. (2019) utilized whey protein concentrate (WPC) and gum arabic as wall components for the nanoencapsulation of fish oil. In this study, the influence of WPC gum arabic on the properties of nanoencapsulated fish oil powder was investigated. The results indicated that the nature and concentration of wall material exhibit significant effect on the characteristics of nanocapsules. The highest encapsulation efficiency and smallest emulsion droplet diameter of about 50 nm were obtained in the sample having 100% gum arabic. It can be established that gum arabic is more appropriate for nanoencapsulation and could properly contain the volatile components within the capsules. The branched structure of gum arabic acts as an emulsifier and decreases the particle size through covalent bonding of the oil droplets and the branches. With the increase in emulsification capacity of the wall, the movement of components to the capsule surface decreased.

In the study of De Campo et al. (2017), chia seed oil was nanoencapsulated with chia seed mucilage as the wall component. The nanocapsules were characterized in terms of encapsulation efficiency, loading capacity, and oxidation stability. The thermal properties of chia seed oil nanoparticles were also performed after preparation. The results showed that nanoencapsulated oil exhibits greater stability against oxidation in comparison to unencapsulated oil. The interaction between oil and mucilage was demonstrated by FT-IR spectroscopy. Moreover, chia seed oil nanoparticles were found to be thermally stable at temperatures up to 300°C.

The use of nanotechnology in the food industry has resulted in the production of food products with better thermal stability, solubility, and higher bioavailability, and thus enhances food quality and safety (Hamad et al., 2018). In this sense, encapsulation of oil is a viable solution for the protection of lipid-soluble components and for improving the solubility and stability of oil (Liu et al., 2008). Oil encapsulation might constitute a physical–chemical barrier against various oxidants like oxygen, free radicals, or ultraviolet radiations and thus inhibit oxidation reactions (Gonnet et al., 2010; Fernández-López and Viuda-Martos; 2018) and also broaden the variety of food commodities intended for enrichment purposes.

6.9 CONCLUSION

Regular consumption of omega-3 fatty acid–rich lipids is said to have a wide range of health benefits including the prevention of inflammation, cardiovascular disease, diabetes, arthritis, and ulcerative colitis. Long-chain PUFAs in the omega-3 form which are said to have particularly high biological activity are abundant in lipids extracted from numerous marine sources. Because of their potential nutritional and health benefits, functional food products aimed to improve human health and well-being are increasingly enriched with these omega-3 PUFAs. Food fortification with PUFAs on the other hand is difficult due to their limited water solubility, quick oxidation, and variable bioavailability. Advanced encapsulation technologies, which often entail integrating omega-3 oils into well-designed colloidal particles produced from food-grade components such as liposomes, emulsion droplets, nanostructured lipid carriers, or microgels, can overcome these problems. These omega-3-enriched colloidal dispersions can be employed in a fluid state or can be spray dried into a powdered form which makes them easier to handle and store while also extending their shelf life.

Oil nanoencapsulation could be a viable option for improving the stability and bioavailability of nanoencapsulated substances. Oil nanoencapsulation is becoming increasingly common particularly in the food business. Conventional nanoencapsulation technologies used in various oils have a direct impact on oil nanoparticle production, influencing factors such as zeta potential, size, and polydispersity index which may limit the use of oils in many industries. Oil nanoencapsulation has seen significant growth in the food business during the last 20 years. Although there are currently no nanoencapsulated oil products in the food sector, various articles in the food science field claim that oil nanoencapsulation will become a market trend. Nonetheless, as indicated by patent applications, nanoencapsulated oils can be used in a variety of applications. Encapsulating and protecting omega-3-rich marine oils has been successfully accomplished using a variety of colloidal delivery methods. However, emulsions and nanoemulsions are usually the most feasible for this purpose since they can be easily synthesized with food-grade ingredients and existing processing activities like mixing and homogenization. Furthermore, they can be made from a variety of food-grade emulsifiers, including proteins, polysaccharides phospholipids, and surfactants giving them a lot of versatility in terms of tailoring their features. Future research will focus on determining the best encapsulation technology for certain food products. Furthermore, before these techniques can be widely used commercially, their economic viability and scalability must be established.

REFERENCES

Abaee, A., Mohammadian, M., & Jafari, S. M. (2017). Whey and soy protein-based hydrogels and nano-hydrogels as bioactive delivery systems. *Trends in Food Science & Technology*, *70*, 69–81.

Ahmad, M. Z., Ahmad, J., Zafar, S., Warsi, M. H., Abdel-Wahab, B. A., Akhter, S., & Alam, M. A. (2020). Omega-3 fatty acids as adjunctive therapeutics: Prospective of nanoparticles in its formulation development. *Therapeutic Delivery*, *11*(1), 851–868.

Arpagaus, C., Collenberg, A., Rütti, D., Assadpour, E., & Jafari, S. M. (2018). Nano spray drying for encapsulation of pharmaceuticals. *International Journal of Pharmaceutics, 546*(1–2), 194–214.

Chau, C. F., Wu, S. H., & Yen, G. C. (2007). The development of regulations for food nanotechnology. *Trends in Food Science & Technology, 18*(5), 269–280.

Chaudhry, Q., Scotter, M., Blackburn, J., Ross, B., Boxall, A., Castle, L., Aitken, R., & Watkins, R. (2008). Applications and implications of nanotechnologies for the food sector. *Food Additives and Contaminants, 25*(3), 241–258.

Choi, M. J., Ruktanonchai, U., Soottitantawat, A., & Min, S. G. (2009). Morphological characterization of encapsulated fish oil with β-cyclodextrin and polycaprolactone. *Food Research International, 42*(8), 989–997.

Considine, T., Patel, H. A., Singh, H., & Creamer, L. K. (2007). Influence of binding conjugated linoleic acid and myristic acid on the heat-and high-pressure-induced unfolding and aggregation of β-lactoglobulin B. *Food Chemistry, 102*(4), 1270–1280.

Das, A. K., Nanda, P. K., Bandyopadhyay, S., Banerjee, R., Biswas, S., & McClements, D. J. (2020). Application of nanoemulsion-based approaches for improving the quality and safety of muscle foods: A comprehensive review. *Comprehensive Reviews in Food Science and Food Safety, 19*(5), 2677–2700.

De Campo, C., Dos Santos, P. P., Costa, T. M. H., Paese, K., Guterres, S. S., de Oliveira Rios, A., & Flôres, S. H. (2017). Nanoencapsulation of chia seed oil with chia mucilage (*Salvia hispanica* L.) as wall material: Characterization and stability evaluation. *Food Chemistry, 234*, 1–9.

De Vos, P., Faas, M. M., Spasojevic, M., & Sikkema, J. (2010). Encapsulation for preservation of functionality and targeted delivery of bioactive food components. *International Dairy Journal, 20*(4), 292–302.

Detsi, A., Kavetsou, E., Kostopoulou, I., Pitterou, I., Pontillo, A. R. N., Tzani, A., Christodoulou, P., Siliachli, A., & Zoumpoulakis, P. (2020). Nanosystems for the encapsulation of natural products: The case of chitosan biopolymer as a matrix. *Pharmaceutics, 12*(5), 669.

Dodds, J. (2013). Techniques to analyse particle size of food powders. In: Bhandari, B. R., Bansal, N., Zhang, M., & Schuck, P. (Eds.), *Handbook of Food Powders* (pp. 309–338). Woodhead Publishing: Sawston, Cambridge.

Eid, M. M. (2022). Characterization of nanoparticles by FTIR and FTIR-microscopy. In: *Handbook of Consumer Nanoproducts* (pp. 1–30). Springer Singapore: Singapore.

Esfanjani, A. F., Jafari, S. M., Assadpoor, E., & Mohammadi, A. (2015). Nano-encapsulation of saffron extract through double-layered multiple emulsions of pectin and whey protein concentrate. *Journal of Food Engineering, 165*, 149–155.

Esquerdo, V. M., Dotto, G. L., & Pinto, L. A. A. (2015). Preparation of nanoemulsions containing unsaturated fatty acid concentrate–chitosan capsules. *Journal of Colloid and Interface Science, 445*, 137–142.

Fernández-López, J., & Viuda-Martos, M. (2018). Introduction to the special issue: Application of essential oils in food systems. *Foods, 7*(4), 56.

Ferreira, C. D., & Nunes, I. L. (2019). Oil nanoencapsulation: Development, application, and incorporation into the food market. *Nanoscale Research Letters, 14*(5), 1–13.

Firtin, B., Yenipazar, H., Saygun, A., & Şahin Yeşilçubuk, N. (2020). Encapsulation of chia seed oil with curcumin and investigation of release behaivour Burcu & antioxidant properties of microcapsules during in vitro digestion studies. *LWT-Food Science and Technology*, 134, 109947.

Frankel, E., Bakhouche, A., Lozano-Sánchez, J., Segura-Carretero, A., & Fernández-Gutiérrez, A. (2013). Literature review on production process to obtain extra virgin olive oil enriched in bioactive compounds. Potential use of byproducts as alternative sources of polyphenols. *Journal of Agricultural and Food Chemistry, 61*(22), 5179–5188.

Ganje, M., Jafari, S. M., Tamadon, A. M., Niakosari, M., & Maghsoudlou, Y. (2019). Mathematical and fuzzy modeling of limonene release from amylose nanostructures and evaluation of its release kinetics. *Food Hydrocolloids, 95*, 186–194.

Ghorbanzade, T., Jafari, S. M., Akhavan, S., & Hadavi, R. (2017). Nano-encapsulation of fish oil in nano-liposomes and its application in fortification of yogurt. *Food Chemistry, 216*, 146–152.

Gómez-Mascaraque, L. G., & López-Rubio, A. (2016). Protein-based emulsion electrosprayed micro-and submicroparticles for the encapsulation and stabilization of thermosensitive hydrophobic bioactives. *Journal of Colloid and Interface Science, 465*, 259–270.

Gonnet, M., Lethuaut, L., & Boury, F. (2010). New trends in encapsulation of liposoluble vitamins. *Journal of Controlled Release, 146*(3), 276–290.

Gordon, V., Marom, G., & Magdassi, S. (2015). Formation of hydrophilic nanofibers from nanoemulsions through electrospinning. *International Journal of Pharmaceutics, 478*(1), 172–179.

Habib, S. M., Amr, A. S., & Hamadneh, I. M. (2012). Nanoencapsulation of alpha-linolenic acid with modified emulsion diffusion method. *Journal of the American Oil Chemists' Society, 89*(4), 695–703.

Hădărugă, D. I., Ünlüsayin, M., Gruia, A. T., Birău, C., Rusu, G., & Hădărugă, N. G. (2016). Thermal and oxidative stability of Atlantic salmon oil (*Salmo salar* L.) and complexation with β-cyclodextrin. *Beilstein Journal of Organic Chemistry*, *12*(1), 179–191.

Hamad, A. F., Han, J. H., Kim, B. C., & Rather, I. A. (2018). The intertwine of nanotechnology with the food industry. *Saudi Journal of Biological Sciences, 25*(1), 27–30.

Henna Lu, F. S., Nielsen, N. S., Timm-Heinrich, M., & Jacobsen, C. (2011). Oxidative stability of marine phospholipids in the liposomal form and their applications. *Lipids*, *46*(1), 3–23.

Herculano, E. D., de Paula, H. C., de Figueiredo, E. A., Dias, F. G., & Pereira, V. D. A. (2015). Physicochemical and antimicrobial properties of nanoencapsulated Eucalyptus staigeriana essential oil. *LWT-Food Science and Technology, 61*(2), 484–491.

Hogan, S. A., O'riordan, E. D., & O'sullivan, M. (2003). Microencapsulation and oxidative stability of spray-dried fish oil emulsions. *Journal of Microencapsulation, 20*(5), 675–688.

Hosseini, S. F., Ramezanzade, L., & McClements, D. J. (2021). Recent advances in nanoencapsulation of hydrophobic marine bioactives: Bioavailability, safety, and sensory attributes of nano-fortified functional foods. *Trends in Food Science & Technology*, *109*, 322–339.

Jacobsen, C., Hartvigsen, K., Lund, P., Adler-Nissen, J., Hølmer, G., & Meyer, A. S. (2000). Oxidation in fish-oil-enriched mayonnaise. *European Food Research and Technology, 210*(4), 242–257.

Jafari, S. M., & Esfanjani, A. F. (2017). Instrumental analysis and characterization of nanocapsules. In: Jafari, S. M. (Ed.), *Nanoencapsulation Technologies for the Food and Nutraceutical Industries* (pp. 524–544). Academic Press: Cambridge, MA.

Jafari, S. M., He, Y., & Bhandari, B. (2007). Optimization of nano-emulsions production by microfluidization. *European Food Research and Technology, 225*(5), 733–741.

Jia, D., Barwal, I., Thakur, S., & Yadav, S. C. (2015). Methodology to nanoencapsulate hepatoprotective components from Picrorhiza kurroa as food supplement. *Food Bioscience, 9*, 28–35.

Joos, A., Weiss, J., & McClements, D. J. (2015). Fabrication of lipophilic nanoparticles by spontaneous emulsification: Stabilization by cosurfactants. *Food Biophysics, 10*(1), 83–93.

Katouzian, I., & Jafari, S. M. (2019). Protein nanotubes as state-of-the-art nanocarriers: Synthesis methods, simulation and applications. *Journal of Controlled Release, 303*, 302–318.

Lammari, N., Louaer, O., Meniai, A. H., Fessi, H., & Elaïssari, A. (2021). Plant oils: From chemical composition to encapsulated form use. *International Journal of Pharmaceutics*, *601*, 120538.

Lee, S. J., & Ying, D. Y. (2008). Encapsulation of fish oils. In: Garti, N. (Ed.) *Delivery and Controlled Release of Bioactives in Foods and Nutraceuticals* (pp. 370–403). CRC Press: Boca Raton, FL.

Liu, Z., Jiao, Y., Wang, Y., Zhou, C., & Zhang, Z. (2008). Polysaccharides-based nanoparticles as drug delivery systems. *Advanced Drug Delivery Reviews, 60*(15), 1650–1662.

Lopez-Huertas, E. (2010). Health effects of oleic acid and long chain omega-3 fatty acids (EPA and DHA) enriched milks: A review of intervention studies. *Pharmacological Research, 61*(3), 200–207.

Martinez, M. L., Curti, M. I., Roccia, P., Llabot, J. M., Penci, M. C., Bodoira, R. M., & Ribotta, P. D. (2015). Oxidative stability of walnut (*Juglans regia* L.) and chia (*Salvia hispanica* L.) oils microencapsulated by spray drying. *Powder Technology, 270*, 271–277.

McClements, D. J. (2021). Advances in edible nanoemulsions: Digestion, bioavailability, and potential toxicity. *Progress in Lipid Research, 81*, 101081.

Mohanraj, V. J., & Chen, Y. (2006). Nanoparticles-a review. *Tropical Journal of Pharmaceutical Research, 5*(1), 561–573.

Montes, C., Villaseñor, M. J., & Rios, A. (2019). Analytical control of nanodelivery lipid-based systems for encapsulation of nutraceuticals: Achievements and challenges. *Trends in Food Science & Technology, 90*, 47–62.

Moomand, K., & Lim, L. T. (2014). Oxidative stability of encapsulated fish oil in electrospun zein fibres. *Food Research International, 62*, 523–532.

Mourdikoudis, S., Pallares, R. M., & Thanh, N. T. (2018). Characterization techniques for nanoparticles: Comparison and complementarity upon studying nanoparticle properties. *Nanoscale, 10*(27), 12871–12934.

Mozafari, M. R., Flanagan, J., Matia-Merino, L., Awati, A., Omri, A., Suntres, Z. E., & Singh, H. (2006). Recent trends in the lipid-based nanoencapsulation of antioxidants and their role in foods. *Journal of the Science of Food and Agriculture, 86*(13), 2038–2045.

Park, Y., Kelleher, S. D., McClements, D. J., & Decker, E. A. (2004). Incorporation and stabilization of omega–3 fatty acids in Surimi made from Cod, Gadus morhua. *Journal of Agricultural and Food Chemistry, 52*(3), 597–601.

Pateiro, M., Gómez, B., Munekata, P. E., Barba, F. J., Putnik, P., Kovačević, D. B., & Lorenzo, J. M. (2021). Nanoencapsulation of promising bioactive compounds to improve their absorption, stability, functionality and the appearance of the final food products. *Molecules, 26*(6), 1547.

Picó, Y. (2016). Challenges in the determination of engineered nanomaterials in foods. *TrAC Trends in Analytical Chemistry, 84*, 149–159.

Rein, M. J., Renouf, M., Cruz-Hernandez, C., Actis-Goretta, L., Thakkar, S. K., & da Silva Pinto, M. (2013). Bioavailability of bioactive food compounds: A challenging journey to bioefficacy. *British Journal of Clinical Pharmacology, 75*(3), 588–602.

Ricaurte, L., de Jesús Perea-Flores, M., Martinez, A., & Quintanilla-Carvajal, M. X. (2016). Production of high-oleic palm oil nanoemulsions by high-shear homogenization (microfluidization). *Innovative Food Science & Emerging Technologies, 35*, 75–85.

Rodríguez, J., Martín, M. J., Ruiz, M. A., & Clares, B. (2016). Current encapsulation strategies for bioactive oils: From alimentary to pharmaceutical perspectives. *Food Research International, 83*, 41–59.

Sharma, S., Jaiswal, S., Duffy, B., & Jaiswal, A. K. (2019). Nanostructured materials for food applications: Spectroscopy, microscopy and physical properties. *Bioengineering, 6*(1), 26.

Silva, H. D., Cerqueira, M. Â. & Vicente, A. A. (2012). Nanoemulsions for food applications: Development and characterization. *Food and Bioprocess Technology, 5(3)*, 854–867.

Tarhan, Ö., & Jafari, S. M. (2020). Spectroscopic and chromatographic analyses of nanoencapsulated food ingredients. In: & Jafari, S. M. (Ed.), *Characterization of Nanoencapsulated Food Ingredients* (pp. 585–615). Academic Press: Cambridge, MA.

Tiede, K., Boxall, A. B., Tear, S. P., Lewis, J., David, H., & Hassellöv, M. (2008). Detection and characterization of engineered nanoparticles in food and the environment. *Food Additives and Contaminants, 25*(7), 795–821.

Titus, D., Samuel, E. J. J., & Roopan, S. M. (2019). Nanoparticle characterization techniques. In: *Green Synthesis, Characterization and Applications of Nanoparticles* (pp. 303–319). Elsevier: Amsterdam, Netherlands.

Tonon, R. V., Grosso, C. R., & Hubinger, M. D. (2011). Influence of emulsion composition and inlet air temperature on the microencapsulation of flaxseed oil by spray drying. *Food Research International, 44*(1), 282–289.

Tontul, İ., Eroglu, E., & Topuz, A. (2017). Nanoencapsulation of fish oil and essential fatty acids. In: Jafari, S. M. (Ed.), *Nanoencapsulation of Food Bioactive Ingredients* (pp. 103–144). Academic Press: Cambridge, MA.

Vahidmoghadam, F., Pourahmad, R., Mortazavi, A., Davoodi, D., & Azizinezhad, R. (2019). Characteristics of freeze-dried nanoencapsulated fish oil with whey protein concentrate and gum arabic as wall materials. *Food Science and Technology, 39*, 475–481.

Venugopalan, V. K., Gopakumar, L. R., Kumaran, A. K., Chatterjee, N. S., Soman, V., Peeralil, S., Mathew, S., McClements, D. J., & Nagarajarao, R. C. (2021). Encapsulation and protection of omega-3-rich fish oils using food-grade delivery systems. *Foods, 10*(7), 1566.

Vieira, S. A., Zhang, G., & Decker, E. A. (2017). Biological implications of lipid oxidation products. *Journal of the American Oil Chemists' Society, 94*(3), 339–351.

Walia, N., Dasgupta, N., Ranjan, S., Chen, L., & Ramalingam, C. (2017). Fish oil based vitamin D nanoencapsulation by ultrasonication and bioaccessibility analysis in simulated gastro-intestinal tract. *Ultrasonics Sonochemistry, 39*, 623–635.

CHAPTER 7

Development and Characterization of Nanocomposite for Organic Acids

Prerna Gupta, Tanu Malik, Rhythm Kalsi, and Jasleen Kaur

CONTENTS

DOI: 10.1201/9781003259183-7

7.1 INTRODUCTION

Organic acids (OA) are a class of natural plant acids that have been neglected due to their lesser-known effects. They contain a functional carboxylic group (R-COOH) attached via an organic group and are commonly known by various names as fatty acid, volatile fatty acid, lipophilic group, or carboxylic acids (Cherrington et al., 1991). These acids are said to play a principal role in maintaining the quality and nutritional value of food and possess anti-oxidative, antimicrobial, and anti-inflammatory properties (Robles et al., 2019). They exert an antimicrobial activity on the gut of microorganisms and improve protein digestibility and amino acid absorption (Akbar et al., 2019). These acids were also recognized as safe for animal feeding by the European Union (EU) (Adil et al., 2010). OA are naturally present as ingredients in foods and can also be either added directly or indirectly to the product after processing. OA are usually the fermentation by-products of microorganisms such as lactic acid bacteria (LAB), which produce lactic acid in turn serving as an important antibacterial and bio-preservative substance (Theron and Lues, 2007).

OA exist in two main forms: pure organic acid or buffered organic acids. Lactic, propionic, acetic, citric, and benzoic acids are the class of pure acids while their respective salts are the buffered acids. These acids have a straight and even number of carbon chains of 4 to 24. They are generally classified based on four categories; (i) On the basis of chain length into three groups as long, medium, and short-chain fatty acids (aromatic, aliphatic, alicyclic, and heterocyclic), (ii) on the basis of a number of functional groups (Table 7.1). The first group involves glycolytic pathway and tricarboxylic acid (biosynthetic pathway from glucose cycle), and the second group involves "small pathway," fundamentally a bioconversion of glucose (Mattey, 1992), (iii) saturation or unsaturation properties; and (iv) substituted or nonsubstituted features. In nature, most organic acids react with glycerol to form triglycerides and phosphoglycerides or their derivatives. For example, one molecule of glycerol can bind up to three molecules of fatty acid, one for each hydroxyl attach site (positions 1, 2, and 3 on the glycerol carbon chain), forming a triacylglycerol. Positions 1 and 3 are said to be chemically equivalent. Once ingested, triglycerides undergo lipolysis in the intestine leading to the formation of diglycerides, monoglycerides, free fatty acids, and free glycerol. Some common examples of organic acids are: butyric acid, acetic acid, proponic acid, caproic acid, linolenic acid, capric acid, lauric acid, etc. (Scitutella et al., 2021). The demand of organic acids is increasing due to their promising effect of inhibition of microorganisms and is introduced as preservatives to food.

Of the various acids present, acetic, citric, proponic, sorbic, and benzoic acids are majorly used and exist between the undissociated and dissociated states. Their versatility lies in the mechanism of their action. These acids decrease the intracellular pH by dissociating inside the cell which is overall required for the control of ATP synthesis, RNA and protein synthesis and DNA replication and cell growth (Anyasi et al., 2017) (Figure 7.1).

OA are more versatile in their role in plant stress tolerance and are more efficient chelating agents than other acids such as amino acids. Root OA exudation is important in soil carbon sequestration. These functions are key processes in combating

Table 7.1 Classification of Organic Acids Based on Chain Length (Self-creation)

Chain Length	Type	Molar Mass (g/mol)	Molecular Formula	Structure
Short chain	Formic acid	46.02	CH_2O_2	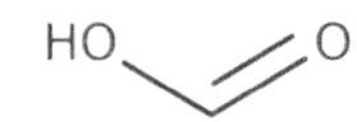
	Acetic acid	60.052	CH_3COOH	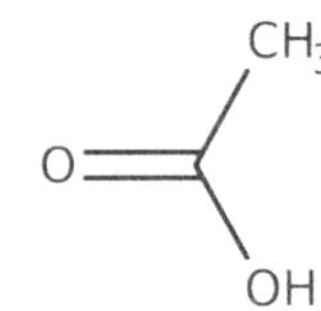
	Proponic acid	74.08	CH_3CH_2COOH	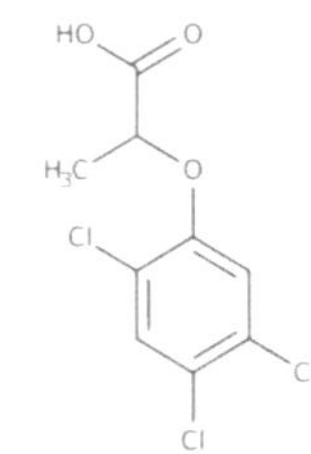
	Butyric acid	88.11	$CH_3(CH_2)_2COOH$	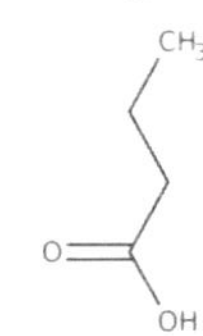

(*Continued*)

Table 7.1 (*Continued*) Classification of Organic Acids Based on Chain Length (Self-creation)

Chain Length	Type	Molar Mass (g/mol)	Molecular Formula	Structure
	Lactic acid	90.1	$C_3H_6O_3$	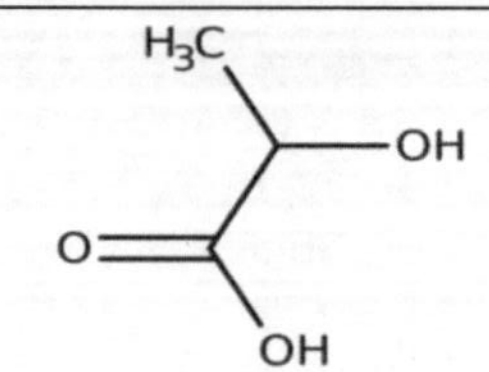
	Valeric acid	102.13	$CH_3(CH_2)_3COOH$	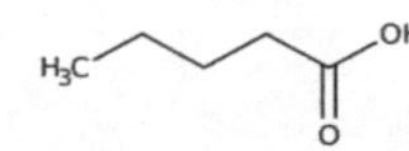
Medium chain	Caproic acid	116.1583	$CH_3(CH_2)4COOH$	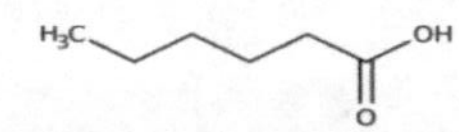
	Benzoic acid	122.1	$C_7H_6O_2$	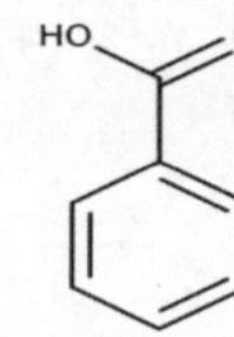
	Enanthic acid	130.1849	$CH_3(CH_2)_5COOH$	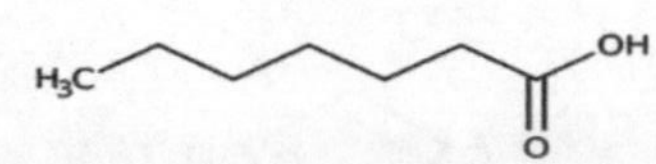
	Caprylic acid	144.21	$CH_3(CH_2)_6COOH$	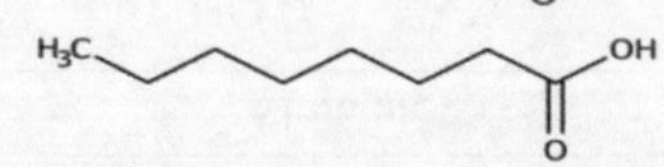

(*Continued*)

Table 7.1 (*Continued*) Classification of Organic Acids Based on Chain Length (Self-creation)

Chain Length	Type	Molar Mass (g/mol)	Molecular Formula	Structure
	Capric acid	172.26	$CH_3(CH_2)_8COOH$	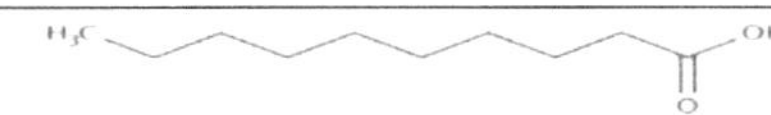
Long chain	Lauric acid	200.3178	$CH_3(CH_2)_{10}COOH$	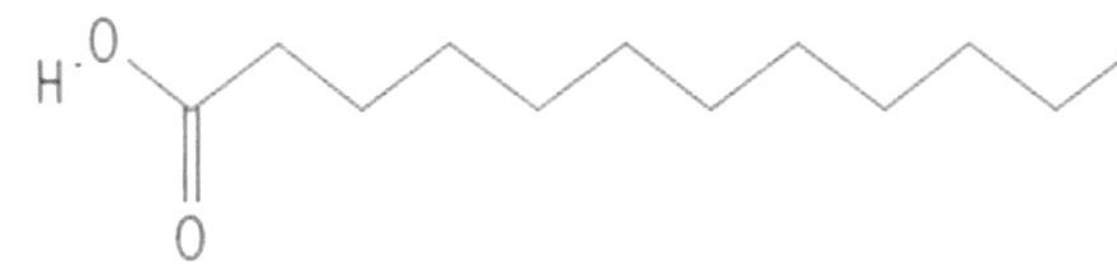
	Myristic acid	228.37	$CH_3(CH_2)_{12}COOH$	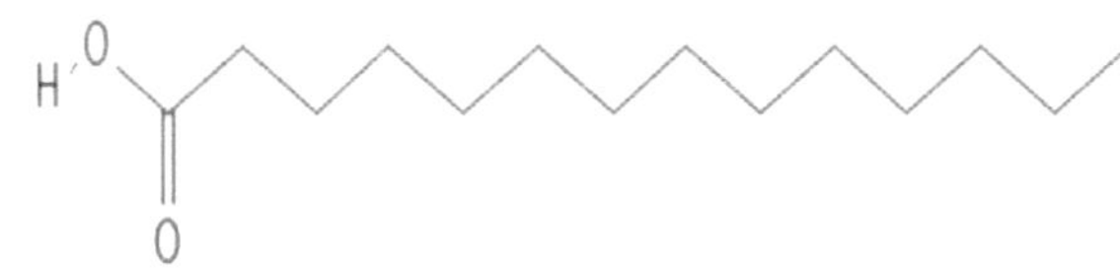
	Palmitic acid	256.4	$CH_3(CH_2)_{14}COOH$	
	Stearic acid	284.48	$CH_3(CH_2)_{16}COOH$	

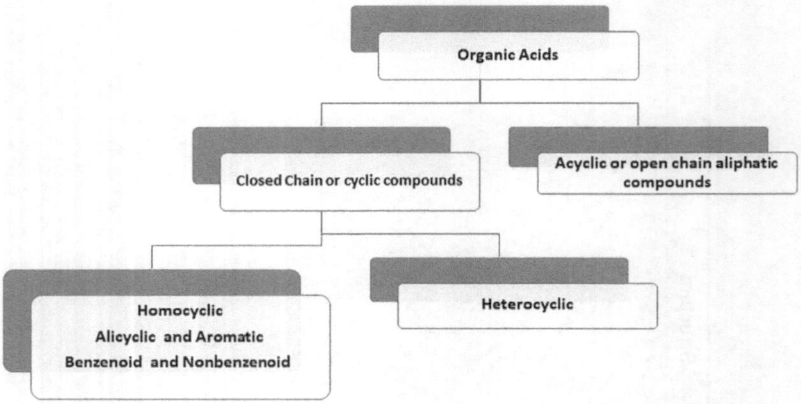

Figure 7.1 Nomenclature of organic acids.

climate change and helping with more sustainable food production. These compounds can be added as acidulants or stabilizers (e.g., citric, ascorbic, benzoic, fumaric, and malic acids) and contribute to the organoleptic (flavor, color, and aroma) properties. In the case of insufficient sterilization and/or microbial contamination during storage, sugar fermentation results in the formation of volatile acids (C_2eC_{12}) and impairs the quality of some products (Ohria et al., 2013). In addition, as natural components, organic acids (antioxidant and antimicrobial activities) are present in all types of food, such as fruits and vegetables, mushrooms and beverages, such as juice, coffee, tea, and wine (Sandín-Espana et al., 2016). Their functions include: "acids" modify pH, "chelators"-binding metals, or as "carbon sources" for microbes (Panchal et al., 2021). Kelebek et al. (2009) have evaluated citric, malic, and ascorbic acids in orange juice and orange wine using HPLC-DAD. Similarly, simultaneous determination of sugars and organic acids in wines and grape juices with HPLC using RID and DAD was described by Monteiro Coelho et al. (2018). UV detection is the most common detection system coupled to HPLC, and these methods were also discussed by Kritsunankul et al. (2009), Cunha et al. (2002) and Zatou et al. (2004). As it has been commented previously, many complex matrices require previous treatments for the chromatographic determination of organic acids. To avoid this problem, Ohira et al. (2014) have reported an electrodialytic transfer of organic acids from wines to ultrapure water using cellulose membranes modified with N, N-dimethylaminoethyl methacrylate before ion exclusion chromatography and direct UV detection.

7.2 FOOD PRESERVATION BY VARIOUS ORGANIC ACIDS

OA help in maintaining the pH of foods and further delaying the microbial growth and spoilage by fungi stats, mold, and rope inhibitors. Microorganisms have an optimum pH level for their growth and any alteration in hydrogen ion concentration leads to their inhibition. Microorganisms are unable to grow under increased acidity conditions and adding an acidulant will further provide a microbiostatic condition for them. Thus to reduce the spoilage of food products, organic acids can be used as a natural, effective, and inexpensive intruder. The effectiveness of these acids depends on the lowering of pH in a product and the pKa (dissociation constant), and the dissociation of the acids lies between 3 and 5 pH and shows their activity in higher acid foods. OA can penetrate the lipid bilayer of the cell membrane in the dissociated form. This dissociation occurs because of the higher pH in the interior of the cell. They are found naturally or can be added directly or indirectly into the foods and can often be produced by microorganisms. OA are said to possess antimicrobial, antimycotic activity (Davidson et al., 2005) and are ubiquitous by-products of organisms like LAB (Theron and Lues, 2007).

Most OA have an advantage because of their simple structure and small molecular size or mass which allow them to move freely throughout cells. OA also have many vital roles and functions in plants including contributing to growth and production of fruit and are therefore applied in many sectors of agriculture. The most important acids possessing strong antimicrobial activity are acetic, citric, lactic, proponic, benzoic, and sorbic acids (Table 7.2.).

7.2.1 Citric Acid

Citric acid is an organic acid that has anti-oxidative, acidulative, sequestrating, buffering, flavoring, and salt-forming properties. Citric acid is used in fruit, soft drinks, laxatives kosher foodstuff, and seasoned meat. Presently *Aspergillus Niger* is found to be an acceptable microorganism for the commercial production of citric acid. The yield could be as high as 57% (Soccol et al., 2006; Dhillon et al., 2011).

7.2.2 Acetic Acid

Acetic acid is one of the oldest known chemicals and is produced naturally during spoilage of fruit and certain other foods by the bacterium, which is fermented both in aerobic and anaerobic conditions. Acetobactum, e.g., is used in the production of vinegar primarily by oxidative fermentation (Molina et al., 2019). Acetic acid is used in fish fingers, butter and margarine, cooking oil, processed cheeses, and curry powder.

7.2.3 Benzoic Acid

Benzoic acid is a food preservative and occurs naturally in certain plants as well as animals. Benzoic acid is used in margarine, certain fish products, fruit juices, and marmalades. Benzoic acid is found naturally at a high level in many fruits like plums, berries, cinnamon, and prunes (Anyasi et al., 2017). Salts of benzoic acid are used as food additives for preservation against bacteria, fungi, and yeast (Bilau et al. 2008). Most of the activity of yeasts and molds are inhibited by addition of 0.002%–0.07% benzoic acid.

Table 7.2 Food Applications of Various Organic Acids along with the Target Microorganisms Involved (Self-Creation)

Organic Acids	Applications	Microorganisms	References
Acetic acid	Bakery and confectionary products, meat and meat products, dairy products, spices and condiments, fats and oils	Bacteria and yeasts	Stratford and Eklund (2003)
Benzoic acid	Fermented and unfermented beverages, margarine, processed fruit products	Yeasts and molds	
Lactic acid	Meat and meat products	Bacteria	Theron and Lues (2007)
Citric acid	Fruit juices, beverages	Bacteria	
Propionic acid	Bakery and dairy products	Molds (*Propionibacterium* sp.)	Ammar et al. (2014) and Wang et al. (2015)
Sorbic acid	Food and drinks	Bacteria, yeast, molds	
Fumaric acid	Animal feed and acidity regulator	Bacteria and yeast (*Rhizopus nigricans*, *Rhizopus arrhizus*, *Rhizopus oryzae*, *Rhizopus formosa*)	Molina et al. (2019)
Malic acid	Food flavoring agent in candy and beverages (vegetable juices, carbonated soft drinks, jams and wines)	Yeast (*Saccharomyces cerevisiae*)	Lee et al. (2019) and Zambaninia et al. (2017)

7.2.4 Propionic Acid

Propionic acid and its salts are already commonly used in the baking industry to extend shelf life. *P. shermanii* and Propionobacteria use a variety of carbohydrate sources such as glucose, fructose, xylose, whey lactose, hemicellulose, and glycerol as carbon sources for the production of propionic acid (Liu et al., 2012).

7.2.5 Lactic Acid

Lactic acid is produced by LAB which are responsible for major biochemical reactions. It is majorly utilized in the production of dairy products such as yogurt, cultured yogurt, and cheese. Major lactic acid production is contributed by bacterial fermentation supporting more yields (Axelsson, 2004). Lactic acid is safe for human consumption by acting as a food additive. Also, it has been majorly used for making polylactic acid polymers for eco-friendly packaging films (Li and Borodina, 2015).

7.2.6 Sorbic Acid

Sorbic acid is a sparingly water-soluble acid, which is used as a preservative in a wide variety of foods. Combination of sorbic acid and lactic acid has also been studied to show the inhibitory properties against *Salmonella* pathogen (Stopforth et al., 2005). Sorbic acid is also widely used to inhibit the growth of yeast and molds in processed cheese, wine, fruit juices, and meat products with nitrites (Zeece, 2020).

7.2.7 Fumaric Acid

Fumaric acid is widely used as a food acidulant in the beverage industry. It provides a fruit-like flavor and has also been used for making baking powders. Fumaric acid was originally isolated from plants belonging to the genus "Fumaria" (Yadav et al., 2022). It is found naturally in fruits such as papaya, pear, and plums. Fumaric acid has been approved by FDA and GRAS in the United States of America; it is mostly applied in fruit juices as a preservative or as an acidulant (Gurtler and Mai, 2014).

7.3 NANOCARRIERS USED IN ENCAPSULATION

Organic acids can be used to improve gut health by direct delivery to the target organ in the form of nano capsules. This will further help in reducing the intestinal pathogenic load without delay. (Dittoe et al., 2018). It has been observed by Hassan et al. (2010) that organic acids especially of one to seven carbon chains are effective enough to improve the digestibility of proteins and minerals such as calcium, phosphorus, magnesium, and zinc in chicken gut. As the pH of the gut is maintained at 3.5–4, the growth of *lactobacilli* is facilitated, whereas growth of *Escherichia coli, Salmonella*, and other pathogenic microorganisms decreases. Due to the emerging growth and adaptability of different pathogens, especially in case of pathogenic microorganisms, blends of encapsulated organic acids have appeared in the market. Encapsulation technique allows for the slow release at the site of action of each acid to the target organ (van den Borne et al., 2015). This allows for an increase in the absorption of nutrients and consequently reduces the amount of the product in the diet. This led to an increase in the maintenance and potentiation of organic acid activity on performance as compared to that of conventional antimicrobials (Galli et al., 2020) or free fatty acids including short-chain fatty acids (C1–C6) and medium-chain fatty acids (C7–C12) (Abdelli et al., 2020). Tungoli et al. (2020) have also used microencapsulation technique as a tool to deliver organic acids along the intestine and allow their positive effects.

7.4 CHARACTERIZATION OF NANOCOMPOSITES

Performance aspects such as mechanical, barrier, optical, thermal, biodegradation, antibacterial, and other functional qualities should be studied for applications of polymer nanocomposites. The microstructure of polymer nanocomposites influences their characteristics. Various polymers/nanofillers and nanocomposites showed a significant increase in characteristics which can be attributed to the increased interfacial area between nanofillers and polymer matrices.

OA are important analytes that can be found in a variety of complex matrices including biological, industrial, environmental, and food samples. The chemical structure, microstructure, and morphology as well as the physical properties of nanocomposites are analyzed using a variety of characterization methods. Many of these techniques are tailored to the characterization of specific features of nanocomposites and the properties of nanocomposites are examined as well. Several characterization techniques are frequently employed to completely comprehend the structure–property connections. The properties of polymer blend nanocomposites are substantially influenced by their content, particle size, interfacial contact, and other factors (Nalwa, 2003). Due to organic acids' excellent sensitivity and precision, high-performance liquid chromatography (HPLC) is often used to determine individual levels of OA (Shui and Leong, 2002; Coelho et al., 2008; Kelebek et al., 2009).

Nanocomposites are typically made up of a matrix material with nano-sized reinforcing components such as particles, whiskers, fibers, nanotubes, and so on (Tai et al., 2003). Various equipment and techniques such as atomic force microscopy (AFM), scanning tunneling microscopy (STM), Fourier transformed infrared spectroscopy (FTIR), X-ray photoelectron spectroscopy (XPS), nuclear magnetic resonance (NMR), differential scanning calorimetry (DSC), scanning and transmission electron microscopies (SEM/TEM), have been used to characterize nanocomposites by various researchers. As evidenced by the work of Veith et al. (2001), AFM is a potent technique for studying the surface down to the nanometer scale.

7.4.1 Acetic Acid

It has the $C_2H_4O_2$ formula and a molecular weight of 60 g/mole making it the second simplest carboxylic acid after formic acid. Because of the high amount of alkyl groups in its structure, acetic acid has higher water solubility than other acids (Vargas, 2016). It is also known as ethanoic acid since it is usually formed by the oxidation of ethanol or acetaldehyde (Li et al., 2015). In the food sector, it has been employed as an emulsifier, stabilizer, flavor enhancer, and firming agent with the European E260 designation in addition to its antibacterial qualities. The annual biological production of acetic acid is expected to be in excess of 190,000 tonnes (Syldatk, 2006).

In situ polymerization with AgNPs was used to create nanocomposites based on indole-3-acetic acid. FTIR and UV spectral analyses were used to completely characterize the parent hydrogel and its nanocomposites. SEM, TEM, and EDAX investigation validated the surface morphology and transport of nanoparticles in polymer hydrogels. TGA–DTA analysis was used to investigate thermal stability. The pH-sensitive swelling behavior of the hydrogel and nanocomposite was induced with increased swelling at pH 3.0 and decreased swelling in pH 10.0 media. The hydrogel and nanocomposites had antifungal properties against *Aspergillus fumigatus*, *Rhizopus oryzae*, and *Candida albicans*. In tests against *Bacillus aureus*, *Staphylococcus aureus*, and *E. coli*, the nanocomposites demonstrated outstanding antibacterial activity. Using the DPPH and NO radicals, the antioxidant properties of the produced hydrogel and nanocomposites' radical scavenging abilities were investigated. The antibacterial, antifungal, and antioxidant capabilities of the nanocomposites as well as their pH-responsive and thermally stable properties can be used in wound healing and burn dressing as well as the treatment of topical fungal infections and other biomedical applications (Chitra et al., 2016).

7.5 TOOL FOR FOOD AND POULTRY INDUSTRY

The meat industry is continually confronted with new waves of changes in consumer nutritional trends, food safety, quality requirements, and legislations resulting in a surge in interest in meat bio-preservation to meet all of these current socio-economic demands. As a result, new strategies for protecting meat products from microbial infection have been developed to replace synthetic and/or expensive additives. OA and their salts have been cited as the most common examples of preservatives that have a variety of benefits when used in the meat business. Over the last decade, there has been an increase in consumer desire for more natural foods which has sparked a lot of interest in food bio-preservation. It is a recent trend that has led to the discovery and development of new antimicrobials from various natural sources (animal, plant, and microbial) (Britton, 2018; Ben Braïek and Smaoui, 2021). OA and their derivative salts are among the more well-studied antimicrobials as bio-preservatives in food systems. Since lactic, acetic, and citric acids along with their salts exist naturally in foods, they have aroused researchers' attention. Furthermore, these preservatives have antimicrobial properties inhibiting the growth of a variety of spoilage and pathogenic microorganisms, improving sensory properties, and having technological functions such as color stabilization, acidity regulation, and the development of distinct flavors in a variety of foods (Hauser et al., 2016). Many organic acids are GRAS and could be employed in the food industry (Theron and Lues, 2010).

The key issues in meat and meat products are the prevention of hazardous bacteria' growth and the preservation of sensory qualities (color, flavor, aroma, texture, and so on). As a result, organic acids and their derivative salts appear to be very good examples of natural preservatives for achieving these goals as well as reacting to present consumer trends and food legislation linked to meat bio-preservation. OA and their salts have recently been shown to have significant utility as natural preservatives of meat and meat products and their use represents a viable biotechnological strategy. In reality, their use is advantageous since they are GRAS-certified natural food components that are produced naturally by a variety of microorganisms (e.g., LAB, yeasts, and fungi) (Hauser et al., 2016). As a result of their widespread and natural distribution, their extraction appears to be considerably easier and straightforward than that of other rare antimicrobials. Another benefit of using organic acids and their salts in meat and meat products is that they have no effect on organoleptic quality when used in the proportions permitted by food legislation.

Furthermore, most of them such as lactic, citric, and acetic acids are not restricted in the ADI for humans (Mani-López et al., 2012). Many of the identified organic acids and salts have demonstrated potent antimicrobial activity against spoilage and pathogenic microorganisms as well as significant improvements in the physicochemical (delay of lipid and protein oxidation) and sensory properties (stabilization of red color, inhibition of off-odor, and enhancement of characteristic flavor or taste or tenderness or juiciness) of treated meats implying that these bioagents coexist. Finally, it is crucial to mention that the use of organic acids and their salts in meat preservation has grown increasingly popular in recent years due to the low cost, ease of use, and speed of application of acid treatment processes (Mani-López et al., 2012). When all of these potential aspects and benefits of organic acids and their salts are considered, it is reasonable to conclude that they are attracting considerable interest which favors their use as innovative natural food preservatives to ensure food safety, improve quality, and extend shelf life of meats.

Antibacterial compounds that are both potent and food-safe and capable of decontaminating carcass surfaces have been studied for a long time because of their importance to meat quality and shelf life. The antibacterial effect of lactic acid (LA) and acetic acid (AA) spray treatments was studied on the aerobic plate count, enterobacteriacae count, coliform count, and *Staphylococcus* count of fresh sheep carcasses' surface after 20 minutes of spraying using three concentrations (1%, 1.5%, and 2%). The results of the investigated bacteriological parameters showed significant reductions after being exposed to organic acids particularly gram-negative bacteria (Enterobacteriacae) which were more sensitive to the used organic acids than gram-positive bacteria (*Staphylococcus*) where higher concentrations resulted in greater bacterial counts reduction. Furthermore, lactic acid spray washing reduced bacterial counts on meat surfaces more effectively than acetic acid. According to the findings, organic acids were found to be a safe, simple, efficient, inexpensive, and highly effective method of meat decontamination. Furthermore, application of lactic acid 2.0% spray had a higher antibacterial effect, so it is recommended to improve the safety of sheep carcasses on a large scale (Saad et al., 2020).

7.5.1 Beef Meat

When 30 g/kg sodium lactate was added to vacuum-packaged raw ground beef, it had an interesting effect on the microbiological and chemical properties (Sallam and Samejima, 2004). In fact, adding sodium lactate to the beef greatly slowed the expansion of aerobic plate counts, psychrotrophic counts, LAB, and Enterobacteriaceae and significantly extended the shelf life of the treated beef to 15 days compared to just 8 days for the control. Furthermore, during the storage trial (21 days at 2°C), this organic salt (sodium lactate) kept the ground beef at a nearly constant pH and did not impair lipid oxidation. Quilo et al. (2009) also found that sprinkling 3% potassium lactate on beef trimmings before grinding could increase sensory (odor and taste) and technological (cooking yield) qualities while maintaining physicochemical (pH and TBARS) parameters.

Another study comparing the effect of sodium and potassium lactates at 2% on the cook yield and tenderness of bovine chuck muscles to noninjected controls found a significant ($P<0.05$) decrease in the WBSF (Warner–Bratzler shear force) and a remarkable improvement in sensory attributes (tenderness and cook yield rating) compared to noninjected controls (Walsh et al., 2010). According to Carpenter et al. (2011), lactic and acetic acids at 2% have substantial ($P<0.05$) antibacterial activity on the decontamination of *E. coli* O157:H7 onto beef meat surfaces. In fact, when compared to control samples (no-wash meats) or water-wash meat samples, these organic acid washes reduced the pathogen by 0.6–1 log/cm^2 and prevented its residual development. The chemical, microbiological, and sensory properties of raw beef meat held at 4°C were investigated using the same organic acids (lactic and acetic acids at 1% and 2%) and their salts (sodium lactate and sodium acetate at 2.5%). The antibacterial activity of these additions was found to be effective against the multiplication of aerobic and psychrotrophic bacteria as well as Enterobacteriaceae according to the findings. The pH of treated beef samples was found to be significantly lower ($P<0.05$) in chemical analysis. OA /salts improved the color, flavor, and texture attributes of treated beef cuts with 1% acetic acid and 1% lactic acid receiving the highest scores (Ben Braïek and Smaoui, 2021).

7.5.2 Sheep Meat

The application of lactic acid at 2% after slaughter and one-day cold storage resulted in significant ($P<0.05$) reduction rates of total viable bacteria, coliforms, and *E. coli*, implying that lactic acid at 2% combined with good sanitation and handling practices could give safer sheep meat (Beyaz and Tayar, 2010). Recently, it was discovered that marinating sheep meat in sodium lactate at 2% enhanced the water-holding capacity, solubility of the protein fraction, and marinade uptake (Vlahova-Vangelova et al., 2017). Furthermore, adding citric acid to sheep "Buchada" reduced microbiological growth (thermotolerant *coliforms*, *Staphylococcus* spp., and *Salmonella* sp.), pH, moisture, and TBARS value while having no effect on sensory qualities (odor, flavor, look, and tenderness) (Albuquerque et al., 2019).

Finally, the antibacterial effect of 1%, 1.5% and 2% lactic and acetic acid spray treatments on the aerobic plate count, *Enterobacteriaceae*, *coliform*, and *Staphylococcus* counts on fresh sheep carcass surfaces was investigated. After exposing sheep flesh to organic acids, the researchers discovered significant reductions in the microbial development of the examined bacteria with a stronger effect of 2% lactic acid spraying wash (Saad et al., 2020).

7.5.3 Poultry Meat Products

They are a type of poultry meat that comes from chickens. The *E. coli* O157:H7, *Listeria monocytogenes*, and *Salmonella typhimurium* populations were significantly ($P<0.05$) reduced by 1–6 log CFU/g in chicken meat by vacuum infusion with acetic, citric, lactic, malic, and tartaric acids (each at 75.0 or 150.0 mM) over 12 days of storage at 4°C and 150.0 mM was the most effective antimicrobial concentration (Over et al., 2009). In addition, the effects of lactic acid at concentrations of 0.2%, 0.3%, 0.5%, 0.6%, 0.8%, and 1% as well as its sodium lactate salt at concentrations of 1%, 1.5%, 2%, 2.5%, and 3% on chemical, microbiological, and sensory properties of marinated chicken thighs stored at 4°C for 15 days were investigated (Smaoui et al., 2011). These additions were effective against the multiplication of spoilage and pathogenic microorganisms such as aerobic and psychrotrophic bacteria, *Pseudomonas* spp., *Salmonella* spp., *Enterobacteriaceae*, and *Staphylococcus aureus* according to the findings.

Chemical analysis revealed that the tested organic acid and its salt lowered the pH value (6.06 at the end of the storage period) and total volatile basic nitrogen (TVBN) contents in treated chicken samples (with 0.2%, 0.5%, and 1% lactic acid and 1%, 2%, and 3% sodium lactate). Furthermore, lactic acid and sodium lactate improved sensory attributes (odor, flavor, color, and texture) with 1% lactic acid achieving the highest overall acceptability score (7.94 on a ten-point hedonic scale note) followed by 0.8% lactic acid (score of 7.38) and 3% sodium lactate (score of 7.38) (score of 6.94). With the general order of the applied organic preservatives as lactic acid > sodium lactate, all of these results allowed for a significant extension of the shelf life of chilled chicken items. Furthermore, 2% potassium lactate was demonstrated to significantly reduce L. monocytogenes and *Salmonella* spp. levels in refrigerated chicken Bologna samples, extending their microbiological stability and shelf life to 49 days (Casco et al., 2015).

According to the results of a recent study (González Sánchez, 2020) on the decontamination efficacy of formic acid (1.5%) and peroxyacetic acid (550 ppm) against *Campylobacter jejuni* growth in chicken wings stored at 4°C for 24 hours by immersion and spraying, acid treatments were effective in reducing *C. jejuni* counts regardless of the antimicrobial application method. Considering all of the abovementioned applications of organic acids (lactic, citric, acetic, formic, malic, and tartaric acids) and their salts as shelf-life extenders and bioagents in the preservation of various meat matrices during refrigerated storage, it has been determined that the weak organic acids and salts were most commonly used. In fact, these are the well-studied agents, more effective than other organic acids/salts, more available and naturally abundant, and finally legally allowed to be used with quantum satis favoring their use in the meat industry to replace commonly used and expensive chemical preservatives ensuring their safety, improving their quality, and safely extending their shelf life.

REFERENCES

Abdelli, N., Perez, J. F., Villarasa, E., Luna, I.C., Duran, D.M., Angelo, M.D., and Orol, D.S. (2020). Targeted-release organic acids and essential oils improve performance and digestive function in broilers under a necrotic enteritis challenge. *Animals, 10*(2), 259. doi: 10.3390/ani10020259.

Adil, S., Banday, T., Bhat, G.A., Mir, M.S., and Rehman, M. (2010). Effect of dietary supplementation of organic acids on performance, intestinal histomorphology, and serum biochemistry of broiler chicken. *Veterinary Medicine International, 10*, 479485.

Akbar, M.A., Tewatia, B., and Kumar, S. (2019). Effect of dietary supplementation of salts of organic acids on growth performance, carcass traits and meat composition of broilers. *International Journal of Chemical Studies, 7*, 2825–2828.

Albuquerque, G.N., Costa, R.G., Barba, F.J., Gómez, B., Ribeiro, N.L., BeltrãoFilho, E.M., and Lorenzo, J.M. (2019). Effect of organic acids on the quality of sheep "buchada": From food safety to physicochemical, nutritional, and sensorial evaluation. *Journal of Food Processing and Preservation, 43*(3), e13877.

Ammar, E.M., Jin, Y., Wang, Z., and Yang, S.-T. (2014). Metabolic engineering of Propionibacterium freudenreichii: Effect of expressing phosphor enolpyruvate carboxylase on propionic acid production. *Applied Microbioogy and Biotechnology, 98*, 7761–7772.

Anyasi, T.A., Jideani, A.I.O., Edokpayi, J.N., and Anokwuru, C.P. (2017). Applications of organic acids in food preservation. In: Vargas, C. (Ed.), *Organic Acids*. Nova Science Publishers: Hauppauge, NY, pp. 1–46.

Axelsson, L. (2004). Lactic acid bacteria: Classification and physiology. In: Salminen, S., Von Wright, A., and Ouwehand, A. (Eds.), *Microbiological and Functional Aspects*. Marcel Dekker, Inc: New York, pp. 1–67.

Ben Braïek, O., and Smaoui, S. (2021). Chemistry, safety, and challenges of the use of organic acids and their derivative salts in meat preservation. *Journal of Food Quality*, 2021, Article ID 6653190.

Beyaz, D., and Tayar, M. (2010). The effect of lactic acid spray application on the microbiological quality of sheep carcasses. *Journal of Animal and Veterinary Advances*, *9*(13), 1858–1863.

Bilau, M., Matthys, C., Vinkx, C., and de Henauw, S. (2008). Intake assessment for benzoates in different subgroups of the Flemish population. *Food Chemistry and Toxicology*, *46*(2), 717–723.

Britton, B.C. (2018). Effects of antimicrobial interventions on food safety and an assessment of the colorado pork supply. Doctoral dissertation, Colorado State University.

Carpenter, C.E., Smith, J.V., and Broadbent, J.R. (2011). Efficacy of washing meat surfaces with 2% levulinic, acetic, or lactic acid for pathogen decontamination and residual growth inhibition. *Meat Science, 88*(2), 256–260.

Casco, G., Taylor, T.M., and Alvarado, C.Z. (2015). Evaluation of novel micronized encapsulated essential oil–containing phosphate and lactate blends for growth inhibition of listeria monocytogenes and salmonella on poultry bologna, pork ham, and roast beef ready-to-eat deli loaves. *Journal of Food Protection, 78*(4), 698–706.

Cherrington, C.A., Hinton, M., Mead, G.C., and Chopra, I. (1991). Organic acids: Chemistry, antibacterial activity and practical applications. *Advances in Microbial Physiology,* 32, 87–108.

Chitra, M., Uthayarani, K., Rajasekaran, N., Neelakandeswari, N., Girija, E. K., Padiyan, D. P., and Mangamma, G. (2016). Enhanced ethanol gas sensing performance of zinc–tin–vanadium oxide nanocomposites at room temperature. *RSC Advances*, *6*(112), 111526–111538.

Coelho, E. M., da Silva Padilha, C. V., Miskinis, G. A., de Sá, A. G. B., Pereira, G. E., de Azevêdo, L. C., and dos Santos Lima, M. (2018). Simultaneous analysis of sugars and organic acids in wine and grape juices by HPLC: Method validation and characterization of products from northeast Brazil. *Journal of Food Composition and Analysis*, *66*, 160–167.

Cunha, S.C., Faria, M.A., Fernandes, J.O., Ferreira, I.M.P.L.V.O., and Ferreira, M.A. (2002). Quantification of organic acids in grape musts and Port wines. *Ciencia Technology Ailment Journal, 3*(4), 212–216.

Davidson, P.M., Sofos, J.N., and Branen, A.L. (2005). *Antimicrobials in Food*, Third Edition. CRC Press, Taylor & Francis Group: Boca Raton, FL.

Dhillon, G.S., Brar, S.K., Verma, M., and Tyagi, R.D. (2011). Recent advances in citric acid bio-production and recovery. *Food and Bioprocess Technology, 4*, 505–529.

Dittoe, D.K., Ricke, S.C., and Kiess, A.S. (2018). Organic acids and potential for modifying the avian gastrointestinal tract and reducing pathogens and disease. *Frontier Veteniary Science*, 5. doi:10.3389/fvets.2018.00216.

Galli, G.M., Gerbet, R.R., Griss, L.G., Fortuoso, B.F., Petrolli, T.G., Boiago, M.M., Souza, C.F., Baldissera, M.D., Mesadri, J., Wagner, R., Da Rosa, G., Mendes, R.E., Gris, A., and Da Silva, A.S. (2020). Combination of herbal components (curcumin, carvacrol, thymol, cinnamaldehyde) in broiler chicken feed: Impacts on response parameters, performance, fatty acid profiles, meat quality and control of coccidia and bacteria. *Microbial Pathology*, 139, 103916. doi: 10.1016/j.micpath.2019.103916.

González Sánchez, S.V. (2020). Efficacy of antimicrobial treatments against Salmonella enterica on pork and Campylobacter jejuni on poultry. Doctoral dissertation, Colorado State University.

Gurtler, J.B., and Mai, T.L. (2014). Preservatives | traditional preservatives: Organic acids. In: Batt, C.A., and Tortorello, M.L. (Eds.), *Encyclopedia of Food Microbiology,* Second Edition. Academic Press: Cambridge, MA, pp. 119–130. ISBN: 780123847331, doi: 10.1016/B978-0-12-384730-0.00260-3.

Hassan, H.M.A., Mohamed, M.A., Amani Youssef, W., and Hassan, E.R. (2010). Effect of using organic acids to substitute antibiotic growth promoters on performance and intestinal microflora of broilers. *Asian-Australian Journal of Animal Sciences, 23*, 1348–1353.

Hauser, C., Thielmann, J., and Muranyi, P. (2016). Organic acids: Usage and potential in antimicrobial packaging. In: Barros-Velazquez, J. (Ed.), Antimicrobial Food Packaging. Academic Press: Cambridge, MA, pp. 563–580.

Kelebek, H., Selli, S., Canbas, A., and Cabaroglu, T. (2009). HPLC determination of organic acids, sugars, phenolic compositions and antioxidant capacity of orange juice and orange wine made from a Turkish cv. Kozan. *Microchemical Journal, 91*(2), 187–192.

Kritsunankul, O., Pramote, P., and Jakmunee, J., (2009). Flow injection on-line dialysis coupled to high performance liquid chromatography for the determination of some organic acids in wine. *Talanta, 79*(4), 1042–1049.

Lee, J.A., Ahn, J.H., and Lee, S.Y. (2019). 3.15- Organic acids: Succinic and malic acids. In: Young, M.M. (Ed.), *Comprehensive Biotechnology*, Third Edition. Pergamon: Oxford, pp. 172–187. ISBN: 9780444640475.

Li, M., and Borodina, I. (2015). Application of synthetic biology for production of chemicals in yeast Saccharomyces cerevisiae. *FEMS Yeast Research*, *15*(1), 1–12.

Li, Y., He, D., Niu, D., and Zhao, Y. (2015). Acetic acid production from food wastes using yeast and acetic acid bacteria micro-aerobic fermentation. *Bioprocess and Biosystems Engineering*, *38*(5), 863–869.

Liu, L., Zhu, Y., Li, J. et al. (2012). Microbial production of propionic acid from propionibacteria: Current state, challenges and perspectives. *Critical Reviews in Biotechnology*, *32*, 374–381.

Mani-López, E., García, H.S., and López-Malo, A. (2012). Organic acids as antimicrobials to control Salmonella in meat and poultry products. *Food Research International*, *45*(2), 713–721.

Mattey, M. (1992). The production of organic acids. *Critical Reviews in Biotechnology*, *12*(1–2), 87–132. doi: 10.3109/07388559209069189.

Molina, G., Gupta, V., Singh, B., and Gathergood, N. (2019). Research and production of organic acids and industrial potential. In: Molina, G., Gupta, V.K., Singh, B.N., and Gathergood, N. (Eds.), *Bioprocessing for Biomolecules Production*. John Wiley & Sons: Hoboken, NJ, pp. 195–209. doi: 10.1002/9781119434436.ch9.

Nalwa, H.S. (2003). *Handbook of Organic-Inorganic Hybrid Materials and Nanocomposites*. American Scientific Publishers: Valencia, CA.

Ohira, S.-I., Kuhara, K., Shigetomi, A., Yamasaki, T., Kodama, Y., Dasgupta, P.K., and Toda, K. (2014). On-line electrodialytic matrix isolation for chromatographic determination of organic acids in wine. *Journal of Chromatography A, 1372,* 18–24. doi: 10.1016/j.chroma.2014.10.077.

Over, K.F., Hettiarachchy, N., Johnson, M.G., and Davis, B. (2009). Effect of organic acids and plant extracts on Escherichia coli O157: H7, Listeria monocytogenes, and Salmonella Typhimurium in broth culture model and chicken meat systems. *Journal of Food Science, 74*(9), M515–M521.

Panchal, P., Miller, A.J., and Giri, J. (2021).Organic acids: Versatile stress response roles in plants. *Journal of Experimental Botany, 72*(11), 4038–4052. doi: 10.1093/jxb/erab019.

Quilo, S.A., Pohlman, F.W., Brown, A.H., Crandall, P.G., Dias-Morse, P.N., Baublits, R.T., and Aparicio, J.L. (2009). Effects of potassium lactate, sodium metasilicate, peroxyacetic acid, and acidified sodium chlorite on physical, chemical, and sensory properties of ground beef patties. *Meat Science, 82*(1), 44–52.

Robles, A., Fabjanowicz, M., Chmiel, T., and Wasylka, J.P. (2019). Determination an identification of organic acids in wine samples. Problems and challenges. *TrAC Trends in Analytical Chemistry, 120,* 115630.

Saad, S.M., Hassanin, F.S., Salem, A.M., and Saleh, E.A.E. (2020). Efficiency of some organic acids as decontaminants in sheep carcasses. *Benha Veterinary Medical Journal, 38*(2), 116–119.

Sallam, K.I., and Samejima, K. (2004). Microbiological and chemical quality of ground beef treated with sodium lactate and sodium chloride during refrigerated storage. *LWT Food Science and Technology, 37*(8), 865–871.

Sandín-Espana, P., Mateo-Miranda, M., Opez-Goti, C.L., De Cal, A., and Alonso- Prados, J.L. (2016). Development of a rapid and direct method for the determination of organic acids in peach fruit using LC e ESI-MS, *Food Chemistry, 192,* 268–273.

Scitutella, F., Manelli, F., Daghio, M., Viti, C., and Buccioni, A. (2021). Polyphenols an organic acids as alternatives to antimicrobials in poultry rearing: A review. *Antibiotics, 10*(8), 1010. doi: 10.3390/antibiotics10081010.

Shui, G., & Leong, L. P. (2002). Separation and determination of organic acids and phenolic compounds in fruit juices and drinks by high-performance liquid chromatography. *Journal of chromatography A, 977*(1), 89–96.

Smaoui, S., Hlima, H.B., Salah, R.B., and Ghorbel, R. (2011). Effects of sodium lactate and lactic acid on chemical, microbiological and sensory characteristics of marinated chicken. *African Journal of Biotechnology, 10*(54), 11317–11326.

Soccol, C.R., Vandenberghe, L.P.S., Rodrigues, C., and Pandey, A. (2006). New perspectives for citric acid production and application. *Food Technology and Biotechnology,* 44, 141–149.

Stopforth, J., Sofos, J., and Busta, F. (2005). Sorbic acid and sorbates. In: Davidson, M., Sofos, J., and Branen, A. (Eds.), *Antimicrobials in Foods.* CRC Press: Boca Raton, FL, pp. 49–90.

Stratford, M., and Eklund, T. (2003). Organic acid and esters. In: Russell, N.J. and Gould, G.W. (Eds.), *Food Preservatives.* Springer: Berlin, Germany, pp. 48–84. ISBN: 978-1-4757-1006-9.

Syldatk, C. (2006). Mikrobielle Stoffproduktion. In: Antranikian, G. (Ed.), *Angewandte Mikrobiologie.* Springer, Berlin, Heidelberg, pp. 305–313.

Tai, W.P., Kim, Y.S., and Kim, J.G. (2003). Fabrication and magnetic properties of Al2O3/Co nanocomposites. *Materials Chemistry and Physics, 82*(2), 396–400.

Theron, M.M., and Lues, J.F.R. (2007). Organic acids and meat preservation: A review. *Food Reviews International, 23*(2), 141–158. doi: 10.1080/87559120701224964.

Theron, M.M., and Lues, J.R. (2010). *Organic Acids and Food Preservation.* CRC Press: Boca Raton, FL.

Tungoli, B., Giovagnoni, G., Piva, A., and Grilli, E. (2020). From acidifiers to intestinal health enhancers: How organic acids can improve growth efficiency of pigs. *Animals*, *10*(1), 134.

van den Borne, J.J.G.C., Heetkamp, M.J.W., Buyse, J., and Niewold, T.A. (2015). Fat coating of Ca butyrate results in extended butyrate release in the gastrointestinal tract of broilers. *Livestock Science*, *175*, 96–10.

Vargas, C. (2016). *Organic Acids: Characteristics, Properties and Synthesis*. Biochemistry Research Trends Series. Nova Science Publishers: Hauppauge, NJ.

Veith, M., Mathur, S., Shen, H., Lecerf, N., Hüfner, S., and Jilavi, M.H. (2001). Single-step preparation of oxide - oxide nanocomposites: Chemical vapor synthesis of LnAlO3/ Al2O3 (Ln=Pr, Nd) thin films. *Chemistry of Materials, 13*(11), 4041–4052.

Vlahova-Vangelova, D.B., Dragoev, S.G., Balev, D.K., Assenova, B.K., and Amirhanov, K.J. (2017). Quality, microstructure, and technological properties of sheep meat marinated in three different ways. *Journal of Food Quality*, 2017, Article ID 5631532.

Walsh, H., Martins, S., O'Neill, E.E., Kerry, J.P., Kenny, T., and Ward, P. (2010). The effect of sodium lactate, potassium lactate, carrageenan, whey protein concentrate, yeast extract and fungal proteinases on the cook yield and tenderness of bovine chuck muscles. *Meat Science*, *85*(2), 230–234.

Wang, Z., Lin, M., Wang, L., et al. (2015). Metabolic engineering of Propionibacterium freudenreichii subsp. shermaniifor enhanced propionic acid fermentation: Effects of overexpressing three biotin-dependent carboxylases. *Process Biochemistry, 50*, 194–204.

Yadav, P., Chauhan, A.K., Singh, S.R., Khan, S., and Halabi, G. (2022). Chapter 22 - Organic acids: Microbial sources, production, and applications. In: Singh, R.B., Watanabe, S., Isaza, A.A. (Eds.), *Functional Foods and Nutraceuticals in Metabolic and Non-Communicable Diseases*. Academic Press: Cambridge, MA, pp. 325–337. ISBN: 9780128198155, doi: 10.1016/B978-0-12-819815-5.00053-7.

Zambaninia, T., Tehrania, H.H., Geisera, E., et al. (2017). Metabolic engineering of Ustilago trichophora TZ1 for improved malic acid production. *Metabolic Engineering Communications, 4*, 12–21.

Zatou, A., Loukou, Z., and Karava, O. (2004). Method development for the determination of seven organic acids in wines by reversed-phase high performance liquid chromatography. *Chromatographia, 60*, 39–44.

Zeece, M. (2020). Chapter seven: Food additives. In: Zeece, M. (Ed.), *Introduction to the Chemistry of Food*. Academic Press: Cambridge, MA, pp. 251–311. ISBN: 9780128094341, doi: 10.1016/B978-0-12-809434-1.00007-4.

CHAPTER 8

An Insight on Nanoencapsulation Techniques and Safety of Bioactives from Microalgae

Renu Sharma, Bharti Mittu, Kawaljeet Kaur, Jasmeet Kour, Mahaldeep Kaur, and Malti Rajput

CONTENTS

DOI: 10.1201/9781003259183-8

8.1 INTRODUCTION

Microalgae are a class of heterogeneous photosynthetic microorganisms with a wide range of biochemical compositions owing to their phylogenetic diversity (Borowitzka, 2013). These can synthesize, secrete, and accumulate a wide range of primary and secondary metabolites as a reaction to alterations in the surrounding environment. Many of these compounds are reported to be extremely valuable with various health benefits and industrial applications (De Morais et al., 2015). The various bioactive compounds include proteins, polyunsaturated fatty acids (PUFAs), polysaccharides, antioxidants, pigments, polyphenols, vitamins, enzymes, and many other high-value compounds (Milledge, 2011). These compounds are reported to have anticancer, antioxidative, antimicrobial, antiviral, anti-inflammatory, and immunomodulatory properties, which are discussed in the chapter later (Mayer and Hamann, 2005). Due to these potential applications, microalgae can be used in medicines, pharmaceuticals, food and cosmetic industries (Markou and Nerantzis, 2013). On the other hand, these bioactive substances are highly unstable, have low solubility and high susceptibility to oxidative degradation in the presence of light, oxygen, moisture, higher temperatures, extreme pH, which significantly limit their application. Degradation caused by oxygen leads to the production of unpleasant flavors and off-odors, which in turn affects stability, sensory properties, and consumer acceptability of the product (Shishir et al., 2018). Gastrointestinal tract internal conditions can also hinder the absorption of bioactive compounds (Rein et al., 2013; Singh, 2016). These technological and developmental challenges address the prominence of research schemes to conserve the functionality of microalgae bioactive compounds from processing till their delivery to the target position. The encapsulation systems, in this regard, are considered as an emerging and auspicious option to overcome these limitations (Vieira et al., 2020).

Encapsulation plays an important role in protecting the bioactive compounds. It involves coating of a substance (active agent) by a carrier material. The carrier substance is termed as capsule, membrane, shell, or external phase, while the encapsulated compound is known as core or internal phase (Shishir et al., 2018). At present, there are two kinds of encapsulation, i.e., microencapsulation and nanoencapsulation. Compounds can be categorized as nanoparticles, in which the particle size ranges from 1 to 100 nm, or microparticles, when the size ranges from 100 nm to 1,000 μm (Nedovic et al., 2011). Nanoencapsulation is preferred over microencapsulation due to its size, as smaller size increases bioavailability and also it can be modified and controlled in a better way. The various functional and physicochemical properties of the final encapsulation system, as well as the properties of the active agent and carrier material, all have a role in the retention and stability of core substance. The molecular weight, chemical behavior, polarity, and solubility are the important considerations for core and carrier materials while permeability, entrapment efficiency, degradability, and release profile must be accounted equally for the final particulate system (Dias et al., 2017; Augustin and Hemar, 2009; Joye and McClements, 2014). The various carrier materials, including proteins, polysaccharides, and lipids, can be used either alone or in combination to improve the efficiency of encapsulation (Khanvilkar et al., 2016). The wall material protects the product from undesirable effects of moisture, light, oxygen, and temperature, therefore resulting in improved stability and shelf life (Fang and Bhandari, 2010). The various encapsulation techniques include spray-drying, coacervation, solvent evaporation, liposomes, electrospray system, and extrusion (Onwulata, 2012; De Vos et al., 2010).

8.2 BIOACTIVE COMPOUNDS FROM MICROALGAE

Microalgae contain an ample amount of biologically active compounds with various health benefits. The natural instability of these bioactive compounds toward abiotic factors like temperature, moisture, and pH limits their application as health-promising substances (Shishir et al., 2018). Therefore, it is important to find ways to keep their properties and benefits intact. Microalgae, such as *Arthrospira*, *Botryococcus braunii*, *Chlorella vulgaris*, *Dunaliella salina*, *Haematococcus pluvialis*, and *Nostoc*, have been studied for their bioactive compounds (Palavra et al., 2011). *Spirulina* (*Arthrospira*) contains PUFAs, phenolic compounds and various micronutrients (such as B1, B2, B12, and E). It also contains pigments (C-phycocyanin) and various minerals (Costa and Morais, 2013). The conditions of hyperlipidemia, cancer, accumulation of fat, HIV, and hypertension can be treated with the help of bioactive compounds. They protect the body against heavy metals and drugs and also help in the reduction of serum levels of glucose and lipids (Colla et al., 2007; Torres-Duran et al., 2007).

The unicellular green microalga, *Chlorella*, is high in protein (about 40%–60%) and mostly comprises of essential amino acids (Vieira et al., 2020). β-1,3 glucan is found to be the most important bioactive compound that reduces free radicals and blood cholesterol. Also, these varieties are high in carotenoids, bioactive compounds, and micronutrients, indicating many benefits to the health system, which includes prevention of atherosclerosis, hyper-cholesterol, and antitumor activities (Costa and Morais, 2013; Kokou et al., 2012).

Nostoc microalgal biomass has been associated with antibacterial, antiviral, and antitumor properties (Deng et al., 2008). Cyanovirin, a protein produced by *Nostoc*, has been shown in studies to be effective against HIV (Boyd et al., 1997) and influenza A (H1N1) (Smee et al., 2008). PUFAs, which include essential fatty acids like linoleic acid, α-linolenic acid, γ-linolenic acid, octadecatetraenoic acid, and eicosapentaenoic acid (EPA) are also present (Wang et al., 2000).

Dunaliella is a type of green single-celled alga that contains pigments, glycerol, fats, as well as other beneficial substances, including enzymes and micronutrients (Preetha et al., 2012). Natural β-carotene is present in a significant amount in this microalga (Costa and Morais, 2013). Biomass extracted from *Dunaliella* exhibits many properties including oxidative, muscle relaxing, antihypertensive, bronchodilatory, pain-relieving, antihepatotoxic, and antiedemal effects (Madkour and Abdel-Daim, 2013).

8.2.1 Bioactive Compounds from Microalgae with Antioxidant Properties

Oxygen is essential for life but it can also cause free radicals at a cellular level via reactive oxygen species (ROS), which can lead to many disorders, including cardiovascular diseases, hardening of the arteries, tumors, and aging (Lane, 2002). Antioxidant substances, such as dimethylsulfoniopropionate and mycosporine amino acids which are powerful chemical ultraviolet blockers, are abundant in algal biomass. Also, antioxidant activity in carotenoids, fats, and polysaccharides has been discovered. Photosynthetic tissues are protected from light and oxygen damage by carotenoid compounds such as β-carotene and astaxanthin (Cazzonelli, 2011). Sulfate esters obtained from microalgal polysaccharides have several medical uses (Mohamed, 2008).

8.2.2 Bioactive Compounds from Microalgae with Antimicrobial Properties

Due to antibiotic resistance in humans, there is a need to develop new compounds with antimicrobial activity which brings microalgae into the limelight. Microalgae have been found to have antibacterial, antifungal, antiprotozoal, and antiplasmodial properties. The presence of bioactive compounds like fatty acids, carbohydrates, glycolipids, phenolics, terpenes, β-diketone, indole alkaloids, acrylic acids, isoprenoid-derived lipids (sterols), halogenated aliphatic compounds, isoprenoids, heterocyclic compounds with sulfur, and phenols contribute to these microorganisms' antimicrobial properties (Prakash et al., 2011).

Antimicrobial characteristics have been discovered in compounds such as α-ionone and β-ionone, β-cyclocitral, neophytadiene, and phytol. γ-linolenic acid, EPA, hexadecatrienoic acid, docosahexaenoic acid (DHA), palmitoleic acid, lauric acid, oleic acid, lactic acid, and arachidonic acid (AA) have all been associated with the suppression of human infections such as *Escherichia coli*, *Pseudomonas aeruginosa*, and *Staphylococcus* (Amaro et al., 2011). Sulfated polysaccharide produced by microalgae with acidic characteristics works as a therapeutic agent. Carrageenan, a sulfated polysaccharide can inhibit the viral adsorption process by directly binding with human papillomavirus (De Jesus Raposo et al., 2014).

8.2.3 Bioactive Compounds from Microalgae with Anti-Inflammatory Properties

The body's natural reaction to unfavorable stimuli such as pathogens, cellular debris, or toxins is inflammation. An inflammatory response causes redness, swelling, pain, and itching. Anti-inflammatory compounds quicken up the healing process and improve the reaction of the body to illness prevention. Microalgal biomass has several anti-inflammatory compounds that protect the body.

To produce scaffolds for the reconstruction process at the cellular and tissue levels, long-chain PUFAs, polysaccharides containing sulfur, and carotenoids are being studied (Steffens et al., 2013; Khan et al., 2007). Some research shows PUFAs like EPA, DHA, and AA can help in treating rheumatism and other skin diseases (Barrow and Shahidi, 2008). Apoptosis in human cancer cells can be stimulated with the help of the fucoxanthin carotenoid (Maeda et al., 2005).

8.2.4 Bioactive Compounds from Microalgae with Other Benefits

Microalgae-derived natural pigments show neuroprotective characteristics (Pangestuti and Kim, 2011). Phenolic compounds from microalgae can fight free radicals (Abd El-Baky et al., 2009). β-Carotene obtained from microalgal biomass has a significant amount of provitamin (A) content that is necessary for eyesight and immunological response. In humans, β-carotene also reduces the chances of getting cancer and degenerative disorders. Lutein can help with cataracts, retinopathy, atherosclerosis, and stroke prevention. Elevated levels of carotenoids with provitamin (A) activity such as α-carotene, β-carotene, and β- cryptoxanthin have been linked to a lower incidence of angina pectoris (Vílchez et al., 2011). Fucoxanthin is a fat-burning vitamin that can also be used to treat obesity (Steffens et al., 2013). Phytosterols have been shown to reduce total cholesterol (LDL) in humans (Francavilla et al., 2010).

Astaxanthin, a natural pigment obtained from *H. pluvialis* is reported to have a wide range of uses in the pharmaceutical, aesthetic, culinary, and feed sectors (Guerin et al., 2003). This pigment is recognized for providing a barrier against peroxidative lipid damage, retinopathy, atherosclerosis, and enhanced immunological responses (De Jesus Raposo et al., 2014). Moreover, some studies recommend that microalgae can even be utilized to treat wastewater and produce bioenergy (Rizwan et al., 2018). Therefore, various bioactive compounds obtained from microalgal biomass can be utilized to prevent many diseases.

8.3 ENCAPSULATION TECHNIQUES

The process of encapsulation consists of the entrapment of bioactive compounds in one or more coating substances using various processes that result in nanoparticles or microparticles (Ye et al., 2018). The encapsulation technique protects the bioactive components throughout production, storing, and circulation. It also helps in masking off unacceptable flavors and improves the solubility and bioavailability of bioactive compounds (Siddiqui and Sanna, 2016). Several techniques for encapsulation processes have been proposed in the literature. However, there is no one methodology that can be considered standardized and is applicable for the various active agents and encapsulated systems. The kind and characteristics of the base and coating materials, as well as the desired characteristics of the completed product are the important parameters before selecting a technique (Shishir et al., 2018; Dias et al., 2017).

Furthermore, the particle size, form, and structural properties differ significantly depending on the manufacturing procedure. Spray-drying, coacervation, fluidized bed, spray-cooling/chilling, liposomes, solvent evaporation, interfacial polymerization, supercritical fluid technology, emulsification, nanoprecipitation, using extruder, and electrospraying systems are some of the frequently used encapsulation methods (Vieira et al., 2020; De Souza Simões et al., 2017).

Bottom-up and top-down techniques are the two most common strategies used in encapsulation systems. Large structures are created from the bottom-up method utilizing atom-by-atom or molecule-by-molecule procedures with the help of various interactions such as van der Waals, ionic, and hydrophobic interactions (Santiago and Castro, 2016). Several factors including pH, heat, and ionic bonding, influence this method (Ezhilarasi et al., 2013). The top-down method on the other hand incorporates physical modification of the base and coating materials that focuses mainly on the reduction of size (De Souza Simões et al., 2017). Bottom-up techniques are more advantageous when compared with the top-down technique. In the bottom-up technique, there is more supervision on the characteristics of the material, less energy consumption, and less risk of sample contamination (Chan and Kwok, 2011; Rehman et al., 2020). The various encapsulation techniques are presented in Figure 8.1.

Spray-drying is the simplest and perhaps the most extensively utilized nanoencapsulation/microencapsulation technology for bioactive compounds. In this technique, the active ingredient is incorporated into an aqueous solution or one made with the coating material. This mix is subsequently atomized, resulting in microcapsules/nanocapsules being produced inside the chamber and dried through hot circulating air, which is then collected by cyclone recovery (Đorđević et al., 2015). The particle size generally ranges from 1 to 50 μm; however by utilizing a nano-spray dryer, it may be decreased to 0.2 μm (Tolve et al., 2016). This technique is simple, flexible, generates fast results and can be operated at low costs. Some of the disadvantages to

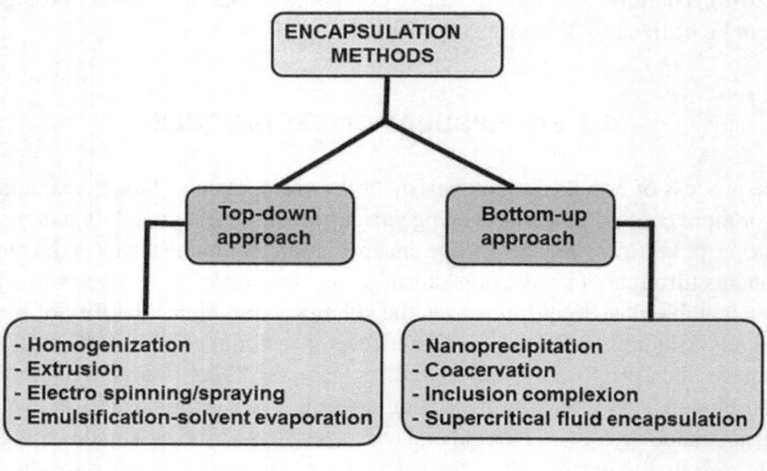

Figure 8.1 Different methods of encapsulation.

be kept in mind while using this technique are that the nanoparticles created may not be homogenous, so there is a possibility of degradation of compounds with a high sensitivity nature as the air used for this technique has a high temperature (Nedovic et al., 2011).

Extrusion is another common technique for the encapsulation of bioactive compounds. In this technique a solution made up of the cladding and an active agent is passed via a nozzle that reaches a gelation atmosphere. Physical or chemical techniques are used to solidify the particles into capsules after particle formation. Physical processes include chilling or heating while chemical techniques include gelatinization (Saifullah et al., 2019). Some of the methods used for the formation of extrusion nanoparticles are electrostatic extrusion, simple dripping, vibrating jet/nozzle, and melt extrusion. Extrusion is a basic and cost-effective method; the only drawback to this technique is its low efficiency (Rodríguez et al., 2016).

Coacervation is the creation of two immiscible phases from a solution having a dispersion polymer. The encapsulated component is disseminated in a polymeric solution that would be used to coat the product. Polymer separation is induced using various methods, resulting in the formation of a new phase (coacervate). Centrifugation or filtration can be used to collect the formed particles, which can then be washed, dried, or hardened. This coacervation technique is further classified as complex and simple. The simple coacervation is aided by a transition in the medium, resulting in the desolvation of the material coating, whereas in the complicated form, mutual neutralization of two polymers with opposing charges will make up the coating (Zhang et al., 2012). This technique can be controlled by altering different processing parameters, which include the types of core and shell materials used and the core/shell ratio. These parameters, together with environmental factors like temperature, ionic strength, and pH, should be taken into consideration while performing this technique. Through this technique, it is feasible to accomplish high encapsulation efficiency and superior particle size control. But, particle agglomeration and high operational costs have been reported as drawbacks of this technique (Vieira et al., 2020; Bosnea et al., 2014; Ach et al., 2015).

Homogenization/emulsification is a process that involves the use of a minimum of two immiscible liquid phases. One of the two immiscible liquids is disseminated in the other as small spherical droplets are encircled by a fine interfacial film of micelles (Bakry et al., 2016). Two types of emulsions can be obtained via this system, i.e., oil-in-water (O/W), in which oil droplets are disseminated in a liquid medium, and water-in-oil (W/O), in which water molecules are dispersed in an oily phase. Different techniques like homogenization at high pressure and emulsification could be employed to obtain emulsions. The procedure used for the preparation of emulsions is easy and cost effective. However, there may be a chance of destabilization physically where emulsions are faced with a variety of storage and processing environments (Zhu et al., 2019).

The emulsification-solvent evaporation technique is another emulsification technique. An emulsion is formed in a volatile organic solvent with an active agent and a polymer that acts as a coating agent. The solvent is evaporated usually under atmospheric conditions (Freitas et al., 2005). Small droplets can be obtained by using this technique. However, because of the large volumes of the organic solvent utilized, production prices may rise (O'Donnell and McGinity, 1997).

In pharmaceutical applications, liposomes are among the most explored encapsulation methods. This method involves forming lipid vacuoles from aqueous dispersions of amphipathic compounds that usually form multilayer structures, such as polar lipids. Liposomes are globular in shape, with varying diameters ranging from nanometers to micrometers. A single or numerous layers of amphiphilic polymolecular membranes may be present in the produced vesicles (Taylor et al., 2005). Its main advantages are the ability to encapsulate compounds having an affinity for lipids (lipophilic) and water (hydrophilic), drug delivery systems on targets, and its adaptability in size and number of layers. The primary operational barriers, on the other hand, include possible toxic effects from organic solvent remnants, expensive on a big scale, and poor stability (Vieira et al., 2020; Mozafari, 2005).

8.4 CARRIERS FOR ENCAPSULATION

The structure of an encapsulation system, which can be generally classified as reservoir or matrix systems, is determined by the arrangement of the core substance and the deposition process of the coating material. The reservoir systems are also termed as microcapsules or nanocapsules, and are further categorized as mono-core (an outer shell surrounding a single core), multicore (distinct cores entrapped in a shell), or multi-shell mono-core (single core surrounded by many shells). On the other hand, the matrix system is obtained when the active agent is evenly distributed throughout the coating material network. Particulate systems are typically obtained as spherical shapes; however, nonspherical or irregular configurations for all systems can be seen (Ye et al., 2018; Saifullah et al., 2019).

In terms of the composition of an encapsulation system, the encapsulated active agents can be lipophilic or hydrophilic and can belong to a variety of classes, including drugs, vitamins, enzymes, minerals, antimicrobials, nutraceuticals, antioxidants, flavors, colorants, and essential oils. Furthermore, the coating material can be chosen from a varied range of natural materials and bio-based substances, as well as synthetic polymers and copolymers. This choice will be influenced by the active agent to be encapsulated and the desired properties for the final system. The other important factors to consider are stability, solubility, safety, and release properties. The substance must be of food grade or pharmaceutical grade, depending upon the application either in the food industry or in pharmaceuticals. It should include criteria such as biocompatibility and biodegradability, among others (Vieira et al., 2020; Mansour et al., 2010).

Natural materials can be categorized as (i) carbohydrates such as starch, pectin, maltodextrin, cyclodextrin, cellulose, and inulin; (ii) waxes or fats mainly glycerides and phospholipids; (iii) proteins including casein, gelatin, whey protein, albumin, bovine serum, and various vegetable sources; and (iv) and gums like arabic, mesquite, and guar. Also, polysaccharides like chitosan and alginate are primarily being researched for food and pharmaceutical applications. The primary benefits of utilizing bio-based materials include their natural abundance, biodegradability, and biocompatibility. However, these natural resources may exhibit unstable properties owing to batch-to-batch variations (Ye et al., 2018; Labuschagne, 2018).

Synthetic polymers (either biodegradable or nonbiodegradable) are another type of coating material used in the encapsulation process. Aliphatic polyesters, such as poly (ϵ-caprolactone), poly (lactic-co-glycolic acid), and poly (lactic acid), are among the biodegradables that are repeatedly investigated in the pharmaceutical industry. The phosphorus-based polymers, polyamides, polyanhydrides, and polyurethanes are also examples of this group. The other widely used nonbiodegradable polymer comprises cellulose derivatives such as ethyl cellulose or cellulose acetate, carboxymethyl cellulose as well as poly (ethylene glycol), polyvinyl alcohol, and poly(N-vinylpyrrolidone). Synthetic polymers differ from natural polymers in terms of chemical and mechanical stability, as well as the ability to alter their properties according to the needs of the final encapsulated system. Nonetheless, biodegradability and biocompatibility are major problems (Ye et al., 2018; Rivas et al., 2017). Furthermore, wall materials can be functionalized, which allows for the creation of encapsulation systems with properties that differ from those observed in the natural world. Increased bio-distribution and additional bio-marker function, for example, have been reported in the literature (Vieira et al., 2020; Nesterenko et al., 2013; Sur et al., 2019).

De Morais et al. (2016) reported the efficiency of the nanoencapsulation process for improving the bioavailability of various bioactive components such as antioxidants, proteins, pigments, biopeptides, and PUFAs obtained from the *Spirulina* biomass. The use of Poly-3-hydroxybutyrate (PHB) biopolymers originating from the *Spirulina* biomass in the nanoencapsulation of active ingredients was observed to be a suitable means of preserving their characteristics. In addition to the nanoencapsulation of the bioactive compounds in *Spirulina*, Deepak et al. (2009) observed that nanoencapsulated nattokinase enzyme coated with PHB biopolymer from microbial sources was found to be highly efficient in preserving enzymatic activity.

Numerous researchers have studied the encapsulation of *H. pluvialis* extracts, especially the carotenoid astaxanthin. The coating material chosen for these encapsulation systems were mainly natural polymeric substances including maltodextrin, chitosan, arabic gum, whey protein, and alginate, as these are usually employed as food additives. The coating layer was found to be highly beneficial in preserving antioxidant activity and also resulted in improved stability even during storage under adverse conditions such as oxygen exposure, elevated temperatures, or higher pH. Moreover, the bioaccessibility and bioavailability of astaxanthin were also reported to be improved (Vieira et al., 2020).

8.5 APPLICATIONS IN THE FOOD INDUSTRY

Several bioactive compounds obtained from microalgae are rich in nutritional components and could be used in the culinary, nutraceutical, and feed sectors. Various microalgal species like *Spirulina, Chlorella*, and *Haematococcus* have been found to produce many vital components like β-carotene, astaxanthin, DHA, EPA, carotenoids, natural dyes, carbohydrates, phytonutrients, and extracts that are known to have health benefits. Microalgae biomass contains almost all the essential vitamins like A, B1, B2, B6, biotin (B7), B12, C, E, nicotinate, pantothenic (B5), folate (B9), minerals (Na, K, Ca, Mg, Fe, Zn), and other micronutrients (Becker, 2004). Microalgae are used industrially, either in the form of paste or as extracts. Microalgal pastes are utilized as protein supplements or as feed for aqueous organisms. On the other hand, extracts obtained from microalgae are used in nutraceuticals and cosmetics. Some of the applications of microalgae include their utilization in food industries, wastewater management, biofuels, and pharmaceutical formulations (Mobin and Alam, 2014).

Microalgae have high food value and can be considered as healthy food for humans. Polysaccharides obtained from microalgae can be utilized for food and fodder additives. Bioactive compounds from microalgae exhibit antioxidant and anti-inflammatory properties. Microalgae can be used in the preparation of biofilms (Hudek et al., 2014) and also as a source of feed for feeding aqueous organisms (Varfolomeev and Wasserman, 2011). Vegetable proteins, oils, fishmeal, and premixes obtained from vitamins and minerals can be used for making feed with better nutritional properties and for the health of animals (Harel and Clayton, 2004). This results in an improved immune system, better growth results, and resistance to diseases.

Microalgae biomass is rich in proteins and various components, including pigments, essential fatty acids, and triglycerides (Pignolet et al., 2013; Borowitzka, 2013; Liu et al., 2016). Studies reveal that polysaccharides and oligosaccharides obtained from microalgae can be utilized as prebiotics for various health benefits. Also, alpha-linolenic acids are found to have a vital contribution to the brain and retinal development of infants (Wroble et al., 2002).

Arthrospira platensis, a microalgal species, contains a high concentration of γ-linolenic acid, an essential fatty acid. It is found to have water-soluble vitamins. *Chlorella vulgaris*, another microalga, is found to be a rich source of dietary antioxidants like lutein, α- and β-carotene, and vitamins C and E. In addition to that, it also contains β-1,3-glucan, which helps in stimulating the human immune system (Becker, 2007). Another form of carotenoid, i.e., astaxanthin, found in *H. pluvialis*, is shown to have various health benefits (Matos et al., 2017). Some of the cyanobacteria are known for their high content of fatty acids. Microalgal oil is a better alternative to fish oil and it also contains high amounts of carotenoids (Ryckebosch et al., 2014).

Spirulina is rich in nutrients and is acknowledged as a "super food" that is high in vitamin B1 (Becker, 2007; Sajilata et al., 2008). It also has a good quantity of γ-linolenic acid (Mobin and Alam, 2017). *D. salina* contains carotenoids which include β-carotene, α-carotene, lutein, zeaxanthin, and cryptoxanthin (Raja et al., 2007). These carotenoids are available in tablets, pills, and fortified products as nutritional supplements. Margarine is colored by using an oil-based extract of β-carotene (Raja and Hemaiswarya, 2010). Beverage coloring is done by using water-soluble extracts of β-carotene. Various nutritional supplements in the form of gels, oils, syrups, and capsules are used commercially, containing astaxanthin (Capelli and Cysewski, 2013).

In a study where spray-dried biomass of *Arthrospira maxima* was used as an ingredient for biscuits, it showed improvement in its overall nutritional quality. In addition to nutritional enhancement, it is found to have improved iron content, high amounts of proteins, and PUFAs (Da Silva et al., 2021). Noodles obtained after combining wheat flour with the flours obtained from *Chlorella vulgaris* and *Arthrospira platensis* in combination with the macroalgae *Eucheuma cottonii* were shown to have a high content of protein, fat, ash, and dietary fiber. However, there was a reduction in carbohydrate content with no changes in moisture.

Microalgal food supplements or fortified foods can be given to children suffering from malnutrition in order to fulfill all their nutritional requirements (Spolaore et al., 2006). To give the required nutrition benefit, they can be integrated into foodstuffs such as spaghetti, cookies, bread, snack items, confectionery, yogurts, and carbonated beverages (Belay et al., 1993).

8.6 CONCLUSIONS

In the last few decades, the interest in utilization of microalgae for food and other industrial applications has been increasing due to the occurrence of ample amount of biologically active compounds with various health benefits. Moreover, the consumers' awareness toward sustainable, healthier, and safer components stimulates the application of natural resources in food and pharmaceuticals. The physicochemical limitations, including high instability, low solubility, and high susceptibility to oxidative degradation, revealed that for the integration of bioactive compounds into products, the emerging encapsulation systems are recognized as a promising option to overcome these limitations owing to the occurrence of a surrounding coating material. Microalgae bioactive compounds can be significantly used in several fields, if these are properly protected during processing and storage. In the present chapter, various types of encapsulation techniques for microalgae biomass, their bioactive compounds for food application, and types of coating materials used in the encapsulation process have been discussed. It was noted that the application of microalgae encapsulation in the food industry is the most studied field that might be due to the significant nutritional composition seen in these microorganisms. The effectiveness of various encapsulation systems has been proven in terms of improved bioavailability and stability. Therefore, merging the benefits of encapsulation techniques and the biological characteristics of microalgal compounds appears to be a promising approach for producing healthier and innovative products.

REFERENCES

Ach, D., Briançon, S., Broze, G., Puel, F., Rivoire, A., Galvan, J. M., & Chevalier, Y. (2015). Formation of microcapsules by complex coacervation. *The Canadian Journal of Chemical Engineering*, *93*(2), 183–191.

Adb El-Baky, H., El-Baz, F. K., & El baroty, G. (2010). Production of phenolic compounds from *Spirulina maxima* microalgae and its protective effects. *African Journal of Biotechnology*, *8*, 7059–7067.

Amaro, H. M., Guedes, A. C., & Malcata, F. X. (2011). Antimicrobial activities of microalgae: an invited review. *Science against Microbial Pathogens: Communicating Current Research and Technological Advances*, *3*, 1272–1284.

Augustin, M. A., & Hemar, Y. (2009). Nano-and micro-structured assemblies for encapsulation of food ingredients. *Chemical Society Reviews*, *38*(4), 902–912.

Bakry, A. M., Abbas, S., Ali, B., Majeed, H., Abouelwafa, M. Y., Mousa, A., & Liang, L. (2016). Microencapsulation of oils: A comprehensive review of benefits, techniques, and applications. *Comprehensive Reviews in Food Science and Food Safety*, *15*(1), 143–182.

Barrow, C., & Shahidi, F. (2008). *Marine Nutraceuticals and Functional Foods*. CRC Press, Taylor & Francis: Boca Raton, FL.

Becker, E. W. (2004). Microalgae in human and animal nutrition. In: Richmond, A., & Hu, Q. (Eds.), *Handbook of Microalgal Culture: Biotechnology and Applied Phycology*. Blackwell Science: Oxford, pp. 312–351.

Becker, E. W. (2007). Micro-algae as a source of protein. *Biotechnology Advances*, *25*(2), 207–210.

Belay, A., Ota, Y., Miyakawa, K., & Shimamatsu, H. (1993). Current knowledge on potential health benefits of Spirulina. *Journal of Applied Phycology*, *5*(2), 235–241.

Borowitzka, M. A. (2013). High-value products from microalgae-their development and commercialisation. *Journal of Applied Phycology*, *25*(3), 743–756.

Bosnea, L. A., Moschakis, T., & Biliaderis, C. G. (2014). Complex coacervation as a novel microencapsulation technique to improve viability of probiotics under different stresses. *Food and Bioprocess Technology*, *7*(10), 2767–2781.

Boyd, M. R., Gustafson, K. R., McMahon, J. B., Shoemaker, R. H., O'Keefe, B. R., Mori, T., Gulakowski, R. J., Wu, L., Rivera, M. I., Laurencot, C. M., Currens, M. J., Cardellina, J. H. 2nd, Buckheit Jr, R. W., Nara, P. L., Pannell, L. K., Sowder, R. C. 2nd, & Henderson, L. E. (1997). Discovery of cyanovirin-N, a novel human immunodeficiency virus-inactivating protein that binds viral surface envelope glycoprotein gp120: potential applications to microbicide development. *Antimicrobial Agents and Chemotherapy*, *41*(7), 1521–1530.

Capelli, B., & Cysewski, G. (2013). *Natural Astaxanthin: The World's Best Kept Health Secret*. Cyanotech Corporation: Kailua-Kona, HI. ISBN: 978-0-9792353-0-6.

Cazzonelli, C. I. (2011). Carotenoids in nature: Insights from plants and beyond. *Functional Plant Biology*, *38*(11), 833–847.

Chan, H. K., & Kwok, P. C. L. (2011). Production methods for nanodrug particles using the bottom-up approach. *Advanced Drug Delivery Reviews*, *63*(6), 406–416.

Colla, L. M., Reinehr, C. O., Reichert, C., & Costa, J. A. V. (2007). Production of biomass and nutraceutical compounds by Spirulina platensis under different temperature and nitrogen regimes. *Bioresource Technology*, *98*(7), 1489–1493.

Costa, J. A. C., & Morais, M. G. (2013). Microalgae for food production. In: Soccol, C. R., & Pandey, A., & Larroche, C. (Eds.), *Fermentation Process Engineering in the Food Industry*. Taylor & Francis: London, p. 486.

Da Silva, S. P., do Valle, A. F., & Perrone, D. (2021). Microencapsulated Spirulina maxima biomass as an ingredient for the production of nutritionally enriched and sensorially well-accepted vegan biscuits. *LWT*, *142*, 110997.

De Morais, M. G., Vaz, B. D. S., de Morais, E. G., & Costa, J. A. V. (2015). Biologically active metabolites synthesized by microalgae. *BioMed Research International*, Article ID 835761, 1–15.

De Morais, M. G., de Morais, E. G., Silva Vaz, B. D., Goncalves, C. F., Lisboa, C., & Vieira Costa, J. A. (2016). Nanoencapsulation of the bioactive compounds of Spirulina with a microalgal biopolymer coating. *Journal of Nanoscience and Nanotechnology*, *16*(1), 81–91.

De Souza Simões, L., Madalena, D. A., Pinheiro, A. C., Teixeira, J. A., Vicente, A. A., & Ramos, O. L. (2017). Micro-and nano bio-based delivery systems for food applications: In vitro behavior. *Advances in Colloid and Interface Science*, *243*, 23–45.

De Vos, P., Faas, M. M., Spasojevic, M., & Sikkema, J. (2010). Encapsulation for preservation of functionality and targeted delivery of bioactive food components. *International Dairy Journal*, *20*(4), 292–302.

Deepak, V., Kalishwaralal, K., & Gurunathan, S. (2009). Purification, immobilization, and characterization of nattokinase on PHB nanoparticles. *Bioresource Technology*, *100*(24), 6644–6646.

Deng, Z., Hu, Q., Lu, F., Liu, G., & Hu, Z. (2008). Colony development and physiological characterization of the edible blue-green alga, Nostocsphaeroides (Nostocaceae, Cyanophyta). *Progress in Natural Science*, *18*(12), 1475–1483.

Dias, D. R., Botrel, D. A., Fernandes, R. V. D. B., & Borges, S. V. (2017). Encapsulation as a tool for bioprocessing of functional foods. *Current Opinion in Food Science*, *13*, 31–37.

Đorđević, V., Balanč, B., Belščak-Cvitanović, A., Lević, S., Trifković, K., Kalušević, A., Kostić, I., Komes, D., Bugarski, B., & Nedović, V. (2015). Trends in encapsulation technologies for delivery of food bioactive compounds. *Food Engineering Reviews*, *7*(4), 452–490.

Ezhilarasi, P. N., Karthik, P., Chhanwal, N., & Anandharamakrishnan, C. (2013). Nanoencapsulation techniques for food bioactive components: A review. *Food and Bioprocess Technology*, *6*(3), 628–647.

Fang, Z., & Bhandari, B. (2010). Encapsulation of polyphenols: A review. *Trends in Food Science & Technology*, *21*(10), 510–523.

Francavilla, M., Trotta, P., & Luque, R. (2010). Phytosterols from Dunaliellatertiolecta and Dunaliellasalina: A potentially novel industrial application. *Bioresource Technology*, *101*(11), 4144–4150.

Freitas, S., Merkle, H. P., & Gander, B. (2005). Microencapsulation by solvent extraction/evaporation: Reviewing the state of the art of microsphere preparation process technology. *Journal of Controlled Release*, *102*(2), 313–332.

Guerin, M., Huntley, M. E., & Olaizola, M. (2003). Haematococcus astaxanthin: Applications for human health and nutrition. *TRENDS in Biotechnology*, *21*(5), 210–216.

Harel, M., & Clayton, D. (2004). Feed formulation for terrestrial and aquatic animals. US Patent 20070082008 (WO/2004/080196).

Hudek, K., Davis, L. C., Ibbini, J., & Erickson, L. (2014). Commercial products from algae. In: Prokop, A., Bajpai, R.K., & Zappi, M.E. (Eds.), *Algal Biorefineries*. Springer: Dordrecht, pp. 275–295.

Joye, I. J., & McClements, D. J. (2014). Biopolymer-based nanoparticles and microparticles: Fabrication, characterization, and application. *Current Opinion in Colloid & Interface Science*, *19*(5), 417–427.

Khan, M. N. A., Cho, J. Y., Lee, M. C., Kang, J. Y., Park, N. G., Fujii, H., & Hong, Y. K. (2007). Isolation of two anti-inflammatory and one pro-inflammatory polyunsaturated fatty acids from the brown seaweed Undariapinnatifida. *Journal of Agricultural and Food Chemistry*, *55*(17), 6984–6988.

Khanvilkar, A. M., Ranveer, R. C., & Sahoo, A. K. (2016). Carrier materials for encapsulation of bio-active components of food. *International Journal of Pharmaceutical Sciences Review and Research*, *40*(1), 62–73.

Kokou, F., Makridis, P., Kentouri, M., & Divanach, P. (2012). Antibacterial activity in microalgae cultures. *Aquaculture Research*, *43*(10), 1520–1527.

Labuschagne, P. (2018). Impact of wall material physicochemical characteristics on the stability of encapsulated phytochemicals: A review. *Food Research International*, *107*, 227–247.

Lane, N. (2002). Elixir of life and death. In: *Oxygen: The Molecule That Made the World*, 2nd ed. Oxford Landmark Science: New York, p. 384.

Liu, L., Pohnert, G., & Wei, D. (2016). Extracellular metabolites from industrial microalgae and their biotechnological potential. *Marine Drugs*, *14*(10), 191.

Madkour, F. F., & Abdel-Daim, M. M. (2013). Hepatoprotective and antioxidant activity of Dunaliellasalina in paracetamol-induced acute toxicity in rats. *Indian Journal of Pharmaceutical Sciences*, *75*(6), 642–648.

Maeda, H., Hosokawa, M., Sashima, T., Funayama, K., & Miyashita, K. (2005). Fucoxanthin from edible seaweed, Undariapinnatifida, shows antiobesity effect through UCP1 expression in white adipose tissues. *Biochemical and Biophysical Research Communications*, *332*(2), 392–397.

Mansour, H. M., Sohn, M., Al-Ghananeem, A., & DeLuca, P. P. (2010). Materials for pharmaceutical dosage forms: Molecular pharmaceutics and controlled release drug delivery aspects. *International Journal of Molecular Sciences*, *11*(9), 3298–3322.

Markou, G., & Nerantzis, E. (2013). Microalgae for high-value compounds and biofuels production: A review with focus on cultivation under stress conditions. *Biotechnology Advances*, *31*(8), 1532–1542.

Matos, J., Cardoso, C., Bandarra, N. M., & Afonso, C. (2017). Microalgae as healthy ingredients for functional food: A review. *Food & Function*, *8*(8), 2672–2685.

Mayer, A. M., & Hamann, M. T. (2005). Marine pharmacology in 2001–2002: Marine compounds with anthelmintic, antibacterial, anticoagulant, antidiabetic, antifungal, anti-inflammatory, antimalarial, antiplatelet, antiprotozoal, antituberculosis, and antiviral activities; affecting the cardiovascular, immune and nervous systems and other miscellaneous mechanisms of action. *Comparative Biochemistry and Physiology Part C: Toxicology & Pharmacology*, *140*(3–4), 265–286.

Milledge, J. J. (2011). Commercial application of microalgae other than as biofuels: A brief review. *Reviews in Environmental Science and Bio/Technology*, *10*(1), 31–41.

Mobin, S., & Alam, F. (2014). Biofuel production from algae utilizing wastewater. In *19th Australasian Fluid Mechanics Conference*, Melbourne, Australia.

Mobin, S., & Alam, F. (2017). Some promising microalgal species for commercial applications: A review. *Energy Procedia*, *110*, 510–517.

Mohamed, Z. A. (2008). Polysaccharides as a protective response against microcystin-induced oxidative stress in Chlorella vulgaris and Scenedesmusquadricauda and their possible significance in the aquatic ecosystem. *Ecotoxicology*, *17*(6), 504–516.

Mozafari, M. R. (2005). Liposomes: An overview of manufacturing techniques. *Cellular and Molecular Biology Letters*, *10*(4), 711–719.

Nedovic, V., Kalusevic, A., Manojlovic, V., Levic, S., & Bugarski, B. (2011). An overview of encapsulation technologies for food applications. *Procedia Food Science*, *1*, 1806–1815.

Nesterenko, A., Alric, I., Silvestre, F., & Durrieu, V. (2013). Vegetable proteins in microencapsulation: A review of recent interventions and their effectiveness. *Industrial Crops and Products*, *42*, 469–479.

O'Donnell, P. B., & McGinity, J. W. (1997). Preparation of microspheres by the solvent evaporation technique. *Advanced Drug Delivery Reviews*, *28*(1), 25–42.

Onwulata, C. I. (2012). Encapsulation of new active ingredients. *Annual Review of Food Science and Technology*, *3*, 183–202.

Palavra, A. M. F., Coelho, J. P., Barroso, J. G., Rauter, A. P., Fareleira, J. M. N. A., Mainar, A., Urieta, J. S., Nobre, B. P., Gouveia, L., Mendes, R. L., Cabral, J. M. S. and Novais, J. M. (2011). Supercritical carbon dioxide extraction of bioactive compounds from microalgae and volatile oils from aromatic plants. *The Journal of Supercritical Fluids*, *60*, 21–27.

Pangestuti, R., & Kim, S. K. (2011). Biological activities and health benefit effects of natural pigments derived from marine algae. *Journal of Functional Foods*, *3*(4), 255–266.

Pignolet, O., Jubeau, S., Vaca-Garcia, C., & Michaud, P. (2013). Highly valuable microalgae: Biochemical and topological aspects. *Journal of Industrial Microbiology and Biotechnology*, *40*(8), 781–796.

Prakash, J. W., Marimuthu, J., & Jeeva, S. (2011). Antimicrobial activity of certain fresh water microalgae from Thamirabarani River, Tamil Nadu, South India. *Asian Pacific Journal of Tropical Biomedicine*, *1*(2), S170–S173.

Preetha, K., John, L., Subin, C. S., & Vijayan, K. K. (2012). Phenotypic and genetic characterization of Dunaliella (Chlorophyta) from Indian salinas and their diversity. *Aquatic Biosystems*, *8*(1), 1–16.

Raja, R., & Hemaiswarya, S. (2010). Microalgae and immune potential. In: Watson, R. R., Zibadi, S., & Preedy, V. R. (Eds), *Dietary Components and Immune Function*. Humana Press: Totowa, NJ, pp. 515–527.

Raja, R., Hemaiswarya, S., & Rengasamy, R. (2007). Exploitation of Dunaliella for β-carotene production. *Applied Microbiology and Biotechnology*, *74*(3), 517–523.

Raposo, M. F. D. J., de Morais, A. M. M. B., & de Morais, R. M. S. C. (2014). Influence of sulphate on the composition and antibacterial and antiviral properties of the exopolysaccharide from Porphyridiumcruentum. *Life Sciences, 101*(1–2), 56–63.

Rehman, A., Tong, Q., Jafari, S. M., Assadpour, E., Shehzad, Q., Aadil, R. M., Iqbal, M. W., Rashed, M., Mushtaq, B. S., & Ashraf, W. (2020). Carotenoid-loaded nanocarriers: A comprehensive review. *Advances in Colloid and Interface Science*, *275*, 102048.

Rein, M. J., Renouf, M., Cruz-Hernandez, C., Actis-Goretta, L., Thakkar, S. K., & da Silva Pinto, M. (2013). Bioavailability of bioactive food compounds: A challenging journey to bioefficacy. *British Journal of Clinical Pharmacology*, *75*(3), 588–602.

Rivas, C. J. M., Tarhini, M., Badri, W., Miladi, K., Greige-Gerges, H., Nazari, Q. A., Galindo Rodríguez, S. A., Román, R. Á., Fessi, H., & Elaissari, A. (2017). Nanoprecipitation process: From encapsulation to drug delivery. *International Journal of Pharmaceutics*, *532*(1), 66–81.

Rizwan, M., Mujtaba, G., Memon, S. A., Lee, K., & Rashid, N. (2018). Exploring the potential of microalgae for new biotechnology applications and beyond: A review. *Renewable and Sustainable Energy Reviews*, *92*, 394–404.

Rodríguez, J., Martín, M. J., Ruiz, M. A., & Clares, B. (2016). Current encapsulation strategies for bioactive oils: From alimentary to pharmaceutical perspectives. *Food Research International*, *83*, 41–59.

Ryckebosch, E., Bruneel, C., Termote-Verhalle, R., Goiris, K., Muylaert, K., & Foubert, I. (2014). Nutritional evaluation of microalgae oils rich in omega-3 long chain polyunsaturated fatty acids as an alternative for fish oil. *Food Chemistry*, *160*, 393–400.

Saifullah, M., Shishir, M. R. I., Ferdowsi, R., Rahman, M. R. T., & Van Vuong, Q. (2019). Micro and nano encapsulation, retention and controlled release of flavor and aroma compounds: A critical review. *Trends in Food Science & Technology*, *86*, 230–251.

Sajilata, M. G., Singhal, R. S., & Kamat, M. Y. (2008). Fractionation of lipids and purification of γ-linolenic acid (GLA) from Spirulina platensis. *Food Chemistry*, *109*(3), 580–586.

Santiago, L. G., & Castro, G. R. (2016). Novel technologies for the encapsulation of bioactive food compounds. *Current Opinion in Food Science*, *7*, 78–85.

Shishir, M. R. I., Xie, L., Sun, C., Zheng, X., & Chen, W. (2018). Advances in micro and nano-encapsulation of bioactive compounds using biopolymer and lipid-based transporters. *Trends in Food Science & Technology*, *78*, 34–60.

Siddiqui, I. A., & Sanna, V. (2016). Impact of nanotechnology on the delivery of natural products for cancer prevention and therapy. *Molecular Nutrition & Food Research*, *60*(6), 1330–1341.

Singh, H. (2016). Nanotechnology applications in functional foods; opportunities and challenges. *Preventive Nutrition and Food Science*, *21*(1), 1–8.

Smee, D. F., Bailey, K. W., Wong, M. H., O'Keefe, B. R., Gustafson, K. R., Mishin, V. P., & Gubareva, L. V. (2008). Treatment of influenza A (H1N1) virus infections in mice and ferrets with cyanovirin-N. *Antiviral Research*, *80*(3), 266–271.

Spolaore, P., Joannis-Cassan, C., Duran, E., & Isambert, A. (2006). Commercial applications of microalgae. *Journal of Bioscience and Bioengineering*, *101*(2), 87–96.

Steffens, D., Lersch, M., Rosa, A., Scher, C., Crestani, T., Morais, M. G., Costa, J. A., & Pranke, P. (2013). A new biomaterial of nanofibers with the microalga Spirulina as scaffolds to cultivate with stem cells for use in tissue engineering. *Journal of Biomedical Nanotechnology*, *9*(4), 710–718.

Sur, S., Rathore, A., Dave, V., Reddy, K. R., Chouhan, R. S., & Sadhu, V. (2019). Recent developments in functionalized polymer nanoparticles for efficient drug delivery system. *Nano-Structures & Nano-Objects*, *20*, 100397.

Taylor, T. M., Weiss, J., Davidson, P. M., & Bruce, B. D. (2005). Liposomal nanocapsules in food science and agriculture. *Critical Reviews in Food Science and Nutrition*, *45*(7–8), 587–605.

Tolve, R., Galgano, F., Caruso, M. C., Tchuenbou-Magaia, F. L., Condelli, N., Favati, F., & Zhang, Z. (2016). Encapsulation of health-promoting ingredients: Applications in foodstuffs. *International Journal of Food Sciences and Nutrition*, *67*(8), 888–918.

Torres-Duran, P. V., Ferreira-Hermosillo, A., & Juarez-Oropeza, M. A. (2007). Antihyperlipemic and antihypertensive effects of *Spirulina maxima* in an open sample of Mexican population: A preliminary report. *Lipids in Health and Disease*, *6*(1), 1–8.

Varfolomeev, S. D., & Wasserman, L. A. (2011). Microalgae as source of biofuel, food, fodder, and medicines. *Applied Biochemistry and Microbiology*, *47*(9), 789–807.

Vieira, M. V., Pastrana, L. M., & Fuciños, P. (2020). Microalgae encapsulation systems for food, pharmaceutical and cosmetics applications. *Marine Drugs*, *18*(12), 644.

Vílchez, C., Forján, E., Cuaresma, M., Bédmar, F., Garbayo, I., & Vega, J. M. (2011). Marine carotenoids: Biological functions and commercial applications. *Marine Drugs*, *9*(3), 319–333.

Wang, M., Xu, Y. N., Jiang, G. Z., Li, L. B., & Kuang, T. Y. (2000). Membrane lipids and their fatty acid composition in Nostocflagelliforme cells. *Acta Botanica Sinica*, *42*, 1263–1266.

Wroble, M., Mash, C., Williams, L., & McCall, R. B. (2002). Should long chain polyunsaturated fatty acids be added to infant formula to promote development? *Journal of Applied Developmental Psychology*, *23*(1), 99–112.

Ye, Q., Georges, N., & Selomulya, C. (2018). Microencapsulation of active ingredients in functional foods: From research stage to commercial food products. *Trends in Food Science & Technology*, *78*, 167–179.

Zhang, K., Zhang, H., Hu, X., Bao, S., & Huang, H. (2012). Synthesis and release studies of microalgal oil-containing microcapsules prepared by complex coacervation. *Colloids and Surfaces B: Biointerfaces*, *89*, 61–66.

Zhu, Q., Pan, Y., Jia, X., Li, J., Zhang, M., & Yin, L. (2019). Review on the stability mechanism and application of water-in-oil emulsions encapsulating various additives. *Comprehensive Reviews in Food Science and Food Safety*, *18*(6), 1660–1675.

CHAPTER 9

Techniques and Processes Involved in Nanoencapsulation of Omega-3, -6, and -9 Fatty Acids

Jessica Pandohee, Yakindra Prasad Timilsena, Jadala Shankaraswamy, and Lisa F. M. Lee Nen That

CONTENTS

DOI: 10.1201/9781003259183-9

9.1 INTRODUCTION

Unsaturated fatty acids such as omega-3, -6, and -9 are found in oils and are identified by the presence of a double bond on the carbon chain (Chang and Nickerson, 2018; Saini and Keum, 2018). Both omega-3 and omega-6 are polyunsaturated fatty acids and are also considered as essential fatty acids for the diet, as they cannot be produced in humans due to the absence of Δ^{12}-desaturase and Δ^{15}-desaturase enzymes (Alagawany et al., 2021; Saini and Keum, 2018). These enzymes are responsible for converting oleic acid to linoleic acid and linoleic acid to linolenic acid (Pandohee, 2022). However, omega-9 is a monounsaturated fatty acid because it has only one double bond and is not essential, as it is synthesized by the body (Alagawany et al., 2021). Fish oil in the form of eicosapentaenoic acid (EPA) and docosahexaenoic acid (DHA), vegetable, seed oils (flaxseed), and *Perilla frutescens* in the form of α-linolenic acid are all good sources of omega-3 fatty acids (Alagawany et al., 2021; Asif, 2011; Saini and Keum, 2018). Omega-6 is present as linoleic acid in many crop seeds (walnuts, Brazil nuts), vegetable oils (corn, soybean), and as arachidonic acid in meat, eggs, and dairy products (Astorg et al., 2004). Docosapentaenoic acid found in fish oil, is an omega-6 fatty acid whereas oleic acid, found primarily in olive oil, is an omega-9 fatty acid (Mousavi et al., 2016).

Obesity, hypertension, and diabetes, as well as autoimmune diseases, cardiovascular diseases, inflammation, cancer, and bone health, have all been linked to unsaturated fatty acids (Chang and Nickerson, 2018; Saini and Keum, 2018). Omega-3 fatty acid has had a protective and therapeutic role in these health disorders, as it has anti-inflammatory and immunomodulatory properties and regulates gene expression and transcription factors in lipid metabolism in different types of cells (Calder, 2010; Calder and Grimble, 2002; Jump and Clarke, 1999). Omega-6, on the other hand, has had a pro-inflammatory effect in the body and many inflammatory mediators such as prostaglandins and leukotrienes are derived from omega-6 (Calder and Grimble, 2002; Dennis and Norris, 2015). It has also been involved in platelet aggregation and vasoconstriction (Saini and Keum, 2018). Omega-3 and -6 fatty acids are precursors to eicosanoids, which regulate inflammation, immunity, blood vessel formation, cell proliferation, and differentiation, as well as serve homeostatic functions by regulating these processes (Berquin et al., 2008; Saini and Keum, 2018). Therefore, a low ratio is necessary between the two fatty acids to reduce the risk of diseases and maintain health (Simopoulos, 2002). The Western diet has contributed to an increase in omega-6 fatty acid consumption, resulting in an omega-6:omge-3 fatty acid ratio imbalance (Simopoulos, 2002). An optimal ratio would be dependent upon the health disorder being diagnosed and a recommended ratio would generally be 1:1 or 2:1 (Saini and Keum, 2018; Simopoulos, 2004).

There have been many hurdles, which have made it difficult to include unsaturated fatty acids in our daily diet. Technical aspects such as storage, processing, and transport have also posed problems (Jurić et al., 2022). These fatty acids are unsaturated and chemically unstable, which makes them prone to lipid oxidation, loss of volatile compounds, and polymerization when exposed to certain environmental conditions (Chang and Nickerson, 2018; Jurić et al., 2022). This could ultimately affect the shelf life and lead to deterioration of nutritional value, sensory characteristics, and the overall acceptability of food products (Chang and Nickerson, 2018; Gökmen et al., 2011; Velasco et al., 2003). Moreover, by-products of lipid oxidation have been linked to cardiovascular disease, cancer, and atherosclerosis (Addis, 1986; Jurić et al., 2022; Yaacoub et al., 2008).

Nanotechnology has been recently introduced in the field of food science and is considered to provide advantages to overcome these issues and prolong the shelf life of unsaturated fatty acids (Hamad et al., 2018). Nanoencapsulation represents a viable method, which would allow the unsaturated fatty acids to be embedded in a matrix (Ferreira and Nunes, 2019b; Jurić et al., 2022). This protects these fatty acids from oxidation and hydrolysis. Moreover, it improves the absorption and bioavailability of the encapsulated fatty acids and carries its content to their targeted delivery with minimal loss of fatty acids before its destination (Ferreira and Nunes, 2019b; Khorasani et al., 2018). No negative impact on the morphology, textural and sensory aspects of the final food product would be observed; instead, there would be additional nutritional value and improved sensory quality (Ferreira and Nunes, 2019b). Different techniques including coacervation, spray-drying, nanoemulsification, and nanoliposome entrapment have been used for encapsulation of unsaturated fatty acids (Saini et al., 2020). This chapter provides an overview of the techniques involved in nanoencapsulation as well as the biodegradability and biocompatibility of the nanoencapsulated fatty acids. With the advent of encapsulated omega-3, -6, and -9, the production of novel enriched foods may encourage a more balanced dietary intake of unsaturated fatty acids.

9.2 BENEFITS OF THE NANOENCAPSULATION PROCESS

9.2.1 Protection from Oxidation

Several experiments have proved that nanoencapsulation protects omega fatty acids against oxidation (Delshadi et al., 2020; Ghorbanzade et al., 2017; Slaven Jurić et al., 2022). Oxygen in contact with food containing omega-3, -6, and -9 causes oxidation of the fatty acids into aldehydes, which results in rancidity, off-flavor, foul odor, texture alterations as well as discoloration of the food. Nanoencapsulation prevents oxidation by surrounding the omega-3, -6, and -9 molecules. Moreover, the nano-sized spheres produced after omega fatty acids have been synthesized have a very low surface oil (<0.5%) at very high payloads (30%–40%) compared to micro-encapsulated fatty acids (Singh, 2016). This further protects omega-3, -6, and -9 from oxidation. Ojagh and Hasani (2018), synthesized nanoliposome fish oil, and tested the properties of bread enriched with the nanoliposomes. The nanoencapsulated fish oil had superior oxidative stability against lipid oxidation than the unencapsulated fish oil with an efficiency of incorporation of 90%.

9.2.2 Increased Solubility, Absorption, and Bioavailability

Omega fatty acids are hydrophobic in nature; they repel water and therefore can be difficult to be absorbed in the gastrointestinal tract, if they are not in the right environment. The hydrophobicity and hydrophilicity of nanoencapsulated omega fatty acids can be controlled. Moreover, they can be manufactured with other bioactive ingredients, such as vitamins or prebiotics which will increase the absorption of omega fatty acids. Shehzad et al. (2021) developed nanoemulsions of tuna fish oil using curcumin and resveratrol as co-antioxidants with an aim to be stabilized by two modified starches (Purity Gum 2000 or Purity Gum Ultra). Tuna fish oil nanoemulsions with the modified starches showed increased storage at 4°C, 25°C, and 40°C when co-encapsulated with curcumin and resveratrol.

The fact that nano-sized particles with omega fatty acids provide a larger surface area to volume ratio compared to omega fatty acids on their own increases their bioavailability. This is due to the increase in favorable interactions between the epithelial cells of the wall of the intestines and the nanoencapsulated particles. It is also believed that nanoencapsulated omega-3, -6, and -9 will stay longer in the intestines. Nanoemulsion carriers have several benefits over common emulsions such as high surface area, relatively high physical stability, better bioavailability and transparent appearance (Jin et al., 2016; Kumar et al., 2019).

9.2.3 Extending Shelf Life and Reducing Food Packaging

The primary purpose of elaborate food packaging is to prevent food spoilage, bacterial infection, or loss of nutrients. Considering how delicate omega-3, -6, and -9 fatty acids are they are at high risk of degrading if not packaged properly or consumed as soon as the food has been processed. Therefore, this calls for extra packaging, which will potentially cause unwanted environmental pollution, depending on the nature of the materials used in packaging. Nanoencapsulation isolates omega fatty acids which minimizes the contact of these fatty acids with any sources of food spoilage. Zimet et al. (2011) used casein, milk proteins to encapsulate DHA; they found that casein and DHA had a high affinity ($K_b = (8.38 \pm 3.12) \times 10^6\ M^{-1}$) and that on average one molecule of casein would bind to three to four molecules of DHA, therefore increasing its bioavailability. More importantly, the casein–DHA nanoparticle showed increased stability and an improved shelf life of 40 days at 4°C.

9.2.4 Sensory Attribute and Customer Perspective

While the benefits of consuming polyunsaturated fatty acid containing food is undeniable, the flavor and taste experience of the consumer is another aspect of nutrition that is important to be considered. Polyunsaturated fatty acids that undergo lipid oxidation produce biochemical compounds with an off-flavor and fish-like smell. Some of these metabolites include aldehydes and carboxylic acids. Lipid oxidation can occur anytime during food processing, storage, and cooking. By reducing the risks of oxidation and hydrolysis of omega-3, -6, and -9 as discussed in Section 9.2.1, the chance for the production of off-flavor biochemical compounds will also be reduced, thus increasing the chances of customers to not dislike food containing nanoencapsulated omega-3, -6, and -9. Therefore, nanoencapsulated omega fatty acids have an improvement in the macroscale level characteristics of food. Interestingly, when comparing bread and milk enriched with either nanoencapsulated omega-3, microencapsulated omega-3, or unencapsulated fish oil (Rasti et al., 2017), no significant difference in perceptible sensory attributes ($p=0.11$) between fish oil and nanoliposomal omega-3 fatty acids was found, while samples enriched with unencapsulated or microencapsulated omega-3 fatty acids were perceived as fishy ($p=0.02$). Higher omega-3 fatty acid recovery was also observed in nanoliposomal samples compared to samples enriched with microencapsulated omega-3 fatty acids.

9.3 TECHNIQUES EMPLOYED IN NANOENCAPSULATION

Omega-3, -6, and -9 fatty acids can be encapsulated in microstructures or nanostructures using a variety of methods. Food ingredient nanoencapsulation techniques are more sophisticated than food ingredient microencapsulation techniques (Anandharamakrishnan, 2014). In general, the techniques of nanoencapsulation are categorized into three main groups (Quintanilla-Carvajal et al., 2010). They are as follows:

- Physical processes; examples include nanoemulsification, spray coating and drying, freeze-drying, and extrusion.
- Simple or sophisticated coacervation and liposome entrapment are examples of physiochemical processes.
- Interfacial polymerization and molecular inclusion are two chemical processes.

Among them, spray-drying, freeze-drying, and coacervation are the most common techniques used to encapsulate polyunsaturated fatty acids and other lipophilic substances (Tontul et al., 2017). Techniques such as nanoprecipitation, emulsion-diffusion, electrospraying, electrospinning, ultrasonication, and supercritical fluid techniques are also used in the production of nanocapsules in food industries (Ferreira and Nunes, 2019a). Some authors classify liquid-based nanoencapsulation techniques as nanoprecipitation, solvent evaporation, coacervation, complex complexation, and the use of supercritical fluid (Walia et al., 2019), while others classify them as bottom-up and top-down approaches (Walia et al., 2019). (Ezhilarasi et al., 2013). In the bottom-up approaches, the nanocapsules are formed by the self-assembly and self-organization of constituents whereas in the case of top-down approaches, the size of the particles is decreased substantially (Tontul et al., 2017). Bottom-up approaches are characterized by low-energy consumption and greater control over particle size and structural morphology, whereas top-down approaches are suitable for both hydrophilic as well as hydrophobic bioactive substances (Basar et al., 2021).

Each method has its merits, demerits, and limitations. The best method to use is determined by the particle size necessary, the physicochemical qualities of the core and carrier substances, the intended use of the encapsulated nanoparticles, and the degree of protection required for the core (Ezhilarasi et al., 2013). This section will give an overview of some of the nanoencapsulation technologies and carriers that are currently being used.

9.3.1 Nanoemulsification

In most nanoencapsulation processes, the first step is the formation of nanoemulsions. In this process, the fatty acids (omega-6 and -9) are emulsified with an aqueous phase in the presence of certain emulsifiers. High-energy processes such as high-shear mixing, high-pressure homogenization, microfluidization, and microchannel homogenization are used to create these emulsions (Liao et al., 2021).

Nanoemulsification is a promising way to incorporate polyunsaturated fatty acids including omega-6 and -9 into food and beverages, salad dressings, sauces, and dips (Walker et al., 2015). Because of the very small droplet size and high surface area, these emulsions offer better stability, high optical clarity, and enhanced bioavailability of the fatty acids (Ezhilarasi et al., 2013).

9.3.2 Nanoencapsulation with Spray-Drying

Spray-drying is a rapid, cost-effective, automated, and high-efficiency method for encapsulating bioactive substances such as omega-6 and -9 fatty acids in nanoforms and microforms (Timilsena et al., 2020). High product yield with substantially high encapsulation efficiency is one of the main advantages of nanoencapsulation techniques by spray-drying. Nano-spray dryers are used to produce nanoparticles of various bioactive substances (Assadpour and Jafari, 2019). In this process, the fatty acids are firstly homogenized at very high shear with the carrier matrix in suitable proportions so that a stable nanoemulsion is achieved. The resultant emulsion is then atomized into a drying chamber and dehydration of atomized particles is carried out. The spray-dried particles contain the lipids in the center, which are covered with a thin exterior layer of carrier matrix (Figure 9.1).

9.3.3 Coacervation

Because of its high payload and encapsulation efficiency, coacervation is becoming a more popular nanoencapsulation technology (up to 99%). In addition, the

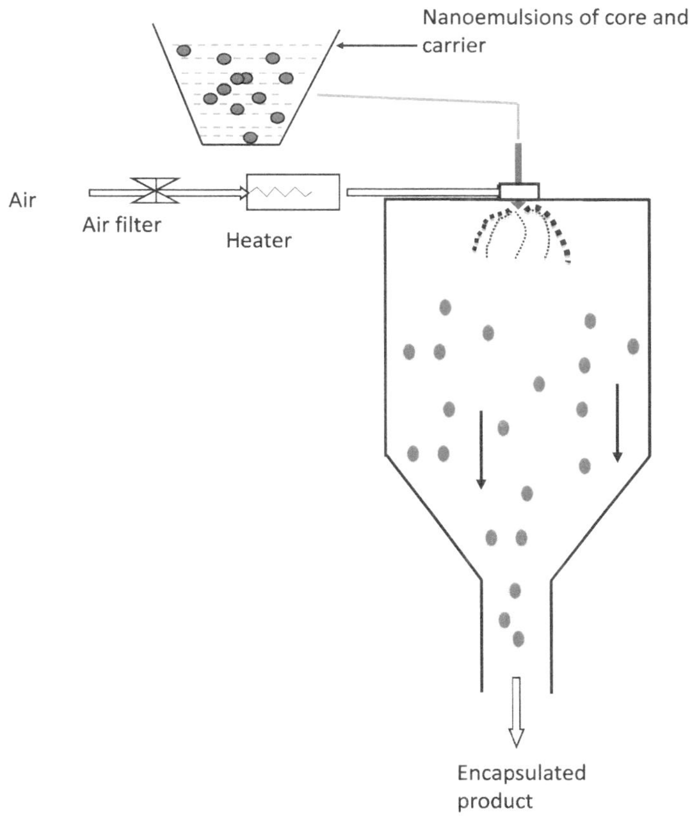

Figure 9.1 A schematic diagram of encapsulation process by nano-spray dryer.

encapsulation is carried out at ambient temperature without the use of any toxic solvents. Furthermore, a large number of substances are available as carriers (Timilsena et al., 2019). In the food and pharmaceutical industries, it's also one of the most efficient micro/nanoencapsulation technologies (Xiao et al., 2014). To make a nanoemulsion, an aqueous solution of two oppositely charged biopolymers, typically proteins and polysaccharides is emulsified with the target lipid. These oppositely charged polymers build a complex shell around the core material when triggered by a change in pH or temperature or the addition of another solvent. Despite its many benefits, the complicated coacervation nanoencapsulation method has certain flaws. The procedure takes a long time. Cross-linking of the newly formed coacervates is essential to make them stable; otherwise, they can dissociate and release the core when the pH and ionic environment is unfavorable (Timilsena et al., 2016). In general, the coacervates are dried (using spray-drying or freeze-drying) and converted to powder form (Figure 9.2).

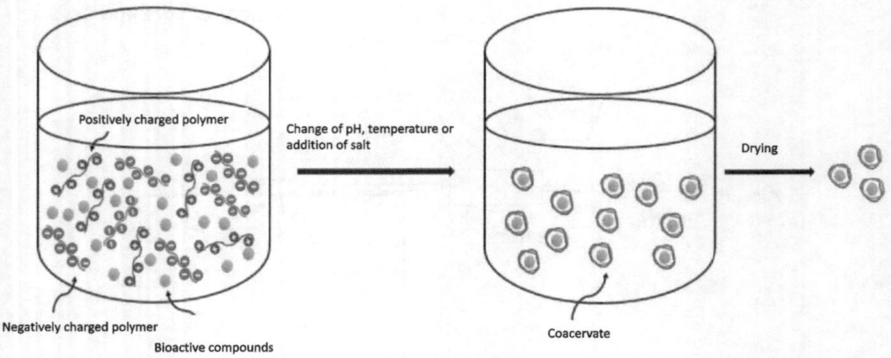

Figure 9.2 General process scheme for capsule preparation by complex coacervation. (Reprinted from Liao et al. (2021) under Creative Commons Attribution license.)

9.3.4 Electrohydrodynamic Technique

Electrohydrodynamic techniques such as electrospinning and electrospraying are emerging methods of nanoencapsulation/microencapsulation of food and pharmaceutical ingredients. In these methods, charged jets of a polymer solution or nanoemulsions are used to rotate or spray, which results in fibers or particles. These are simple, low-cost, flexible, and single-step process of preparation of monodispersed oil-based nanoparticles or fibers in microscale or nanoscale (Trotta et al., 2010). These technologies produce fiber nanoparticles with a manageable size and substantial surface area. Similarly, heat-sensitive bioactives such as polyunsaturated fatty acids are well-suited to these approaches. During this encapsulation technique, very high encapsulation efficiency (up to 96.4%) can be achieved (Ghayempour and Mortazavi, 2014).

9.3.5 Nanoliposome Entrapment

Nanoliposomes are lipid-based carriers for encapsulation of bioactive ingredients. Liposomes are very promising delivery systems owing to their flexible biophysical and chemical characteristics. These liposomes can be easily modified to match the delivery system's requirements. As a result, they are the most commonly used and researched nanocarriers for delivering bioactive compounds to specified sites (Sercombe et al., 2015). Liposomes are phospholipids encapsulating different aqueous areas with one or more concentric lipid bilayers (Sercombe et al., 2015). As a result, they can encapsulate both hydrophilic and lipophilic compounds making them versatile nanoencapsulation carriers. Nanoliposome entrapment is used to encapsulate omega-6 and -9 fatty acids and similar lipophilic substances. For this purpose, the fatty acids and the phospholipids are dissolved in a suitable solvent. An external energy (e.g., ultrasonication) is applied to the system making the phospholipids aggregate outside of the fatty acids forming a double-layered capsule (Abbasi et al., 2019a; Shukla et al., 2017). Finally, the dried nanocapsules are recovered by lyophilization, freezing, spray-drying, or the supercritical fluid extraction method (Fathi et al., 2012).

9.3.6 Nanoprecipitation

Nanoprecipitation is a simple and reproducible technique of developing nanoparticles with particle size ranging from 50 to 300 nm (Barreras-Urbina et al., 2016). This method is known for its simplicity, rapidity, reduced energy consumption, and reproducibility (Barreras-Urbina et al., 2016; Walia et al., 2019).

Nanoprecipitation involves three steps: particle nucleation, growth, and aggregation. Nucleation is initiated by the addition of a polymeric solution to an aqueous/solvent phase containing the bioactive compound. The growth of the particles takes place through coagulation or condensation process.

9.4 CARRIERS USED IN NANOENCAPSULATION

Carriers are inert materials that surround the core material. Appropriate carrier material is chosen from a large pool based on its physicochemical attributes, cost, availability and scale; the degree of protection it can provide to the core; the site; and the rate of release of the core and environmental aspects of its production and disposal (Assadpour and Jafari, 2019). Broadly speaking, carriers used in the nanoencapsulation of essential fatty acids are grouped into lipid-based materials and biopolymer-based materials. In addition, fiber-like nanostructures are also formed by electrospinning or electrospraying processes. Therefore, some authors classify the nanocarriers into three groups (Chen et al., 2006). Nanoemulsions, nanoliposomes, solid lipid nanoparticles, and nanostructured lipid carriers (NLCs) are among the first class of materials, whereas biopolymer-based carriers primarily consist of proteins (e.g., albumin, lactalbumin, gelatin, collagen, casein, whey proteins, and soy proteins) and polysaccharides (e.g., alginate, starch, chitosan, cyclodextrin, pectin, and gum Arabic). Biopolymer-based carriers are obtained from various sources. Both lipid-based and polymer-based carriers have long been used in the formulation of nanodelivery systems for fatty acids and other lipophilic substances (Ezhilarasi et al., 2013).

9.4.1 Nanoemulsions

Oil-in-water nanoemulsions are formed by dispersing small emulsifier-coated oil droplets having a diameter of <200 nm in water (McClements and Rao, 2011). Before or after the nanoemulsion is formed the bioactive lipids are normally retained within the oil droplets. Low-energy or high-energy techniques can be used to create this type of nanoemulsion from food-grade oil, water, and an emulsifier. Low-energy approaches rely on the spontaneous formation of microscopic oil droplets when the system composition or ambient factors are changed in a specified way. In high-energy procedures, mechanical devices called homogenizers are employed to break up the oil and water phases and form minute oil droplets coated with emulsifiers. Food-grade emulsifiers include synthetic surfactants, biosurfactants, phospholipids, proteins, and polysaccharides (Shu et al., 2018; Ozturk and McClements, 2016; Komaiko and McClements, 2015; Qian and McClements, 2011). To prevent the emulsifier-coated oil droplets from aggregating, significant electrostatic and/or steric repulsive forces are typically generated between them (flocculating or coalescing). Because of the small size of the droplets, nanoemulsions have a great resistance to gravitational separation (McClements and Rao, 2011). When the oil phase used has a significant solubility in water such as essential or flavor oils, nanoemulsions are susceptible to droplet growth through Ostwald ripening. To reduce this form of instability, they must be developed properly in this situation, such as by adding a ripening inhibitor. Microemulsions and nanoemulsions should be distinguished because they are both used as bioactive delivery methods and share some compositional and structural characteristics. Thermodynamically, nanoemulsions are unstable, whereas microemulsions are stable. As a result, while microemulsions can be made simply by mixing bioactives, surfactants, and water, nanoemulsions necessitate specialized preparation techniques that can be time-consuming and energy-intensive. Microemulsions have the disadvantage of requiring a lot of synthetic surfactants which makes them unsuitable for a lot of culinary uses (McClements, 2012).

9.4.2 Solid Lipid Nanoparticles and Nanostructured Lipid Carriers

Solid lipid nanoparticles (SLN) contain small emulsifier-coated lipid particles (d <200 nm), yet the molecules within the lipid phase are organized into extremely regular crystalline structures rather than chaotic like in a fluid (Weiss et al., 2020).

The necessary fatty acids or lipids are once again discovered in the lipid phase. If they are successfully trapped within the crystal structure formed by the lipid molecules, they may have better chemical resistance and a longer release rate. When the lipid phase crystallizes, they are sometimes ejected to the surfaces of the SLNs which can compromise their chemical stability. Furthermore, due to their huge surface area the form of lipid nanoparticles may change from spherical to nonspherical after production making them more prone to aggregation.

Both the type of lipid phase and the emulsifier used in SLNs must be carefully chosen to ensure optimal performance. To address the issues with SLNs, NLCs were developed. They are made up of emulsifier-coated lipid particles having a diameter of <200 nm that are disseminated in water. This time a mixture of solid and liquid lipids is used to create a less ordered lipid phase which allows for more bioactive lipid solubilization and retention as well as fewer particle shape changes during the lipid phase crystallization process (Tamjidi et al., 2013; Katouzian et al., 2017).

9.4.3 Nanoliposomes

Nanoliposomes are colloidal dispersions that contain lipid nanoparticles with diameters of <200 nm and are made up of one or more concentric lipid bilayers produced by phospholipid molecules self-associating in water (Khorasani et al., 2018). At the particle surfaces, the phospholipids' polar head groups protrude into the surrounding water phase.

Within the concentric bilayers, the nonpolar phospholipid tails face each other allowing bioactive lipids to be contained within these hydrophobic regions (Akhavan et al., 2018; Assadpour and Jafari, 2019; Tan et al., 2014). The two ways for generating nanoliposomes are coating-solvent evaporation and homogenization operations, with the former being more ideal for large-scale continuous commercial production and the latter being more suitable for laboratory use (Akhavan et al., 2018).

9.4.4 Nanogels and Nanofibers Made of Biopolymer

Biopolymer nanogels and nanofibers have been developed to encapsulate bioactive lipids (Assadpour and Jafari, 2019). Nanogels are made up of small biopolymer-rich particles suspended in water with a diameter of <200 nm. The particles form a 3D network of cross-linked biopolymer molecules that retain a large quantity of water due to hydration and capillary pressures. Biopolymer molecules can be cross-linked by physical or chemical interactions, and the pore size of the resulting network can be changed by modifying biopolymer or cross-linking concentrations as well as preparation conditions.

Before being added to the biopolymer network, bioactive lipids can be enclosed in lipophilic nanocarriers (such as nanoemulsions or nanoliposomes). Biopolymer nanogels can be utilized to increase the stability and dispersion of bioactive lipids. Biopolymer nanofibers are made up of thin filaments of cross-linked biopolymer molecules and electrospinning is a common method for manufacturing them. Bioactive lipids can be trapped directly or indirectly in the biopolymer solution used to make the nanofibers prior to the electrospinning process. Nanochitin, nanocellulose, and nano-starch delivery systems for food processing and packaging have all been developed using this method (Kumar et al., 2019; Rostamabadi et al., 2020).

9.4.5 Pickering Emulsions Stabilized with Nanoparticles

Oil-in-water colloidal particles that have been covered with oil droplets make up pickering emulsions (Mwangi et al., 2020). These types of structured emulsions can be created entirely of food-grade ingredients and used to encapsulate bioactive lipids. Food-grade nanoparticles such as nanocellulose, nanochitin, nano-starch, protein nanoparticles, and SLNs can be used as emulsifiers in pickering emulsions (Zhou et al., 2020). Oil droplets can be used to encapsulate bioactive lipids or nanoparticles can be used to stabilize them.

9.5 BIODEGRADABILITY OF NANOPARTICLES

Nanotechnology can be used to encapsulate these lipophilic molecules in biocompatible materials, reducing oxygen exposure and preventing omega-6 and omega-9 fatty acid oxidation. By enhancing bioavailability and as a result, therapeutic efficacy, nanoencapsulation of these unsaturated fatty acids promotes the maintenance of their functional properties. According to Cheong et al. (2017), nanoencapsulation is an effective technique for increasing bioactive chemical intake through food by delaying oxidation, increasing solubility, and preventing the unpleasant taste of components.

Biodegradable nanomaterials are nanoscale materials that can disintegrate naturally in living environments (Hans and Lowman, 2002). Biodegradable nanoparticles can be made with proteins, polysaccharides, and synthetic biodegradable polymers among other constituents. Based on a range of designs and end-use criteria, the base polymer is chosen. It is governed by a variety of criteria, including the ideal nanoparticle's size, the chemical and physical features of omega fatty acids (such as solubility in water and stability), the surface qualities, and functionality and the level of biodegradability and biocompatibility. Depending on the desired parameters for nanoparticle production, the methods can be classed as (i) prepared polymer dispersion, (ii) monomer polymerization, and (iii) method of ionic gelation for hydrophilic polymers.

Nanoscale materials that degrade naturally in living settings are known as biodegradable nanomaterials (Wiwanitkit, 2019). Because optimal biodegradable nanoparticles must decay efficiently at the target site while remaining stable at off-target regions, their ability to degrade can be a useful attribute for managing release (Wiwanitkit, 2019). Because they have a regulated release profile and are stable in the circulation system, biodegradable nanoparticles have a lot of potential in the drug delivery system. They are also nontoxic, nonimmunogenic, and can escape the reticuloendothelial system which is a portion of the body's immune system that collects and eliminates foreign things, allowing them to circulate longer (Wiwanitkit, 2019). Agents of interest can be encapsulated in nanoparticles or adsorbed on the nanomaterials' surfaces, and biodegradable nanoparticles, similar to their nonbiodegradable counterparts can be classified based on the structure and arrangement of the nanomaterials which can be achieved by encapsulating the agents of interest in a nanocapsule or merging those agents into a nanosphere (Kumari et al., 2010).

Because biodegradable nanomaterials reduce cytotoxicity in the body, biodegradable nanoparticles retain the general advantages of nanoparticles in nanomedicine, such as controlled cargo release and targeted delivery, resulting in improved therapeutic effects and fewer side effects, especially for certain cytotoxic drugs. To increase the drug release profile and targeting efficiency, more surface changes can be made. Some examples of the advantages of biodegradable nanoencapsulated omega fatty acids include excellent biocompatibility, low immunogenicity, delivery of omega fatty acids in a slow and controlled way, improved treatment outcomes, reduced cytotoxicity, and adverse effects and malleability.

Limitation of biodegradable nanoparticles:

1. If not modified, they will become unstable.
2. Particle creation is expensive.
3. Particles of consistent shape and size are challenging to create.
4. The encapsulation of drugs is fairly poor.
5. Production and sterilizing on a large scale are difficult.

Polymers that allow for controlled drug release of core components are desirable and polymeric nanoparticles for cancer treatment and vaccine delivery have become increasingly popular (Masood, 2016). Polymeric nanoparticles' chemical properties and flexibility make them excellent for integrating with biomaterials (such as genetic material and growth hormones) and delivering tailored tissue regeneration (Saravanan et al., 2017). Biodegradable nanoparticles have been used to deliver drugs, vaccinations, and other biomolecules such as omega-6 and omega-9 fatty acids to specified regions.

Polymeric nanoparticles are made from biocompatible and biodegradable polymers acquired from natural and manmade sources. The most frequent wall materials used in nanoencapsulating omega-6 and -9 fatty acids are biodegradable polymers. Polycaprolactone (Choi et al., 2010; Prieto and Calvo, 2017), whey protein (Jafari et al., 2008), sodium caseinate (Xue et al., 2015), maltodextrin, or modified starch (Jafari et al., 2008) are some common wall materials utilized in oil nanoencapsulation. Because of the spontaneous emulsification process used, some authors did not publish the wall material used in their study (Bernardi et al., 2011). Synthetic biodegradable polymers frequently use poly(lactic acid), polyglycolic acid, poly(lactic-co-glycolic acid), poly(-caprolactone), polymethyl methacrylate, and poly (lactic-co-glycolic acid) (amino acid). Biodegradable polymers can also be made from natural polymers such agarose, sodium alginate, chitosan, collagen, and fibrin (Li and Tuan, 2005) as well as chia seed mucilage (de Campo et al., 2017).

9.6 APPLICATION IN THE DEVELOPMENT OF NOVEL FOODS

Foods that are supplemented with encapsulated essential fatty acids such as omega-3, -6, and -9 are functional foods that have a higher nutritional value (Delshadi et al., 2020). Several research on the addition of encapsulated fatty acids to foods have yielded positive findings, since they boost nutritional content and confer other beneficial effects.

9.6.1 Dairy Products

Nanoencapsulation of unsaturated fatty acids has been applied to the production of fortified dairy products (Pudtikajorn et al., 2021; Slewa and Al-Mowsowy, 2018). Addition of nanoencapsulated fish oil using gum Arabic to low fat probiotic fermented milk had a beneficial effect on the probiotic bacteria *Lactobacillus plantarum* (Moghadam et al., 2019). The viability of *L. plantarum* was enhanced considerably in the presence of nanoencapsulated fish oil. Furthermore, the levels of EPA and DHA increased and the addition of nanoencapsulated fish oil had no influence on the total acceptability score.

Skim milk was fortified with shrimp oil rich in polyunsaturated fatty acids enclosed in nanoliposomes and characteristics of the fortified skim milk were analyzed (Gulzar and Benjakul, 2020). There was a color change from white to yellow and red as concentration of the nanoliposomes increased. Viscosity also increased as rheological parameters were modified. The sensory evaluation revealed that the participants did not report any fishy odor as nanoliposomes masked the fishy smell and was deemed generally acceptable. Storage of the fortified milk over a period of 15 days at 4°C limited the microbial load to <2.54 log CFU mL^{-1}. Although there were minor changes in pH and acidity, they were still within acceptable limits. The oxidative stability as well as the concentration of unsaturated fatty acids retained was maintained due to encapsulation by the nanoliposome.

A study by Marsanasco et al. (2015a) showed that chocolate milk could be enriched with liposomes containing omega-3, -6, vitamin E, and folic acid. The carrier was made of soy phosphatidylcholine (SPC), while stearic acid or calcium stearate was used as stabilizers. The liposomes were stable when morphological parameters such as size, shape, rheology, and membrane packing were tested. Furthermore, the carrier agent protected the thermolabile folic acid after pasteurization. The encapsulation efficiency was 48.5% and 50.4% for SPC: calcium stearate and SPC:stearic acid systems, respectively. There were significant differences in the sensory perception between chocolate milk only and enriched chocolate milk with either stabilizer. Fortified milk with calcium stearate had higher acceptability than with stearic acid. Chocolate milk with only liposomes did not affect on the acceptability and did not have any difference from commercial milk. Another study by Marsanasco et al. (2016) demonstrated the fortification of chocolate milk with omega-3, -6, and vitamins C and E. The results were similar to the previous study. The liposomes also protected vitamin C known for its thermolability after pasteurization. Their texture, morphology, membrane packing, and oxidative stability were stable and they exhibited low lipid peroxidation. Sensory evaluations revealed that the chocolate milk with liposome and stearic acid complex did not affect the sensory perception of participants although they knew what they were judging.

Olive, corn, and linseed oils are high in omega-3, -6, and -9 fatty acids and they were integrated into the crystal structure of anhydrous milk fat (AMF) (Viriato et al., 2020). Several parameters such as fatty acid composition, crystallization kinetics, thermal behavior, microstructure, and lipid compatibility have been assessed. Concentration of saturated fatty acids (41.37%) decreased as both monounsaturated (77.39%) and polyunsaturated fatty acids (808.70%) increased when compared with control AMF. Atherogenicity and thrombogenicity indices are lipid quality indices that quantify the role of saturated and unsaturated fatty acids in the progression of cardiovascular disorders. When vegetable oils were added to AMF, both indices decreased indicating a high concentration of unsaturated fatty acids. The inclusion of the vegetable oils changed the kinetics of nucleation and crystallization produced by minor chemicals in olive oil. The crystalline shape and polymorphism of AMF, on the other hand, remained unchanged. Furthermore, at 20°C, AMF–vegetable oil blends included 15%–20% solid fats which may be used as confectionery fats and form stable emulsions.

In other dairy products, such as yogurt, unsaturated fatty acids, namely, omega-3, -6, and -9 were microencapsulated with maltodextrin, whey protein concentrate, gum Arabic, and lactoferrin (Ramadan et al., 2021). Concentrations of Gum Arabic and whey protein concentrate in the carriers varied. The most stable carrier agents were those with the highest concentration of gum Arabic (25 g). Yogurt which had been flavored by orange essential oil was then fortified with the microcapsules. Encapsulation efficiency ranged from 52.63% to 88.61%. The fortified yogurt was also tested in a sensory evaluation and consumers found it acceptable, especially with the addition of orange essential oil as it camouflaged undesirable taste. Moreover, the flavored yogurt fortified with microcapsules protected rats suffering from neurotoxicity. The consumption of fortified yogurt alleviated oxidative stress by reducing malondialdehyde, catalase activity, and nitric oxide, and increased superoxide dismutase and glutathione levels. Butyrylcholinesterase and acetylcholinesterase are two enzymes that promote the breakdown of cholinesterase in the brain and could lead to Alzheimer's disease. Their levels were reduced in rats that had consumed fortified yogurt.

T. Ghorbanzade et al. (2017a) investigated the incorporation of fish oil into nanoliposomes and yogurt. The physicochemical properties of fortified yogurt were assessed, including pH, acidity, syneresis, fatty acid profile, and peroxide value. The centrifugal test revealed that the nanoliposomes were stable at 70.03% and had a size of 300–500 nm. Syneresis which is an indicator of the yogurt quality revealed that there were changes over time due to pH change. However, the changes were similar for control yogurt and fortified yogurt. The peroxide value of fortified yogurt remained constant over the 21 days of storage. Chromatogram profile also showed that there was an increase in DHA and EPA, indicating that the carrier could be protecting the fatty acids from deteriorating. Participants did not find any significant differences in the texture, color, taste, and aroma of control yogurt and fortified yogurt when evaluating the sensory characteristics. Moreover, there was also no significant difference in the overall acceptance between control and fortified yogurt with nanoliposomes.

A bioavailability study demonstrated that the ingestion of yogurt drink fortified with water and nanoemulsion of omega-3 led to a significant decline in the ratio of omega-6 and -3. The ratio was lower when compared to the consumption of strawberry milk with the bulk algae oil at similar amount. (Lane and Derbyshire, 2013; Lane et al., 2014). Omega-3 (EPA) to omega-6 (arachidonic acid) ratio is crucial since it is a health indicator and a vegetarian diet is typically high in omega-6 and low in omega-3. This shows that consuming an omega-3 nanoemulsion enhanced the ratio since it had a higher bioavailability and could be eaten by vegetarians. Nanoemulsions were able to pass through cell membranes in the body due to their small size.

9.6.2 Meat Products

The fortification of meat products by adding nanoencapsulated unsaturated fatty acids has had a beneficial effect. Animal fat usually contains saturated fatty acids and encapsulated unsaturated fatty acids have been used to enrich their nutritional value. In order to reduce saturated fatty acids in burgers, Heck et al. (2019) used a hydrogelled emulsion combining chia and linseed oils with high omega-3 content. Although the addition of the hydrogelled emulsion increased the burgers' hardness and chewiness, it had no effect on cooking loss, moisture retention, or diameter decrease. Furthermore, participants found that replacing 60% of the pork fat with the hydrogelled emulsion was acceptable. The fatty acid profile improved as well, with a lower ratio of omega-6 to omega-3 and lower lipid index scores. This shows that adding the hydrogelled emulsion to the burger made it healthier.

Ojha et al. (2017) used an ultrasonic device to encapsulate fish oil in nanovesicles which were then integrated into hog flesh. The nanovesicles ranged in size from 345.9 ± 5.1 to 352.8 ± 8.5 nm. When the pork meat was subjected to ultrasound treatment over long periods of time, the level of omega-3 increased indicating the positive effect of the treatment. The thrombogenic index is a measure of the likelihood of blood clots forming in blood arteries and both ultrasonography treatment and more time reduced the thrombogenic index, implying that saturated and unsaturated fatty acids have a strong beneficial relationship. After ultrasonic treatment, the omega-6/omega-3 ratio fell as well.

For the enrichment of broiler chicken meats, flaxseed oil, which is high in omega-3 fatty acids, was encapsulated in nanoemulsions using the ultrasonication process (Abbasi et al., 2019b). The nanoemulsions provided protection in simulated gastric conditions (10% released oil) and released their contents in simulated intestinal conditions (44.66% released oil), indicating that they provide targeted delivery. The malondialdehyde levels were also lowered in thigh and breast meat when slaughtered and after one-month storage at freezing temperature.

Broilers fed with nanoencapsulated flaxseed oil had increased body weight and improved lipid profile as levels of omega-3 were significantly higher in thigh and breast meat of broilers. In broilers fed with nanoencapsulated flaxseed oil, the omega-6:omega-3 ratio was considerably lower. The addition of nanoencapsulated flaxseed oil enhanced the expression of desaturase and elongase enzymes involved in the manufacture and metabolism of unsaturated fatty acids (Zhang et al., 2016).

9.6.3 Drinks and Other Beverages

Studies have demonstrated that nanoencapsulated omega-3 and -6 enrich beverages that do not have or lack fat content. Omega-3, -6, and vitamins E and C were encapsulated in liposomes made of SPC (Marsanasco et al., 2015b). The liposomes were analyzed for their morphology, size, structure, rheology, membrane packing, and oxidative stability, and demonstrated stability before and after pasteurization. The antioxidant activity of vitamin C which is thermolabile did not change after pasteurization. The liposomes were also incorporated in orange juice and sensory evaluation was tested using overall acceptability and triangular tests in a group of 78 evaluators. The acceptance of the product was unaffected by the addition of liposomes to orange juice.

Using a biopolymer called sodium caseinate–gum Arabic complex, nanoencapsulation for omega-3 polyunsaturated fatty acids from fish oil was also developed (Ilyasoglu and Nehir El, 2014). At a pH of 4, the biopolymer was stable and the optimal concentration was 0.1 g sodium caseinate and 0.2 g gum Arabic per 100 mL. The encapsulation effectiveness of the nanoencapsulated fish oil was found to be 78.88 ± 2.89%, and the size was found to be 232.3 nm. Fruit juice was also supplemented with nanoencapsulated omega-3 (40, 50, and 60 mg/100 mL fruit juice), and EPA and DHA bioaccessibility was 47.37%. These findings prove promising for incorporating nanoencapsulated essential fatty acids into beverages.

9.6.4 Bakery and Pastries

Nanoencapsulated unsaturated fatty acids have also been investigated for use in baking products. Indeed, Gökmen et al. (2011) described the encapsulation of flax seed oil, a key source of omega-3 fatty acids and the synthesis of nanocomplexes utilizing high amylose maize starch. Spray-drying was also employed to transform the flax seed oil into powder microparticles with a flax seed oil concentration of 11.6±0.6 g per 100 g powder.

Bread was then enriched with nanoencapsulated flax seed oil and the texture was affected by incorporation of nanocomplexes. Although, texture parameters such as hardness, chewiness, and gumminess increased as the concentration of nanocomplexes increased, cohesiveness and springiness remained constant. There was also a decline in the lipid peroxidation and the nanocomplexes prevented the production of harmful Maillard reaction compounds in bread during baking.

Nanoliposomes containing omega-3 were added in bread and its application was evaluated (Rasti et al., 2017). Fortified bread with nanoliposomes had higher % recovery of DHA and EPA than bread enriched with fish oil and also had lower peroxidation value. The anisidine value which is a measure of secondary oxidation of lipids was also reduced (Tompkins and Perkins, 1999). Sensory evaluation demonstrated that the addition of nanoliposomes did not have any significant difference from control bread samples and no fishy flavor was detected by participants.

9.7 CONCLUSIONS

Nanotechnology is being researched because it has the ability to protect bioactive lipid molecules that are susceptible to environmental conditions, such as unsaturated fatty acids, which are essential for human health. As a result, nanoencapsulation of monounsaturated and/or polyunsaturated fatty acids is a promising method for conserving nutraceutical properties, reducing oxidation, and enhancing bioavailability, thereby improving therapeutic efficacy. However, numerous ways are being researched in the literature making it difficult to establish which is the most successful in terms of preventing oxidation, high temperatures, and shelf life. Depending on the type of nanoparticle, particle size, stability duration in encapsulating oils, the number of surfactants and emulsifiers, the volume of encapsulated oil, and the stability period vary greatly.

REFERENCES

Abbasi, F., Samadi, F., Jafari, S. M., Ramezanpour, S., & Shams-Shargh, M. (2019a). Production of omega-3 fatty acid-enriched broiler chicken meat by the application of nanoencapsulated flaxseed oil prepared via ultrasonication. *Journal of Functional Foods, 57*, 373–381. doi: 10.1016/j.jff.2019.04.030.

Abbasi, F., Samadi, F., Jafari, S. M., Ramezanpour, S., & Shargh, M. S. (2019b). Ultrasound-assisted preparation of flaxseed oil nanoemulsions coated with alginate-whey protein for targeted delivery of omega-3 fatty acids into the lower sections of gastrointestinal tract to enrich broiler meat. *Ultrasonics Sonochemistry, 50*, 208–217.

Addis, P. B. (1986). Occurrence of lipid oxidation products in foods. *Food and Chemical Toxicology, 24*(10/11), 1021–1030.

Alagawany, M., Elnesr, S. S., Farag, M. R., El-Sabrout, K., Alqaisi, O., Dawood, M. A. O., Soomro, H., & Abdelnour, S. A. (2021). Nutritional significance and health benefits of omega-3, -6 and -9 fatty acids in animals. *Animal Biotechnology*, 1–13.

Anandharamakrishnan, C. (2014). *Techniques for nanoencapsulation of food ingredients* (Vol. 8, pp. 65–67). New York: Springer.

Asif, M. (2011). Health effects of omega-3,6,9 fatty acids: *Perilla frutescens* is a good example of plant oils. *Oriental Pharmacy and Experimental Medicine, 11*, 51–59.

Assadpour, E., & Jafari, S. M. (2019). Advances in spray-drying encapsulation of food bioactive ingredients: From microcapsules to nanocapsules. *Annual Review of Food Science and Technology, 10*, 103–131.

Astorg, P., Arnault, N., Czernichow, S., Noisette, N., Galan, P., & Hercberg, S. (2004). Dietary intakes and food sources of n-6 and n-3 PUFA in French adult men and women. *Lipids, 39*(6), 527–535.

Barreras-Urbina, C. G., Ramírez-Wong, B., López-Ahumada, G. A., Burruel-Ibarra, S. E., Martínez-Cruz, O., Tapia-Hernández, J. A., & Rodriguez Felix, F. (2016). Nano-and micro-particles by nanoprecipitation: Possible application in the food and agricultural industries. *International Journal of Food Properties, 19*(9), 1912–1923.

Basar, A. O., Prieto, C., & Lagarón, J. M. (2021). Chapter 4: Novel encapsulation of bioactives: Use of electrohydrodynamic processing and applications. In: *Importance & Applications of Nanotechnology* (pp. 24–46). MedDocs Publishers: Reno, NV.

Bernardi, D. S., Pereira, T. A., Maciel, N. R., Bortoloto, J., Viera, G. S., Oliveira, G. C., & Rocha-Filho, P. A. (2011). Formation and stability of oil-in-water nanoemulsions containing rice bran oil: In vitro and in vivo assessments. *Journal of Nanobiotechnology, 9*(1), 1–9.

Berquin, I. M., Edwards, I. J., & Chen, Y. Q. (2008). Multi-targeted therapy of cancer by omega-3 fatty acids. *Cancer Letters, 269*(2), 363–377.

Calder, P. C. (2010). Omega-3 fatty acids and inflammatory processes. *Nutrients, 2*(3), 355–374.

Calder, P. C., & Grimble, R. F. (2002). Polyunsaturated fatty acids, inflammation and immunity. *European Journal of Clinical Nutrition, 56*, S14–S19.

Chang, C., & Nickerson, M. T. (2018). Encapsulation of omega 3–6–9 fatty acids-rich oils using protein-based emulsions with spray drying. *Journal of Food Science and Technology, 55*(8), 2850–2861. doi: 10.1007/s13197-018-3257-0.

Chen, L., Remondetto, G. E., & Subirade, M. (2006). Food protein-based materials as nutraceutical delivery systems. *Trends in Food Science & Technology, 17*(5), 272–283.

Cheong, A. M., Tan, C. P., & Nyam, K. L. (2017). Physicochemical, oxidative and anti-oxidant stabilities of kenaf seed oil-in-water nanoemulsions under different storage temperatures. *Industrial Crops and Products, 95*, 374–382.

Choi, M.-J., Ruktanonchai, U., Min, S.-G., Chun, J.-Y., & Soottitantawat, A. (2010). Physical characteristics of fish oil encapsulated by β-cyclodextrin using an aggregation method or polycaprolactone using an emulsion–diffusion method. *Food Chemistry, 119*(4), 1694–1703.

Delshadi, R., Bahrami, A., Tafti, A. G., Barba, F. J., & Williams, L. L. (2020). Micro and nano-encapsulation of vegetable and essential oils to develop functional food products with improved nutritional profiles. *Trends in Food Science & Technology, 104*, 72–83.

Dennis, E. A., & Norris, P. C. (2015). Eicosanoid storm in infection and inflammation. *Nature Reviews Immunology, 15*, 511–523.

Ezhilarasi, P., Karthik, P., Chhanwal, N., & Anandharamakrishnan, C. (2013). Nanoencapsulation techniques for food bioactive components: A review. *Food and Bioprocess Technology, 6*(3), 628–647.

Fathi, M., Mozafari, M. R., & Mohebbi, M. (2012). Nanoencapsulation of food ingredients using lipid based delivery systems. *Trends in Food Science & Technology, 23*(1), 13–27.

Ferreira, C. D., & Nunes, I. L. (2019a). Oil nanoencapsulation: development, application, and incorporation into the food market. *Nanoscale Research Letters,* 14, 13. doi: 10.1186/s11671-018-2829-2.

Ferreira, C. D., & Nunes, I. L. (2019b). Oil nanoencapsulation: Development, application, and incorporation into the food market. *Nanoscale Research Letters, 14*, 9. doi: 10.1186/s11671-018-2829-2.

Ghayempour, S., & Mortazavi, S. M. (2013). Fabrication of micro–nanocapsules by a new electrospraying method using coaxial jets and examination of effective parameters on their production. *Journal of Electrostatics, 71*(4), 717–727.

Ghorbanzade, T., Jafari, S. M., Akhavan, S., & Hadavi, R. (2017). Nano-encapsulation of fish oil in nano-liposomes and its application in fortification of yogurt. *Food Chemistry, 216*, 146–152.

Gökmen, V., Mogol, B. A., Lumaga, R. B., Fogliano, V., Kaplun, Z., & Shimoni, E. (2011). Development of functional bread containing nanoencapsulated omega-3 fatty acids. *Journal of Food Engineering, 105*, 585–591. doi: 10.1016/j.jfoodeng.2011.03.021.

Gulzar, S., & Benjakul, S. (2020). Fortification of skim milk with nanoliposomes loaded with shrimp oil: Properties and storage stability. *Journal of the American Oil Chemists' Society, 97*, 929–940. doi: 10.1002/aocs.12371.

Hamad, A. F., Han, J.-H., Kim, B.-C., & Rather, I. A. (2018). The intertwine of nanotechnology with the food industry. *Saudi Journal of Biological Sciences, 25*, 27–30.

Hans, M. L., & Lowman, A. M. (2002). Biodegradable nanoparticles for drug delivery and targeting. *Current Opinion in Solid State and Materials Science, 6*(4), 319–327.

Heck, R. T., Saldaña, E., Lorenzo, J. M., Correa, L. P., Fagundes, M. B., Cichoski, A. J., Menezes, C. R. D., Wagner, R., & Campagnol, P. C. B. (2019). Hydrogelled emulsion from chia and linseed oils: A promising strategy to produce low-fat burgers with a healthier lipid profile. *Meat Science, 156*, 174–182. doi: 10.1016/j.meatsci.2019.05.034.

Ilyasoglu, H., & El, S. N. (2014). Nanoencapsulation of EPA/DHA with sodium caseinate–gum arabic complex and its usage in the enrichment of fruit juice. *LWT-Food Science and Technology, 56*(2), 461–468.

Jafari, S. M., Assadpoor, E., Bhandari, B., & He, Y. (2008). Nano-particle encapsulation of fish oil by spray drying. *Food Research International, 41*(2), 172–183.

Jin, W., Liang, H., Li, Y., Liu, S., & Li, B. (2016). Nanoemulsions for food: Properties, production, characterization, and applications. *Emulsions, 3*, 1–36.

Jump, D. B., & Clarke, S. D. (1999). Regulation of gene expression by dietary fat. *Annual Review of Nutrition, 19*, 63–90.

Jurić, S., Jurić, M., Siddique, M. A. B., & Fathi, M. (2022). Vegetable oils rich in polyunsaturated fatty acids: Nanoencapsulation methods and stability enhancement. *Food Reviews International, 38*(1), 32–69.

Katouzian, I., Esfanjani, A. F., Jafari, S. M., & Akhavan, S. (2017). Formulation and application of a new generation of lipid nano-carriers for the food bioactive ingredients. *Trends in Food Science & Technology*, *68*, 14–25.

Khorasani, S., Danaei, M., & Mozafari, M. R. (2018). Nanoliposome technology for the food and nutraceutical industries. *Trends in Food Science & Technology, 79*, 106–115.

Komaiko, J., & McClements, D. J. (2015). Low-energy formation of edible nanoemulsions by spontaneous emulsification: Factors influencing particle size. Journal of food engineering, 146, 122-–128.

Kumar, M., Bishnoi, R. S., Shukla, A. K., & Jain, C. P. (2019). Techniques for formulation of nanoemulsion drug delivery system: A review. *Preventive Nutrition and Food Science, 24*(3), 225–234.

Lane, K. E., & Derbyshire, E. J. (2013). Functional foods enriched with an omega-3 nanoemulsion: Potential to improve the long-term health of vegetarians? *Proceedings of the Nutrition Society*, *72*(OCE4), E239. doi: 10.1017/S0029665113002644.

Lane, K. E., Li, W., Smith, C., & Derbyshire, E. (2014). The bioavailability of an omega-3-rich algal oil is improved by nanoemulsion technology using yoghurt as a food vehicle. *International Journal of Food Science and Technology, 49*, 1264–1271. doi: 10.1111/ijfs.12455.

Liao, W., Badri, W., Dumas, E., Ghnimi, S., Elaissari, A., Saurel, R., & Gharsallaoui, A. (2021). Nanoencapsulation of essential oils as natural food antimicrobial agents: An overview. *Applied Sciences, 11*(13), 5778.

Marsanasco, M., Márquez, A. L., Wagner, J. R., Chiaramoni, N. S., & Alonso, S. D. V. (2015a). Bioactive compounds as functional food ingredients: Characterization in model system and sensory evaluation in chocolate milk. *Journal of Food Engineering*, *166*, 55–63. doi: 10.1016/j.jfoodeng.2015.05.007.

Marsanasco, M., Piotrkowski, B., Calabró, V., Alonso, S. D. V., & Chiaramoni, N. S. (2015b). Bioactive constituents in liposomes incorporated in orange juice as new functional food: Thermal stability, rheological and organoleptic properties. *Journal of Food Science and Technology, 52*(12), 7828–7838. doi: 10.1007/s13197-015-1924-y.

Marsanasco, M., Calabró, V., Piotrkowski, B., Chiaramoni, N. S., & Alonso, S. D. V. (2016). Fortification of chocolate milk with omega-3, omega-6, and vitamins E and C by using liposomes. *European Journal of Lipid Science and Technology, 118*, 1271–1281. doi: 10.1002/ejlt.201400663.

McClements, D. J., & Rao, J. (2011). Food-grade nanoemulsions: formulation, fabrication, properties, performance, biological fate, and potential toxicity. *Critical Reviews in Food Science and Nutrition*, *51*(4), 285–330.

Moghadam, F. V., Pourahmad, R., Mortazavi, A., Davoodi, D., & Azizinezhad, R. (2019). Use of fish oil nanoencapsulated with gum arabic carrier in low fat probiotic fermented milk. *Food Science of Animal Resources, 39*(2), 309–323. doi: 10.5851/kosfa.2019.e25.

Mousavi, S. N., Koohdani, F., Shidfar, F., & Shafiei-Neek, L. (2016). Effects of diets enriched in omega-9 or omega-6 fatty acids on reproductive process. *Journal of Family and Reproductive Health, 10*(2), 85–91.

Ojagh, S. M., & Hasani, S. (2018). Characteristics and oxidative stability of fish oil nanoliposomes and its application in functional bread. *Journal of Food Measurement and Characterization, 12*(2), 1084–1092.

Ozturk, B., & McClements, D. J. (2016). Progress in natural emulsifiers for utilization in food emulsions. *Current Opinion in Food Science, 7*, 1–6.

Pandohee, J. (2022). Alpha-linolenic acid. In: Kour, J., & Nayik, G. A. (Eds.), *Nutraceuticals and Health Care* (pp. 279–288). Academic Press: Cambridge, MA. doi: 10.1016/B978-0-323-89779-2.00003-X.

Prieto, C., & Calvo, L. (2017). The encapsulation of low viscosity omega-3 rich fish oil in polycaprolactone by supercritical fluid extraction of emulsions. *The Journal of Supercritical Fluids, 128*, 227–234.

Pudtikajorn, K., Sae-leaw, T., & Benjakul, S. (2021). Characterization of fortified pasteurized cow milk with nanoliposome loaded with skipjack tuna eyeball oil. *International Journal of Food Science and Technology, 56*, 5893–5903. doi: 10.1111/ijfs.15196.

Quintanilla-Carvajal, M. X., Camacho-Díaz, B. H., Meraz-Torres, L. S., Chanona-Pérez, J. J., Alamilla-Beltrán, L., Jimenéz-Aparicio, A., & Gutiérrez-López, G. F. (2010). Nanoencapsulation: A new trend in food engineering processing. *Food Engineering Reviews, 2*(1), 39–50.

Ramadan, M. M., El-Said, M. M., El-Messery, T. M., & Mohamed, R. S. (2021). Development of flavored yoghurt fortified with microcapsules of triple omega 3-6-9 for preventing neurotoxicity induced by aluminum chloride in rats. *Journal of Food Processing and Preservation, 45*, e15759. doi: 10.1111/jfpp.15759.

Rasti, B., Erfanian, A., & Selamat, J. (2017). Novel nanoliposomal encapsulated omega-3 fatty acids and their applications in food. *Food Chemistry, 230*, 690–696. doi: 10.1016/j.foodchem.2017.03.089.

Saini, R. K., & Keum, Y. S. (2018). Omega-3 and omega-6 polyunsaturated fatty acids: Dietary sources, metabolism, and significance: A review. *Life Sciences, 203*, 255–267.

Saini, A., Panwar, D., Panesar, P. S., & Bera, M. B. (2020). Encapsulation of functional ingredients in lipidic nanocarriers and antimicrobial applications: A review. *Environmental Chemistry Letters, 19*, 1107–1134.

Sercombe, L., Veerati, T., Moheimani, F., Wu, S. Y., Sood, A. K., & Hua, S. (2015). Advances and challenges of liposome assisted drug delivery. *Frontiers in Pharmacology, 6*, 286.

Shehzad, Q., Rehman, A., Jafari, S. M., Zuo, M., Khan, M. A., Ali, A., Khan, S., Karim, A., Usman, M., & Hussain, A. (2021). Improving the oxidative stability of fish oil nanoemulsions by co-encapsulation with curcumin and resveratrol. *Colloids and Surfaces B: Biointerfaces, 199*, 111481.

Shu, G., Khalid, N., Tan, T. B., Zhao, Y., Neves, M. A., Kobayashi, I., & Nakajima, M. (2018). In vitro bioaccessibility of ergocalciferol in nanoemulsion-based delivery system: the influence of food-grade emulsifiers with different stabilising mechanisms. *International Journal of Food Science & Technology, 53*(2), 430–440.

Shukla, S., Haldorai, Y., Hwang, S. K., Bajpai, V. K., Huh, Y. S., & Han, Y.-K. (2017). Current demands for food-approved liposome nanoparticles in food and safety sector. *Frontiers in Microbiology, 8*, 2398.

Simopoulos, A. P. (2002). The importance of the ratio of omega-6/omega-3 essential fatty acids. *Biomedicine & Pharmacotherapy, 56*, 365–379.

Simopoulos, A. P. (2004). Omega-6/Omega-3 essential fatty acid ratio and chronic diseases. *Food Reviews International, 20*(1), 77–90.

Singh, H. (2016). Nanotechnology applications in functional foods; opportunities and challenges. *Preventive Nutrition and Food Science, 21*(1), 1.

Slewa, E. K., & Al-Mowsowy, A. J. (2018). Use nanotechnology in capsulation omega-3 fatty acid to improve its thermal stability and use it to enrich pasteurized milk. *Pakistan Journal of Biotechnology, 15*(1), 33–43.

Tamjidi, F., Shahedi, M., Varshosaz, J., & Nasirpour, A. (2013). Nanostructured lipid carriers (NLC): A potential delivery system for bioactive food molecules. *Innovative Food Science & Emerging Technologies, 19*, 29–43.

Timilsena, Y. P., Wang, B., Adhikari, R., & Adhikari, B. (2016). Preparation and characterization of chia seed protein isolate–chia seed gum complex coacervates. *Food Hydrocolloids, 52*, 554–563.

Timilsena, Y. P., Akanbi, T. O., Khalid, N., Adhikari, B., & Barrow, C. J. (2019). Complex coacervation: Principles, mechanisms and applications in microencapsulation. *International Journal of Biological Macromolecules, 121*, 1276–1286.

Timilsena, Y. P., Haque, M. A., & Adhikari, B. (2020). Encapsulation in the food industry: A brief historical overview to recent developments. *Food and Nutrition Sciences, 11*(06), 481.

Tompkins, C., & Perkins, E. G. (1999). The evaluation of frying oils with the *p*-Anisidine value. *Journal of the American Oil Chemists' Society, 76*(8), 945–947.

Tontul, İ., Eroglu, E., & Topuz, A. (2017). Chapter 3: Nanoencapsulation of fish oil and essential fatty acids A2. In: Jafari, S. M. (Ed.), *Nanoencapsulation of Food Bioactive Ingredients* (pp. 103–144). Academic Press: Cambridge, MA.

Trotta, M., Cavalli, R., Trotta, C., Bussano, R., & Costa, L. (2010). Electrospray technique for solid lipid-based particle production. *Drug Development and Industrial Pharmacy, 36*(4), 431–438.

Velasco, J., Dobarganes, C., & Márquez-Ruiz, G. (2003). Variables affecting lipid oxidation in dried microencapsulated oils. *Grasas y Aceites, 54*, 304–314.

Walia, N., Dasgupta, N., Ranjan, S., Ramalingam, C., & Gandhi, M. (2019). Methods for nanoemulsion and nanoencapsulation of food bioactives. *Environmental Chemistry Letters, 17*(4), 1471–1483.

Walker, R., Decker, E. A., & McClements, D. J. (2015). Development of food-grade nanoemulsions and emulsions for delivery of omega-3 fatty acids: Opportunities and obstacles in the food industry. *Food & Function, 6*(1), 41–54.

Xue, J., Davidson, P. M., & Zhong, Q. (2015). Antimicrobial activity of thyme oil co-nanoemulsified with sodium caseinate and lecithin. *International Journal of Food Microbiology, 210*, 1–8.

Yaacoub, R., Saliba, R., Nsouli, B., Khalaf, G., & Birlouez-Aragon, I. (2008). Formation of lipid oxidation and isomerization products during processing of nuts and sesame seeds. *Journal of Agricultural and Food Chemistry, 56*, 7082–7090.

Zimet, P., Rosenberg, D., & Livney, Y. D. (2011). Re-assembled casein micelles and casein nanoparticles as nano-vehicles for ω-3 polyunsaturated fatty acids. *Food Hydrocolloids, 25*(5), 1270–1276.

CHAPTER 10

Mineral Nanoencapsulation

Sangeeta, Hitesh Chopra, and Gagandeep Garg

CONTENTS

DOI: 10.1201/9781003259183-10

10.1 INTRODUCTION

Minerals are important components of nutrition because of their involvement in metabolism of the cellular system and various biological functions but plants and animals are not able to synthesize these components (GharibZahedi and Jafari, 2017). Minerals mainly exist in soil and are absorbed through the roots of plants; are present in drinking water in some cases (Watzke, 1998); and are found in fruits and vegetables as the main dietary component (Singh et al., 2016). Although minerals are required by the body only in small amounts, the risk of mineral deficiency has become crucial in some cases like pregnancy, adolescence, and patients having dietary restrictions like low-calorie diets, geriatric patients, vegetarians etc. So, a balanced diet is highly recommended, as it can fulfill the demand of all body's essential minerals (Berginc, 2015; Lukaski, 2004; Singh et al., 2016). The common application that helps in fulfilling micronutrient demand and controls the mineral-related deficiencies is fortification of food. The main objective of food fortification processes is to deliver and release these functional ingredients at a proper location in the body and at a proper time. In spite of all these applications mentioned, direct addition of minerals in food formulation has a lot of limitations like solubility and stability restrictions, interaction with other food components, reduction in sensory evaluation scores (unfavorable flavor/taste/color), oxidative degradation of many minerals, etc. So protection of minerals in supplemented/fortified food products during storage is necessary and also crucial (Akhtar et al., 2010). Nanoencapsulation or microencapsulation is one of the most successful techniques among all the strategies used, as this can effectively overcome the abovementioned challenges and protect and deliver the minerals as well as other biologically active ingredients for different food uses (Faridi, Esfanjani and Jafari, 2016; Katouzian and Jafari, 2016). Nanoencapsulation controls the minerals' release by providing stable conditions as well as the separation of reactive or unstable ingredients at the final stage of formulation of different foods (Mozafari et al., 2008). The minerals' incorporation, dispersion, or absorption can be prompted or facilitated by capsule formation having submicron diameters (Oehlke et al., 2014); this is possible because of the strong relationship between particle size and delivery rate of bioactive constituents (Hughes, 2005). Nanocapsules based on minerals and/or other trace elements have been reported to increase the maintenance time of gastrointestinal tract and afterward the bioadhesiveness of poorly soluble minerals in the mucosal layer of the intestinal epithelium. Other than the qualities mentioned here, minerals' nanoencapsulation has been reported for various positive attributes like improved stability, antioxidation, and improved sensory quality as well as controlled mineral release and so on (Hurrell, 2002). Consequently, the main emphasis of this chapter is to highlight the benefits of mineral nanoencapsulation; the recent techniques/instruments of microencapsulation/nanoencapsulation of minerals; carrier used (their availability and choice), and their bioavailability and bioaccessibility.

10.2 BENEFITS OF MINERAL NANOENCAPSULATION

In terms of the nutritional and industrial perspective, there are innumerable benefits of encapsulation for minerals:

- Mineral prevention from several environmental circumstances, for instance acids, heat, moisture, oxygen etc.
- Biological engineering of utilitarian parts not only safeguarded minerals from handling misfortunes and can be well released when exposed to explicit circumstances.
- Reducing dissatisfactory flavors as well as odors connected with certain minerals can notably further enhance consumer choice and the desire to utilize this technology.
- Giving an easy dealing control for free-streaming and dry constituents.
- Capability in estimation and conveyance of the most desirable mineral in exact levels due to stability acquired by the encapsulation process.
- This process has high effectiveness for the manufacturing of functional food varieties with incredible bioavailability, stability as well as delivery viewpoints (Ghosh, 2006; Gouin, 2004).

10.3 CHOICE AND AVAILABILITY OF CARRIERS USED IN ENCAPSULATION

10.3.1 Carbohydrates

In the spray-drying encapsulation technique, carbohydrates are widely utilized as carrier or wall material (Reineccius, 1991; Kenyon, 1995; McNamee et al., 1998). The carbohydrates (like acacia gums, starches, maltodextrins (MDs), and corn syrup solids) have the capacity to develop flavors owing to its diversification, low cost and broad food utilization thus making them a better encapsulation option (Mutka and Nelson, 1988; Dziezak, 1988). Furthermore, there are numerous properties of these carriers like good solubility and low viscosities at high solid content. These properties make it desirable to use carbohydrates as encapsulating agents. Starch as well as other products obtained from it such as cyclodextrin and MDs are known for the encapsulation of aroma compounds thus rendering these as a significant topic of reasonable research (Reineccius, 1991; Goubet et al., 1998).

10.3.2 Starch

For the protection, maintenance, and stability of volatile compounds, starch as well as ingredients based on starch (modified form of starches, MDs, b-cyclodextrins) is broadly utilized in the food processing industry. These can be used as carriers, stabilizers as well as fat replacers in aroma encapsulation emulsions. Numerous scientists have produced new microporous materials based on starch for enhancement in aroma retention (Golovnya et al., 1998; Glenn and Stern, 1999; Thomas and Atwell, 1999). Few starch granules categories have occurred naturally with surface pores of size 1–3 μm diameter. The tiny granules of starch have the potentiality to integrate into porous spheres, which is interesting as well as potentially helpful during being spray-dried (Zhao and Whistler, 1994). Hence, handling the granules of starch with enzyme amylase can form a structure, which is longer and highly porous in nature (Karathanos and Saravacos, 1993; Fannon et al., 1993; Yamada et al., 1995; Zeller et al., 1999).

10.3.3 Maltodextrin

MD is obtained by the partial hydrolyzation of corn flour with acids as well as enzymes. To form matrices, it has the ability that is essential in the formation of wall material (Shahidi and Han, 1993; Kenyon and Anderson, 1988). While choosing the wall material, MD is best suitable because of its viable cost as well as effective nature for encapsulation. It has low viscosity at high solid ratio and also presents in varying molecular weights (Apintanapong and Noomhorm, 2003). The main limitation is the virtual absence of capacity to emulsify and volatile components with short retention (Reineccius, 1991).

10.3.4 Gums

In general, gums as well as thickeners are tasteless; however, they have a noticeable effect on flavor and taste of food. Generally, hydrocolloids reduce sweetness and a high amount of gum mostly acts as a flavor-encapsulating material (Godshall, 1997). Gums have various characteristics like solubility, emulsification, low viscosity, and better volatile retention that make these components extremely versatile for encapsulation. Furthermore, these are ideal wall materials for lipid droplets' encapsulation because of their drying matrix and surface active role; hence they protect from volatiles loss during atmospheric contact. Apart from this, its use in the food industry is minimal, that's why in comparison to MD, gum Arabic (GA) is costly (Kenyon, 1995; Shiga et al., 2001). In addition, their availability as well as price is subjected to fluctuations; thus the evaluation of their alternative is necessary. In the studies related to cardamom oil microencapsulation through spray-drying, GA and MD mixtures have been reported to be elevated solid carriers with allowable viscosity (Sankarikutty et al., 1988). The release of flavor from spray-dried soy matrices or MD/GA and the reduction in the release of ethyl butyrate with increased concentration of MD in feed liquid has been reported (Yoshii et al., 2001).

10.3.5 Proteins

Food proteins like whey protein and sodium caseinate as well as protein isolates have not been used extensively for flavor encapsulation but food hydrocolloids are used at a broad spectrum (Kim and Morr, 1996). The disparate chemical groups, capability to self-associate, amphiphilic properties, and the molecular chain flexibility of these proteins result in various functional properties such as emulsification, viscosity, solubility, film-formation and encapsulation etc. Throughout the formation of emulsion, the protein molecules are speedily absorbed on the interface of freshly formed oil-water emulsion. The formation of final steric stabilization prevents the recoalescence of oil droplets instantly and after that during storage it gives physical stability to the emulsion (Dalgleish, 1997; Dickinson, 2001).

10.3.6 Whey Proteins

Numeral functional properties are present in whey proteins that are required in a wall material for encapsulation (Amrita et al., 1999). As a microencapsulating agent, various applications of whey proteins have been observed by Young, Sarda and Rosenberg (1993a, b), Moreau and Rosenberg (1993, 1996, 1998, 1999), Rosenberg and Sheu (1996), and Rosenberg (1997). Whey proteins are present in the form of concentrated whey protein powders (WPC-50, WPC-70) and isolates of whey protein (95%–96% protein).

Whey protein isolates provide a good barrier from oxygen and prevents from oxidation as reported in the encapsulated orange oil (Kim and Morr, 1996). Moreover, they also provide an effectual base for volatiles microencapsulation using the spray-drying method. Surface properties that are required to stabilize the emulsions were offered by WPC-70. Faldt and Bergenståhl (1996) reported lower encapsulating capacity of WPC than sodium caseinate using soybean oil as an encapsulating material.

10.3.7 Other Proteins

Materials based on proteins like soy protein, gelatin derivatives, or polypeptone are efficient in the production of volatiles-based stable emulsions (Bangs and Reineccius, 1988). Gelatin is a hydrolyzed collagen material and is utilized broadly in complex coacervation (Ducel et al., 2004). This type of material is soluble in water and has the capability to produce walls by spray-drying of aromatic volatile compounds, wall material, and water mixture (Lee et al., 1999). The flavored gelatin capsules formed by this method are being used in various types of materials such as seasonings as well as other flavorings for consumption in foods (Gourdel and Tronel, 2001). The encapsulation of orange oil by initial triple emulsion (o/w/o/w) and then the evaporation of the outer continuous aqueous phase by spray-drying using sodium caseinate as well as lactose as the wall material has been reported by Edris and Bergmtahl (2001). The encapsulation of liquid oils has been reported to be more efficient using sodium caseinate as compared to whey proteins as the encapsulant (Keogh et al., 2001; Millqvist-Fureby et al., 1999). The solution mixture of carbohydrates like corn syrup solids and MD with casein has been found to be the core encapsulating material having cost-effective and potential functional properties (Hogan et al., 2001).

10.4 TECHNIQUE/INSTRUMENTS USED IN ENCAPSULATION

In encapsulation, small-sized coating agents are gaseous, liquid, and solid in nature. They are mostly used for covering the ingredients. These developed microcapsules/nanocapsules have diverse morphological forms like spherical and irregular comprising a core area enclosed by a continuous wall which envelopes various fine droplets/particles. Waxes, fats, and biopolymers are the examples of capsule wall materials (Jafari et al., 2015). In the present days, numerous techniques that are used to microencapsulate minerals have been used especially Fe, Zn, and I. These techniques can be divided into two categories:

1. Physical/mechanical techniques.
2. Chemical techniques.

10.4.1 Mechanical Methods

10.4.1.1 Freeze-Drying

Freeze-drying (lyophilization) is used for the microencapsulation of components that are intolerable toward high temperature necessary to prepare their powder. As compared to spray-drying, this method has low oxidation rate and exhibits a more porous structure; however, it takes longer time of up to 20 hours considering the used components as well as the loading rate (Desai and Park, 2005). Gupta, Chawla and Arora (2015) observed that to encapsulate Fe salt, modified starch (Ms), and Na-alginate, Na-alginate was individually mixed at 500 rpm stirring rate into distilled water for 0.5 hours. For degassing the mixture, a sonicator was used for 15 minutes, followed by the addition of Fe salt and vitamin C at a ratio of 15:1 to prepare solutions under specific dissolving conditions. For the solidification of obtained microcapsules, the mixture is sprinkled into chilled calcium chloride ($CaCl_2$) solution and after that it is stored for 3–4 hours. Then, vacuum was attained to remove $CaCl_2$ by filtration. Freezing of the obtained solution was done at –20°C and the powder was obtained by lyophilization using the freeze-drying technique. The increase in encapsulation efficiency (EE) of Fe microencapsules by lowering the amount of $CaCl_2$ (1.0 to 0.1 M) has been reported. On the basis of Na-alginate-Ms, the EE of the microcapsules produced at 0.1 M $CaCl_2$ in combination with Na-alginate-Ms becomes a great deal in comparison with Na-alginate alone. Consequently, it is crystal clear that in the expansion of EE, Ms played a vital role (Gupta, Chawla and Arora, 2015).

10.4.1.2 Fluidized-Bed Coating

In the fluidized-bed coating (FBC) method, a bundle of particles slings into the air stream as well as the particles enveloped with coating solution equally by spraying. The coating solution carries single or ample number of biomacromolecules like lipids, polysaccharides, proteins, and emulsifiers. Eventually, drying operation after cycling the freshly coated solidified particles is conducted using cooling or solvent extraction into a specific level. Basically, three types of fluidized beds contain tangential spray, top spray as well as bottom spray, depending on the nozzle's position/arrangements. The FBC process gives appropriate mass rates and transfer of heat; however, it is only used for microcapsules of 0.10 μm (Gouin, 2004). Actually, this technique is used to prepare premixes as well as supplements made up of vitamins (including vitamins B and C group) and minerals (ferrous sulfate ($FeSO_4$), ferrous fumarate ($C_4H_2FeO_4$), Na-ascorbate, potassium iodide (KI) etc.,) (Turchiuli et al., 2005). It has been reported that the FBC system is hard to control and for this reason particles bunch jointly by the addition of wall material solution (Diosady et al., 2002).

Yadava et al. (2012) improved the numerous factors and concluded that for fluidizing, 65°C–75°C is the optimum temperature with a flow rate of 1.5 mL/min of coating solution, 35%–45% flow rate of fluidizing air, 1.8–2.2 bar nozzle air pressure along with 30 cm vertical nozzle, resulting in a better encapsulation process to produce small-sized Fe-premix into fortified salt. Li et al. (2011) produced high porosity and low-density double fortified salt (DFS) particles premix to fortify salt by using fluidized-bed agglomeration.

10.4.1.3 Spray-Drying

Spray-drying is often used for the microencapsulation of functional food ingredients (1–150 μm) due to its low cost and simple scale-up procedure. Romita et al. (2011) used mini spray drier for microencapsulation of Fe (<20 μm) by using aqueous and suspended solution of $C_4H_2FeO_4$ (9% w/v) into Hydroxypropyl methylcellulose (HPMC) (6% w/v) with Na-fumarate (63%) and TiO_2 (22%). So, this technique is scaled easily and the process is a single step used for $C_4H_2FeO_4$ premix production. However, this method has few disadvantages. First and foremost, due to the high water solubility of carriers applied, microcapsules can quickly release the core compounds. So, it is difficult to choose the appropriate wall material. Another limitation is the limited applications of organic solvents because of flammability, low payload rates (<40%), and toxicity (Madene et al., 2006). On the contrary, enteric-coated iron microparticles have been developed using Eudragit EPO and chitosan (CS) as wall material using a mini spray dryer (Dueik and Diosady, 2016). The mixture containing CS had 120°C inlet temperature, 667 L/hour air flow rate, 618 kPa atomizer pressure, and 24.5 Pa aspirator rate, whereas the solution containing Eudragit EPO had an inlet temperature of 105°C and a feed flow rate of 2.7 mL/minute. The polysaccharides of CS were found to be a superior aspirant than EPO in the production of iron-premix for the preparation of stable DFS fortified with Fe and iodine. The round-shaped microcapsules of zinc chlorophyll (from pandan leaf) have been developed by using spray-drying (Porrarud and Pranee, 2010).

10.4.1.4 Spray Chilling

Spray chilling is alike to spray-drying; melted wall material and a solution of bioactive constituents are sprayed into a cavity cooled by cold air or liquid nitrogen. The carrier that is used for the cooling medium is solidified to form powder with improved stability. Okuro et al. (2013) investigated that the limitation of this technique is a presence of unprotected core compounds on the surface of particles that significantly decreases the EE. Iodine, iron (40% w/w ferric pyrophosphate), and vitamin A containing microcapsules have been prepared by spray chilling by using 1% w/w lecithin and hydrogenated palm fat (HPF) as the wall material (Wegmuller et al., 2006). A two-liner tube system facilitating the transport of HPF into stainless-steel spraying was examined, although in the external tube, hot water at about 90°C was flowed to prevent blockage due to the solidification of fat at normal temperature. The prepared suspension was sprayed into the pre-cooled tower with liquid nitrogen using two fluid nozzles having air at 2.5 bar and 90°C temperature. This method has been used to prepare 100 μm microcapsules containing $Fe_4O_{21}P_6$, potassium iodate (KIO_3), and retinyl palmitate (Zimmermann et al., 2004). Hence, 94°C temperature has been reported to be the optimum to stop obstruction for achieving a 2-acetylpyridine (2-APri)-zinc chloride ($ZnCl_2$) of 5% loading. Encapsulation using spray chilling is reported to be the most helpful technique in pharmaceutical as well as food sectors, and it prevents the formation of less stable nutritional complexes (Moran, 2014).

10.4.1.5 Extrusion

The plasticized combine matrices such as fat starch combination either polyethylene glycol (PEG) is utilized for encapsulation of core compounds in extrusion technique. After the addition of 20% w/w water for dry mixture, to attain 500–1,000 μm portions, the wet paste is shaped by the use of extruder and air dried finally. Extrusion gives a chance to utilize glassy polymers for lowering the oxygen permeability rate and increasing the storage time period of finished food products; a serious obstacle is the production of big microparticles for practical use (Li, 2009). Li et al. (2010) prepared a fresh design of extrusion-based products to produce $C_4H_2FeO_4$ microencapsules to fortify salt. Moretti et al. (2005) utilized two extrusion premix tactics: cold

(30°C and 40°C) and hot (50°C and 70°C) using the Brabender single screw extruder for the fortification of rice flour with $Fe_4O_{21}P_6$ with no discoloration.

10.4.2 Chemical Methods

10.4.2.1 Liposome Entrapment

Initially, the liposomal encapsulation technique was used by pharmaceutical industries to establish the drug-based delivery systems, whereas cosmetic as well as food industries moderately applied this technique for the production of particles having fine size containing dissolved bioactive components. The monolayer or multilayer of liposomal particles can have amphiphilic polymolecular films, which are the same as the cell membrane material of plants (Taylor et al., 2005). Numerous researchers have concentrated on the liposome growth using fundamentals and producing methods of emulsion fabrication (Gharibzahedi, Razavi and Mousavi, 2015). Li (2009) concluded that even if the liposomal structures give rise to high stability for water-soluble ingredients because of their intrinsic aqueous solubility, there is a significant limitation so long as the coated ingredients have been required to be under a dry environment. Hirotsuka et al. (1984) sonicated soy lecithin with aqueous solutions with different concentrations of ($C_6H_{10}CaO_6$) calcium lactate to make a liposomal structure that comprises Ca_{21}. They didn't detect any occurrence of coagulation and/or precipitation by incorporating liposomal structure containing 60 m MCa^{2+} to soy protein solution. However, they observed that soymilk can be systematically fortified up to 120 mg/100 g based on Ca-lecithin. Boccio et al. (1997) investigated the fortified dairy items with enhanced Fe bioavailability by the use of encapsulation of $FeSO_4$ in lecithin with no hurdles. Without any off-flavor attributes, lecithin was used for beverage fortification by liposome encapsulation with nutritious Fe composition (Mehansho and Mannar, 1999).

Xia and Xu (2005) performed four particular techniques containing thin-film, freeze-thawing, sonication, and reverse phase evaporation (RPE) to produce liposomes loaded with $FeSO_4$ in hydrating form of deionized water. Initially, they noticed that the optimum cholesterol and tween 80% can notably expand physical stability (electrostatic and steric) of vacant liposomes to encapsulate Fe^{3+}. Furthermore, an appreciable effect was observed on the quantity of encapsulated Fe^{3+} by studying the concentration of deionized water, $FeSO_4$, and sonication. The formation of liposome using the RPE technique had most EE about 67% of $FESO_4$. This resulted that the concentration of Fe can be leveled likely by 15 mg/L fortified milk with $FeSO_4$ filled liposomes. Likewise, the RPE method was also used to produce liposomes of ferrous glycinate as Fe fortifier (Ding et al., 2009). To encapsulate ferrous gluconate ($C_{12}H_{24}FeO_{14}$) and Fe forms of $Fe_4O_{21}P_6$, liposomes are also used. An in vivo study also showed equivalent and favorable bioavailability rate (Navas-Carretero et al., 2007). But, the milk fortified with liposomal microcapsules exhibits lesser score of appearance, color, odor, mouthfeel, taste as well as total acceptability in sensory evaluation as compared to prepared microcapsules with further methods (Gupta, Chawla and Arora, 2015). Xu et al. (2014) evaluated the reduction of anemia in rats and observed that Fe liposome is the best supplement that can notably reduce the deficiency of Fe with negligible side effects.

10.4.2.2 Niosome Entrapment

Niosome is nonionic surfactant vesicles that are able to capture lipophilic as well as hydrophilic compounds in the aqueous form (Devaraj et al., 2002). It is used in food industries, bio, pharma as well as cosmetics due to its long storage stability and low cost (Marianecci et al., 2014). Because of easily scaled and simple without any use of hazardous material, conventional ethanol injection (CEI) method is included in the most important techniques (Pham et al., 2012). Gutierrez et al. (2016) recently observed the preparation of Fe microcapsules with the use of niosomes produced by CEI to improve yogurt. Wagner et al. (2016) studied and reported that concurrent encapsulation involving a mixture of lipophilic $FeSO_4$ and vitamin D_3 into niosomes was the major application of latest supercritical CO_2 technique. They observed that this technique makes capsules with 7.21 and 1.44 μm with vitamin D_3-EE of ~96% and $FeSO_4$-EE of ~25%. A superior shelf life for encapsulation of niosomes at 20°C was noticed as compared to 4°C for 21 days.

10.4.2.3 Fatty Acid Esters

The fatty acid ester technique is generally used to produce microcapsules comprising water-soluble mineral by the use of FAE. In this technique, coating agent solution, PGMS as a FAE, distilled water, and a mineral salt is prepared, followed by the heating and agitating process under specific conditions. After that, the obtained solution is nebulized by surfactant (like tween 60), which is dispersed in distilled water using a sprayer. After centrifugation, the separated phases are pulled out. Eventually, the collection of the microcapsules and storage at cool temperature were done till they were used. To microencapsulate Fe salts ($FeSO_4$, $FeC_{13,}$ $NH_4Fe(SO_4)_2$), the FAE technique has been used by various scientists (Jackson and Lee, 1992a,b; Kwak, Ihm and Ahn, 2001; Kim et al., 2003; Kwak et al., 2003). The study reported by Abbasi and Azari (2011) stated that, the efficiency of the FAE method for the encapsulation of Fe salt to fortify milk acquired-EE (~82%) using the FAE method and has an optimum ratio of salt PGMS:Fe (15:1). However, Gupta, Chawla and Arora (2015) observed that microcapsules produced with the FAE technique had less EE in samples prepared by other techniques of encapsulation. Abbasi and Azari (2011) observed that the obtained microcapsules by liposome entrapment and the FAE method had nonuniform round shape with even surface morphology. The finest sensory element of pasteurized milk was control samples and fortified by FAE-microencapsulated Fe with down level of 7 mg/L. They highlight that fatty acid esters encapsulation method can be observed as a simple, low-cost, appropriate, efficient, and fast approach for encapsulation of distinctive Fe salts and afterward production of Fe-fortified pasteurized milk (Abbasi and Azari, 2011; Jafari, 2017).

10.4.2.4 Coacervation

The coacervation method is the technique based on separation of phase (diluted colloidal and concentrated phases), and it possesses insufficient desolvation of entirely solvated polymers. If the solubility of macromolecules was altered with environmental changes for instance, addition of polymer or salt with reverse-charge may suggestively start the chemical process (Kruif et al., 2004). According to the number of polymers and mechanism of phase separation, coacervation may be divided into simple as well as complex categories. In simple process, only one macromolecule is participated and phase separation is performed by salt incorporation (like Ca salt to alginate polysaccharide), whereas the complex group occurs by anion-cation interactivities, mainly GA as well as gelatin biopolymers (Shewan and Stokes, 2013). This method was firstly applied in zinc sulfate ($ZnSO_4$) macromolecules by ethyl cellulose as wall material in ratios of 1:1 and 1:2 of two walls (Oner et al., 1988). Khosroyar and Arastehnodeh (2007) prepared microcapsules involving ferric saccharate together with the coating material substance of alginate by coacervation technique at 30:70 wall ratios. In the existence of $CaCl_2$ 400 µm, particles are obtained that had a short Fe release rate of <0.04% at harsh and mild temperature as well as RHs. Nonetheless, by raising the amount of RH, a high-rise Fe release was acquired. Using coacervation, spray-drying, and emulsification, encapsulation of ferric saccharate into Na-algiante wall material has been studied by Khosroyar et al. (2013).

10.4.2.5 Modified Solvent Evaporation

The present method was designed by Gupta et al. (2015) to produce $FeSO_4$ microcapsules with the help of the following coating agents, i.e., Ms, MD, and GA mixture at a ratio of 4:1:1.6 g carbohydrates dissolved in 10 mL distilled water at 60°C; this blend was maintained at a temperature of 4°C–7°C for 12 hours for rehydration. Antioxidant agent of vitamin C and $FeSO_4$ in the ratio 1:15 was dissolved in 10 mL water and then added to rehydrated mixture and stirred well and sonicated in a water bath at 5°C for 15 minutes. The carboxylic groups that are present in GA can productively interact with Fe^{2+} by powerful electrostatic bonds. The acquired solution was sprayed on chilled alcohol with the help of an airless paint sprayer and after 5 minutes the solution was filtered under vacuum conditions with Whatman filter paper no. 1. The rest of alcohol left in the microcapsules (EE = 91.58%, 6.8–33.4 µm) that was transferred on the filter paper and kept at 5°C for 13 hours. The effective technique for microencapsulation is modified solvent evaporation technique as compared to other methods because the recycling of solvent is easy by distillation of filtered alcohol. On the basis of these Fe micromolecules, microcapsules containing fortified milk were prepared, their comparison of oxidative stability as well as sensory components was done. Due to longer bioavailability, the values of thiobarbituric acid (TBA) were listed for the Fe microcapsules fortified milk by modified solvent evaporation (Gupta et al., 2015).

10.4.2.6 Emulsification

The encapsulation of ferrous bis-glycinate ($C_4H_8FeN_2O_4$, 15% wt) using multiple emulsions stabilized with 5% wt of Panodan SDK (mono- and di-glyceride esters of di-acetyl tartaric acid) and polyglycerol polyricinoleate (PGPR) in 4:6 ratio has been performed by Jimenez-Alvarado et al. (2009). For producing the last multiple emulsions, the preliminary emulsions were re-emulsified and stabilized with one of the polysaccharides of mesquite gum (MG), GA, low methoxyl pectin, and whey protein concentrate. The favorable Fe-EE, Fe bioavailability, and physicochemical stability of emulsion in terms of the lower coalescence rate, oxidation rate, release of $C_4H_8FeN_2O_4$, and droplet size has been observed from the results. The two-step emulsion method to produce Fe capsules (2–12 μm) to formulate into milk has been studied in another study (Chang et al., 2016). Initially, $FeSO_4$ had been dissolved in water (distilled) and added into the palm oil generated medium chain triglyceride (MCT) as the lipid phase, and the hydrophobic emulsifier of PGPR were also contained in it. Firstly, the core agent, i.e., iron solution has been mixed with preliminary wall agent, i.e., MTC (ratio 4:6). The emulsion thus obtained has been mixed with whey protein isolate solution (30%) (ratio 2.5:7.5), and with polyoxyethylene sorbitan monolaurate (hydrophilic emulsifier) also. The resulting solution was homogenized to create a fine droplets-based emulsion (W/O/W). The obtained spherical microcapsules have a smooth and serrated surface having high EE (93.63%) and good stability in simulated gastric conditions (during 0.5–2 hours of incubation – almost 1.2%–1.9% Fe release) (Chang et al., 2016). For the fortification of soymilk by fabricating Ca microcapsules, an analogous process has been designed f. It has been reported that during pasteurization, the fortified products have high physical stability. The optimized EE and payload of Ca microcapsules by response-surface methodology (RSM) has been reported by various researchers. An increment of protective oil and wall components concentration resulted in a decline in Ca-release rate (Saeidy, Keramat and Nasirpour, 2014). A formulated emulsion with gelatin (3%), agar (0.5%), W_1/O (48.58%), and W_2 (47.92%) resulted in the maximum EE values and Ca payload (Saeidy, Keramat and Nasirpour, 2014).

10.4.2.7 Salt-Induced Cold Gelation

The gelled microcapsules having functional components can be produced by gelation through cross-linking by the use of cooling, Ca-alginate beads, and chemical reactions. Gelation process that is used for designing of microcapsules is costly as well as a tough-scaled technique, which resulted in liquids with uniformity in porous structure and controlled diffusion ability (Li, 2009). The higher Fe-EE and the low color variations have been achieved by optimizing the process for the production of Fe-encapsulated cold-set WPI gels using the salt-induced cold gelation method (Bagci and Gunasekaran, 2017). The WPI Fe particles of 1 μm produced by using optimum conditions have been reported to be an outstanding mean for site-specific delivery of Fe particles by fortification into various dairy products as evident from in vitro gastrointestinal studies (Bagci and Gunasekaran, 2017). The development of Fe microparticles has been reported by the entrapment of $FeSO_4$ into Whey Protein Isolate (WPI) matrix using the cold-set gelation process (Martin and de Jong, 2012). The strength optimization of protein gel has been studied by evaluating the effect of heating intensities (slight/middle/vigorous) and amount of Fe loading on EE and effectiveness. The brilliant Fe-EE as well as bonding ability by WPI has been reported due to high and low Fe release rates at pH values of neutral and acidic range, respectively. Therefore, an increment in the Fe release level has been reported by an increase in the intensity of heating, and this fact can be considered to be effective for the successful utilization of Fe WPI particles for the fortification of food products, which are processed at high temperatures (pasteurized/sterilized) (Jafari, 2017).

10.5 BIOAVAILABILITY AND BIOACCESSIBILITY

The creation of functionally enhanced meals necessitates the integration of several variables under consideration. Finding the right source, discovering bioactive chemicals using separation and recovery procedures, conducting toxicity evaluations as well as determining bioaccessibility measures are all part of this process. Gastrointesinal (GI) digestion, absorption, metabolism, distribution, and bioactivity are all components of bioavailability. The term might indicate many things depending on the field of study. When it comes to drugs, bioavailability refers to how quickly and efficiently the therapeutic component is absorbed and made accessible for use at the drug's target location. Bioavailability refers to the percentage of a nutrient that is stored or accessible for use by the body's systems. It's an important phrase in nutritional efficacy since the body doesn't use all of the bioactive chemicals that are given to it. When different foods come into touch with the mouth or digestive system, e.g., multiple interactions may take place that impact the bioavailability of phytochemicals (e.g., fat enhances quercetin bioavailability in meals). Therefore, bioavailability expresses the percentage of nutrients or bioactive compounds that are absorbed into the bloodstream and ultimately utilized by the body after they are consumed.

Bioactive chemicals must first be liberated from the food matrix and changed in the gastrointestinal system before they can be absorbed into the body. Consequently, the phrase bioavailability incorporates the word bioaccessibility. The digestion of bioactive compounds and the stability of those compounds must be examined before any possible health effects can be concluded. The amount of a substance that is liberated from its matrix and made accessible for absorption in the gastrointestinal system is known as its "bioaccessibility" (e.g., enters the blood stream). Presystemic, intestinal, and hepatic metabolisms are all included in this phrase, which encompasses digestion, absorption, and excretion of nutrients from the digestive tract. Definitions based on absorption, on the other hand, ignore the positive effects of nutrients that are not absorbed, such as calcium binding of bile salts in the tract. It is common practice to conduct in vitro digesting methods that mimic stomach and small intestine digestion, followed by Caco-2 cell absorption, in order to assess bioaccessibility.

Biological activity is the impact that a substance has on the body when it is exposed to it. Tissue uptake and physiological response are included in this definition (antioxidant, anti-inflammatory, etc.). Also, transport and target tissue interactions, biomolecule interactivity, metabolism and biotransformation properties as well as the development of the biomarkers and the resulting physiological reactions contain this information. For the purposes of this definition, "digestibility" refers to the proportion of a food's constituent parts that are converted into material that may be used by the body's lumen by physical and chemical means. The process of transepithelial absorption on the other hand is referred to as assimilation and describes the uptake by the epithelium of bioaccessible material. Nondigestible polysaccharides, oligosaccharides, and dietary fiber are some examples where only bioactivity applies. Despite the fact that these chemicals are not absorbed, they provide several health advantages.

Events that occur during the time an active component interacts with biomolecules are used to quantify bioactivity in vivo, ex vivo as well as in vitro. Until the systemic reaction is formed, this interaction creates metabolites, signals, or responses that are modulated and amplified (health benefit). Measurement methodologies for bioactivity must be tailored to each specific health benefit claim. Bioactivity data are used to back up claims about a food's health benefits or the ability to lower one's risk of illness. Bioaccessibility may be used to make claims about a food's nutritional content or nutritional value in contrast to other foods without the requirement for bioactivity investigations.

In order to enclose bioactive chemicals into a matrix smaller than 1,000 nm, nanoencapsulation is thought of as a revolutionary technique. With this technique, new mineral and other functional component delivery systems might be created that are more stable, water-soluble, bioavailable, and readily absorbed by the body. There has been a recent emphasis on developing practical methods for nanoencapsulating minerals for the purpose of fortifying food items, since doing so may significantly accelerate manufacturing of new functional food compositions. Rather than using physicomechanical processes, studies observed that the development of mineral ions/salts-nanocapsules has relied on chemical processes (such as nanoliposomes, nanoemulsions, and cyclodextrin inclusions) rather than physical ones. For the production of mineral nanoparticles, we found the chemical methods to be more practical (by >95%) than the physicomechanical ones.

As a result of the current drawbacks of using mineral salts to fortify food products, the incorporation of nanoencapsulated minerals into food and drug articulation has unique benefits in the development of new useful products having enhanced physicochemical as well as sensory properties. There are several advantages to using nanoencapsulated mineral salts, including the ability to inhibit interactivity with other materials already available in the matrix, avoid discoloration, reduce off-flavors through the use of flavor-masking agents, and have a more controlled release of the mineral components.

10.6 CONCLUSION

Nanoencapsulation of minerals using different carriers and techniques is a promising approach to enhance their functionality, bioavailability, and stability. Nanoencapsulation of minerals enhances their use in food as well as the pharmaceutical industry. The nanotechnology used for the encapsulation of minerals has much importance in food processing as well as packaging. Nanoencapsulation of minerals is useful for the improvement in stability and bioavailability of minerals. This controls the release of bioactive compounds and acts as a defensive mechanism by protecting them in different adverse conditions which in turn increase the usefulness of minerals in an effective way. Minerals are the important components of nutrition because of their involvement in cellular metabolism and different biological functions, but they are unable to be synthesized by plants and animals. So nanoencapsulation of minerals is important and carrier as well as techniques used for their encapsulation is equally important for their better bioavailability and bioaccessability. Emulsification, nanoprecipitation, coacervation, spray-drying, solvent evaporation, and freeze-drying are some of the widely used techniques for mineral nanoencapsulation. Furthermore, the safety- and toxicity-related issues of nanoencapsulation resulting components and their understanding require further research and work. The global legislation rules need to be implemented to facilitate their safe utilization and marketing.

REFERENCES

Abbasi, S., & Azari, S. (2011). Efficiency of novel iron microencapsulation techniques: Fortification of milk. *International Journal of Food Science & Technology*, *46*(9), 1927–1933. doi: 10.1111/j.1365-2621.2011.02703.x.

Akhtar, S., Anjum, F. M., & Sheikh, M. A. (2010). Effect of storage and baking on mineral contents of fortified whole wheat flour. *Journal of Food Processing and Preservation*, *34*(2), 335–349. doi: 10.1111/j.1745-4549.2009.00405.x.

Amrita, K., Dabur, R. S., & Kapoor, C. M. (1999). Application of microencapsulation technology in dairy industry. *Indian Food Industry*, *18*, 57–64.

Apintanapong, M., & Noomhorm, A. (2003). The use of spray drying to microencapsulate 2-acetyl-1-pyrroline, a major flavour component of aromatic rice. *International Journal of Food Science & Technology*, *38*(2), 95–102. doi: 10.1046/j.1365-2621.2003.00649.x.

Bagci, P. O., & Gunasekaran, S. (2017). Iron-encapsulated cold-set whey protein isolate gel powder-Part 1: Optimisation of preparation conditions and in vitro evaluation. *International Journal of Dairy Technology*, *70*(1), 127–136. doi: 10.1111/1471-0307.12317.

Bangs, W. E. & Reineccius, G. A. (1988). Corn starch derivatives. In: *Flavor Encapsulation* (edited by S. J. Risch & G. A. Reineccius). ACS Symposium Series 370. Washington, DC: American Chemical Society, pp. 12–28.

Berginc, K. (2015). Pharmacokinetic interactions between drugs and dietary supplements: Carbohydrate, protein, vitamin and mineral supplements. In: *Dietary Supplements* (edited by K. Berginc & S. Kreft). Sawston: Woodhead Publishing, pp. 85–126. doi: 10.1533/9781782420811.2.85.

Boccio, J. R., Zubillaga, M. B., Caro, R. A., Gotelli, C. A., Gotelli, M. J., & Weill, R. (1997). A new procedure to fortify fluid milk and dairy products with high-bioavailable ferrous sulfate. *Nutrition Reviews*, *55*(6), 240–246. doi: 10.1111/j.1753-4887.1997.tb01611.x.

Chang, Y. H., Lee, S. Y., & Kwak, H. S. (2016). Physicochemical and sensory properties of milk fortified with iron microcapsules prepared with water-in-oil-in-water emulsion during storage. *International Journal of Dairy Technology*, *69*(3), 452–459.

Dalgleish, D. G. (1997). Adsorption of protein and the stability of emulsions. *Trends in Food Science & Technology*, *8*(1), 1–6. doi: 10.1016/S0924-2244(97)01001-7.

Desai, K. G. H., & Jin Park, H. (2005). Recent developments in microencapsulation of food ingredients. *Drying Technology*, *23*(7), 1361–1394. doi: 10.1081/DRT-200063478.

Devaraj, G. N., Parakh, S. R., Devraj, R., Apte, S. S., Rao, B. R., & Rambhau, D. (2002). Release studies on niosomes containing fatty alcohols as bilayer stabilizers instead of cholesterol. *Journal of Colloid and Interface Science*, *251*(2), 360–365. doi: 10.1006/jcis.2002.8399.

Dickinson, E. (2001). Milk protein interfacial layers and the relationship to emulsion stability and rheology. *Colloids and Surfaces B: Biointerfaces*, *20*(3), 197–210. doi: 10.1016/S0927-7765(00)00204-6.

Ding, B., Xia, S., Hayat, K., & Zhang, X. (2009). Preparation and pH stability of ferrous glycinate liposomes. *Journal of Agricultural and Food Chemistry*, *57*(7), 2938–2944. doi: 10.1021/jf8031205.

Diosady, L. L., Alberti, J. O., & Mannar, M. V. (2002). Microencapsulation for iodine stability in salt fortified with ferrous fumarate and potassium iodide. *Food Research International*, *35*(7), 635–642. doi: 10.1016/S0963-9969(01)00166-1.

Ducel, V., Richard, J., Saulnier, P., Popineau, Y., & Boury, F. (2004). Evidence and characterization of complex coacervates containing plant proteins: Application to the microencapsulation of oil droplets. *Colloids and Surfaces A: Physicochemical and Engineering Aspects*, *232*(2–3), 239–247. doi: 10.1016/j.colsurfa.2003.11.001.

Dueik, V., & Diosady, L. L. (2016). Microencapsulation of iron in a reversed enteric coating using spray drying technology for double fortification of salt with iodine and iron. *Journal of Food Process Engineering*, *40*(2), e12376. doi: 10.1111/jfpe.12376.

Dziezak, J. D. (1988). Microencapsulation and encapsulated ingredients. *Food Technology (Chicago), 42*(4), 136–153.

Edris, A., & Bergmtahl, B. (2001). Encapsulation of oil orange in a spray dried double emulsion. *Nahrung/Food*, *45*, 133–137.Esfanjani, A. F., & Jafari, S. M. (2016). Biopolymer nano-particles and natural nano-carriers for nano-encapsulation of phenolic compounds. *Colloids and Surfaces B: Biointerfaces*, *146*, 532–543. doi: 10.1016/j.colsurfb.2016.06.053.

Faldt, P., & Bergenståhl, B. (1996). Spray-dried whey protein/lactose/soybean oil emulsions. 1. Surface composition and particle structure. *Food Hydrocolloids*, *10*(4), 421–429. doi: 10.1016/S0268-005X(96)80020-8.

Fannon, J. E., Shull, J. M., & Bemiller, J. N. (1993). Interior channels of starch granules. *Cereal Chemistry, 70*, 611–611.

Gharibzahedi, S. M. T., & Jafari, S. M. (2017). The importance of minerals in human nutrition: Bioavailability, food fortification, processing effects and nanoencapsulation. *Trends in Food Science & Technology*, *62*, 119–132. doi: 10.1016/j.tifs.2017.02.017.

Gharibzahedi, S. M., Razavi, S. H., & Mousavi, M. (2015). Optimal development of a new stable nutraceutical nanoemulsion based on the inclusion complex of 2-hydroxypropyl-β-cyclodextrin with canthaxanthin accumulated by Dietzia natronolimnaea HS-1 using ultrasound-assisted emulsification. *Journal of Dispersion Science and Technology*, *36*(5), 614–625. doi: 10.1080/01932691.2014.921188.

Ghosh, S. K. (2006). Functional coatings and microencapsulation: A general perspective. In: *Functional Coatings: By Polymer Microencapsulation*. Weinheim: Wiley-VCH, Verlag GmbH&Co. KgaA, pp. 1–28.

Glenn, G. M., & Stern, D. J. (1999). Starch-based microcellular foams. United States Department of Agriculture patents 1999 no. 5,958,589.

Godshall, M. A. (1997). How carbohydrates influence food flavor. *Journal of Food Technology, 51*, 63–67.

Golovnya, R. V., Misharina, T. A., & Terenina, M. B. (1998). GC evaluation of flavour compound sorption from water solutions by corn starch cryotextures obtained by freezing. *Food/Nahrung, 42*(06), 380–384. doi: 10.1002/(SICI)1521-3803(199812)42:06%3C380::AID-FOOD380%3E3.0.CO;2-V.

Goubet, I., Le Quere, J. L., & Voilley, A. J. (1998). Retention of aroma compounds by carbohydrates: Influence of their physicochemical characteristics and of their physical state: A review. *Journal of Agricultural and Food Chemistry, 46*(5), 1981–1990. doi: 10.1021/jf970709y.

Gouin, S. (2004). Microencapsulation: Industrial appraisal of existing technologies and trends. *Trends in Food Science & Technology, 15*(7–8), 330–347. doi: 10.1016/j.tifs.2003.10.005.

Gourdel, Y., & Tronel, J. (2001). Flavoured gelatin capsules capable of containing various substances used in the food, animal feed, dietetic and pharmaceutical sectors. *French Patent Application FR, 2*(807), 677.

Gupta, C., Chawla, P., & Arora, S. (2015). Development and evaluation of iron microencapsules for milk fortification. *CyTA-Journal of Food, 13*(1), 116–123. doi: 10.1080/19476337.2014.918179.

Gutierrez, G., Matos, M., Barrero, P., Pando, D., Iglesias, O., & Pazos, C. (2016). Iron-entrapped niosomes and their potential application for yogurt fortification. *LWT, 74*, 550–556. doi: 10.1016/j.lwt.2016.08.025.

Hirotsuka, M., Taniguchi, H., Narita, H., & Kito, M. (1984). Calcium fortification of soy milk with calcium-lecithin liposome system. *Journal of Food Science, 49*(4), 1111–1112. doi: 10.1111/j.1365-2621.1984.tb10405.x.

Hogan, S. A., McNamee, B. F., O'riordan, E. D., & O'sullivan, M. (2001). Microencapsulating properties of whey protein concentrate 75. *Journal of Food Science, 66*(5), 675–680. doi: 10.1111/j.1365-2621.2001.tb04620.x.

Hughes, G. A. (2005). Nanostructure-mediated drug delivery. *Nanomedicine: Nanotechnology, Biology and Medicine, 1*(1), 22–30. doi: 10.1016/j.nano.2004.11.009.

Hurrell, R. (2002). How to ensure adequate iron absorption from iron-fortified food. *Nutrition Reviews, 60*, S7–S15. doi: 10.3929/ethz-b-000422926.

Jackson, L. S., & Lee, K. (1992a). The effect of dairy products on iron availability. *Critical Reviews in Food Science & Nutrition, 31*(4), 259–270. doi: 10.1080/10408399209527573.

Jackson, L. S., & Lee, K. (1992b). Fortification of cheese with microencapsulated iron. *Cultured Dairy Products Journal (USA), 27*(2), 4–7.

Jafari, S. M. (Ed.). (2017). *Nanoencapsulation of Food Bioactive Ingredients: Principles and Applications*. Cambridge, MA: Academic Press.

Jimenez-Alvarado, R., Beristain, C. I., Medina-Torres, L., Román-Guerrero, A., & Vernon-Carter, E. J. (2009). Ferrous bisglycinate content and release in W1/O/W2 multiple emulsions stabilized by protein–polysaccharide complexes. *Food Hydrocolloids, 23*(8), 2425–2433. doi: 10.1016/j.foodhyd.2009.06.022.

Karathanos, V. T., & Saravacos, G. D. (1993). Porosity and pore size distribution of starch materials. *Journal of Food Engineering, 18*(3), 259–280. doi: 10.1016/0260-8774(93)90090-7.

Katouzian, I., & Jafari, S. M. (2016). Nano-encapsulation as a promising approach for targeted delivery and controlled release of vitamins. *Trends in Food Science & Technology, 53*, 34–48. doi: 10.1016/j.tifs.2016.05.002.

Kenyon, M. M. (1995). Modified starch, maltodextrin, and corn syrup solids as wall materials for food encapsulation. In: *Encapsulation and Controlled Release of Food Ingredients* (edited by S. J. Risch & G. A. Reineccius). ASC Symposium Series 590. Washington, DC: American Chemical Society, pp. 43–50.

Kenyon, M. M., & Anderson, R. J. (1988). Maltodextrins and low-dextrose-equivalence corn syrup solids: Production and technology for the flavor industry.

Keogh, M. K., O'kennedy, B. T., Kelly, J., Auty, M. A., Kelly, P. M., Fureby, A., & Haahr, A. M. (2001). Stability to oxidation of spray-dried fish oil powder microencapsulated using milk ingredients. *Journal of Food Science, 66*(2), 217–224. doi: 10.1111/j.1365-2621.2001.tb11320.x.

Khosroyar, S., & Arastehnodeh, A. (2007). Ferric saccharate microencapsulation with alginate coating using the coacervation method. *African Journal of Microbiology Research, 6*(10), 2455–2461.

Khosroyar, S., Akbarzade, A., Mortazavi, M., Arjomand, M., Safekordi, A. A., & Mortazavi, S. A. (2013). The study of dispersion method in the formation of ferric saccharate microcapsules with alginate coating. *International Journal of Life Science and Pharma Reviews, 2*, 325–334.

Kim, Y. D., & Morr, C. V. (1996). Microencapsulation properties of gum arabic and several food proteins: Spray-dried orange oil emulsion particles. *Journal of Agricultural and Food Chemistry, 44*(5), 1314–1320. doi: 10.1021/jf9503927.

Kim, S. J., Ahn, J., Seok, J. S., & Kwak, H. S. (2003). Microencapsulated iron for drink yogurt fortification. *Asian-Australasian Journal of Animal Sciences, 16*(4), 581–587. doi: 10.5713/ajas.2003.581.

Kruif, C. G., Weinbreck, F., & de Vries, R. (2004). Complex coacervation of proteins and anionic polysaccharides. *Current Opinion in Colloid & Interface Science, 9*(5), 340–349. doi: 10.1111/j.1365–2621.2011.02703.x.

Kwak, H. S., Ihm, M. R., & Ahn, J. (2001). Microencapsulation of β-galactosidase with fatty acid esters. *Journal of Dairy Science, 84*(7), 1576–1582. doi: 10.3168/jds.S0022-0302(01)74590-0.

Kwak, H. S., Ju, Y. S., Ahn, H. J., Ahn, J., & Lee, S. (2003). Microencapsulated iron fortification and flavor development in Cheddar cheese. *Asian-Australasian Journal of Animal Sciences, 16*(8), 1205–1211. doi: 10.5713/ajas.2003.1205.

Lee, S. W., Kim, M. H., & Kim, C. K. (1999). Encapsulation of ethanol by spray drying technique: Effects of sodium lauryl sulfate. *International Journal of Pharmaceutics, 187*(2), 193–198. doi: 10.1016/S0378-5173(99)00185-4.

Li, Y. O. (2009). Development of microencapsulation-based technologies for micronutrient fortification in staple foods for developing countries. University of Toronto (Canada).

Li, Y. O., Diosady, L. L., & Wesley, A. S. (2010). Iodine stability in iodized salt dual fortified with microencapsulated ferrous fumarate made by an extrusion-based encapsulation process. *Journal of Food Engineering, 99*(2), 232–238. doi: 10.1016/j.jfoodeng.2010.03.007.

Li, Y. O., Yadava, D., Lo, K. L., Diosady, L. L., & Wesley, A. S. (2011). Feasibility and optimization study of using cold-forming extrusion process for agglomerating and microencapsulating ferrous fumarate for salt double fortification with iodine and iron. *Journal of Microencapsulation, 28*(7), 639–649. doi: 10.3109/02652048.2011.604434.

Lukaski, H. C. (2004). Vitamin and mineral status: Effects on physical performance. *Nutrition, 20*(7–8), 632–644. doi: 10.1016/j.nut.2004.04.001.

Madene, A., Jacquot, M., Scher, J., & Desobry, S. (2006). Flavour encapsulation and controlled release–a review. *International Journal of Food Science & Technology*, *41*(1), 1–21.

Marianecci, C., Di Marzio, L., Rinaldi, F., Celia, C., Paolino, D., Alhaique, F., ... & Carafa, M. (2014). Niosomes from 80s to present: The state of the art. *Advances in Colloid and Interface Science*, *205*, 187–206. doi: 10.1016/j.cis.2013.11.018.

Martin, A. H., & De Jong, G. A. H. (2012). Impact of protein pre-treatment conditions on the iron encapsulation efficiency of whey protein cold-set gel particles. *European Food Research and Technology*, *234*(6), 995–1003. doi: 10.1007/s00217-012-1717-8.

McNamee, B. F., O'Riorda, E. D., & O'Sullivan, M. (1998). Emulsification and microencapsulation properties of gum arabic. *Journal of Agricultural and Food Chemistry*, *46*(11), 4551–4555. doi: 10.1021/jf9803740.

Mehansho, H., & Mannar, M. G. V. (1999). The mineral fortification in developing countries. In: *The Mineral for Tification of Foods* (edited by R. Hurrell). Leatherhead, England: Leatherhead International Ltd, pp. 210–228.

Millqvist-Fureby, A., Burns, N., Landstrom, K., Faldt, P. & Bergenstahl, B. (1999). Surface activity at the air-water interface in relation to surface composition of spray-dried milk protein-stabilizes emulsions. In: *Food Emulsions and Foams* (edited by E. Dickinson & J. M. Rodriguez-Patino). London: Royal Society of Chemistry, pp. 236–245.

Moran, L. (2014). Feasibility of spray chilling encapsulation for protection and stabilization of flavor compound-zinc ion complexes.

Moreau, D. L., & Rosenberg, M. (1993). Microstructure and fat extractability in microcapsules based on whey proteins or mixtures of whey proteins and lactose. *Food Structure, 12*(4), 6.

Moreau, D. L., & Rosenberg, M. (1996). Oxidative stability of anhydrous milkfat microencapsulated in whey proteins. *Journal of Food Science*, *61*(1), 39–43. doi: 10.1111/j.1365-2621.1996.tb14721.x.

Moreau, D. L., & Rosenberg, M. (1998). Porosity of whey protein-based microcapsules containing anhydrous milkfat measured by gas displacement pycnometry. *Journal of Food Science*, *63*(5), 819–823. doi: 10.1111/j.1365-2621.1998.tb17907.x.

Moreau, D. L., & Rosenberg, M. (1999). Porosity of microcapsules with wall systems consisting of whey proteins and lactose measured by gas displacement pycnometry. *Journal of Food Science*, *64*(3), 405–409. doi: 10.1111/j.1365-2621.1999.tb15052.x.

Moretti, D., Lee, T. C., Zimmermann, M. B., Nuessli, J., & Hurrell, R. F. (2005). Development and evaluation of iron-fortified extruded rice grains. *Journal of Food Science*, *70*(5), S330-S336. doi: 10.1111/j.1365-2621.2005.tb09987.x.

Mozafari, M. R., Khosravi-Darani, K., Borazan, G. G., Cui, J., Pardakhty, A., & Yurdugul, S. (2008). Encapsulation of food ingredients using nanoliposome technology. *International Journal of Food Properties*, *11*(4), 833–844. doi: 10.1080/10942910701648115.

Mutka, J. R., & Nelson, D. B. (1988). Preparation of encapsulated flavors with high flavor level. *Food Technology (Chicago), 42*(4), 154–157.

Navas-Carretero, S., Perez-Granados, A. M., Sarria, B., Schoppen, S., & Vaquero, M. P. (2007). Iron bioavailability from pate enriched with encapsulated ferric pyrophosphate or ferrous gluconate in rats. *Food Science and Technology International*, *13*(2), 159–163. doi: 10.1177%2F1082013207077931.

Oehlke, K., Adamiuk, M., Behsnilian, D., Gräf, V., Mayer-Miebach, E., Walz, E., & Greiner, R. (2014). Potential bioavailability enhancement of bioactive compounds using food-grade engineered nanomaterials: A review of the existing evidence. *Food & Function*, *5*(7), 1341–1359. doi: 10.1039/C3FO60067J.

Okuro, P. K., de Matos Junior, F. E., & Favaro-Trindade, C. S. (2013). Technological challenges for spray chilling encapsulation of functional food ingredients. *Food Technology and Biotechnology, 51*(2), 171–182.

Oner, L., Kaş, H. S., & Hincal, A. A. (1988). Studies on zinc sulphate microcapsules (1): Microencapsulation and in vitro dissolution kinetics. *Journal of Microencapsulation*, *5*(3), 219–223. doi: 10.3109/02652048809064166.

Pham, T. T., Jaafar-Maalej, C., Charcosset, C., & Fessi, H. (2012). Liposome and niosome preparation using a membrane contactor for scale-up. *Colloids and Surfaces B: Biointerfaces*, *94*, 15–21. doi: 10.1016/j.colsurfb.2011.12.036.

Porrarud, S., & Pranee, A. (2010). Microencapsulation of Zn-chlorophyll pigment from Pandan leaf by spray drying and its characteristic. *International Food Research Journal, 17*(4), 1031–1042.

Reineccius, G. A. (1991). Carbohydrates for flavor encapsulation. *Food Technology, 45,* 144–147.

Romita, D., Cheng, Y. L., & Diosady, L. L. (2011). Microencapsulation of ferrous fumarate for the production of salt double fortified with iron and iodine. *International Journal of Food Engineering*, *7*(3). doi: 10.2202/1556-3758.2122.

Rosenberg, M. (1997). U.S. Patent No. 5,601,760. Washington, DC: U.S. Patent and Trademark Office.

Rosenberg, M., & Sheu, T. Y. (1996). Microencapsulation of volatiles by spray-drying in whey protein-based wall systems. *International Dairy Journal*, *6*(3), 273–284. doi: 10.1016/0958-6946(95)00020-8.

Saeidy, S., Keramat, J., & Nasirpour, A. (2014). Microencapsulation of calcium using water-in-oil-in-water double emulsion method. *Journal of Dispersion Science and Technology*, *35*(3), 370–379. doi: 10.1080/01932691.2013.788453.

Sankarikutty, B., Sreekumar, M. M., Narayanan, C. S. & Mathew, A. G. (1988). Studies on encapsulation of cardamom oil by spray-drying technique. *Journal of Food Science and Technology, 25*, 352–356.

Shahidi, F., & Han, X. Q. (1993). Encapsulation of food ingredients. *Critical Reviews in Food Science & Nutrition*, *33*(6), 501–547. doi: 10.1080/10408399309527645.

Shewan, H. M., & Stokes, J. R. (2013). Review of techniques to manufacture micro-hydrogel particles for the food industry and their applications. *Journal of Food Engineering*, *119*(4), 781–792. doi: 10.1016/j.jfoodeng.2013.06.046.

Shiga, H., Yoshii, H., Nishiyama, T., Furuta, T., Forssele, P., Poutanen, K., & Linko, P. (2001). Flavor encapsulation and release characteristics of spray-dried powder by the blended encapsulant of cyclodextrin and gum arabic. *Drying Technology*, *19*(7), 1385–1395.

Singh, S., Rathi, N., Angal, A., Parida, P., & Rautaray, D. (2016). Biofortification of food with minerals and vitamins encapsulated in silica. In: *Nanoscience in Food and Agriculture* (edited by S. Ranjan, N. Dasgupta, & E. Lichtfouse). Cham: Springer, pp. 157–206. doi: 10.1007/978-3-319-39306-3_5.

Taylor, T. M., Weiss, J., Davidson, P. M., & Bruce, B. D. (2005). Liposomal nanocapsules in food science and agriculture. *Critical Reviews in Food Science and Nutrition*, *45*(7–8), 587–605. doi: 10.1080/10408390591001135.

Thomas, D. J. & Atwell, W. A. (1999). *Starches*. St Paul, MN: Eagan Press.

Turchiuli, C., Fuchs, M., Bohin, M., Cuvelier, M. E., Ordonnaud, C., Peyrat-Maillard, M. N., & Dumoulin, E. (2005). Oil encapsulation by spray drying and fluidised bed agglomeration. *Innovative Food Science & Emerging Technologies*, *6*(1), 29–35. doi: 10.1016/j.ifset.2004.11.005.

Wagner, M. E., Spoth, K. A., Kourkoutis, L. F., & Rizvi, S. S. (2016). Stability of niosomes with encapsulated vitamin D3 and ferrous sulfate generated using a novel supercritical carbon dioxide method. *Journal of Liposome Research, 26*(4), 261–268. doi: 10.3109/08982104.2015.1088868.

Watzke, H. J. (1998). Impact of processing on bioavailability examples of minerals in foods. *Trends in Food Science & Technology, 9*(8–9), 320–327. doi: 10.1016/S0924-2244(98)00060-0.

Wegmuller, R., Zimmermann, M. B., Bühr, V. G., Windhab, E. J., & Hurrell, R. F. (2006). Development, stability, and sensory testing of microcapsules containing iron, iodine, and vitamin A for use in food fortification. *Journal of Food Science, 71*(2), S181–S187. doi: 10.1111/j.1365-2621.2006.tb08923.x.

Xia, S., & Xu, S. (2005). Ferrous sulfate liposomes: Preparation, stability and application in fluid milk. *Food Research International, 38*(3), 289–296. doi: 10.1016/j.foodres.2004.04.010.

Xu, Z., Liu, S., Wang, H., Gao, G., Yu, P., & Chang, Y. (2014). Encapsulation of iron in liposomes significantly improved the efficiency of iron supplementation in strenuously exercised rats. *Biological Trace Element Research, 162*(1), 181–188. doi: 10.1007/s12011-014-0143-0.

Yadava, D., Olive Li, Y., Diosady, L. L., & Wesley, A. S. (2012). Optimisation of polymer coating process for microencapsulating ferrous fumarate for salt double fortification with iodine and iron. *Journal of Microencapsulation, 29*(8), 729–738. doi: 10.3109/02652048.2011.651493.

Yamada, T., Hisamatsu, M., Teranishi, K., Katsuro, K., Hasegawa, N., & Hayashi, M. (1995). Components of the porous maize starch granule prepared by amylase treatment. *Starch-Stärke, 47*(9), 358–361. doi: 10.1002/star.19950470907.

Yoshii, H., Soottitantawat, A., Liu, X. D., Atarashi, T., Furuta, T., Aishima, S., & Linko, P. (2001). Flavor release from spray-dried maltodextrin/gum arabic or soy matrices as a function of storage relative humidity. *Innovative Food Science & Emerging Technologies, 2*(1), 55–61. doi: 10.1016/S1466-8564(01)00019-4.

Young, S. L., Sarda, X., & Rosenberg, M. (1993a). Microencapsulating properties of whey proteins. 1. Microencapsulation of anhydrous milk fat. *Journal of Dairy Science, 76*(10), 2868–2877. doi: 10.3168/jds.S0022-0302(93)77625-0.

Young, S. L., Sarda, X., & Rosenberg, M. (1993b). Microencapsulating properties of whey proteins. 2. Combination of whey proteins with carbohydrates. *Journal of Dairy Science, 76*(10), 2878–2885. doi: 10.3168/jds.S0022-0302(93)77626-2.

Zeller, B. L., Saleeb, F. Z., & Ludescher, R. D. (1999). Trends in development of porous carbohydrate food ingredients for use in flavor encapsulation. *Trends in Food Science & Technology, 9*(11–12), 389–394. doi: 10.1016/S0924-2244(99)00007-2.

Zhao, J. & Whistler, R. L. (1994). Spherical aggregates of starch granules as flavor carrier. *Food Technology, 48,* 104–105.

Zimmermann, M. B., Wegmueller, R., Zeder, C., Chaouki, N., Biebinger, R., Hurrell, R. F., & Windhab, E. (2004). Triple fortification of salt with microcapsules of iodine, iron, and vitamin A. *The American Journal of Clinical Nutrition, 80*(5), 1283–1290. doi: 10.1093/ajcn/80.5.1283.

CHAPTER 11

Nanoencapsulation of Different Bioactive Isoprenoids

Syeda Saniya Zahra and Kawaljeet Kaur

CONTENTS

DOI: 10.1201/9781003259183-11

11.1 INTRODUCTION

Isoprenoids (aka, terpenoids) encompass numerous applications in the fields of medicine, nutraceuticals, cosmeceuticals, and biofuel. Embellished with a variety of substituents, they are basically composed of either single or multiple isoprene subunits ranging to more than 40,000 compounds (Immethun et al., 2013). In plants, these compounds are responsible for maintaining fluidity of the cell membrane, regulating phytohormones immunity, and assisting in pollination (Leisegang, 2021).

Isoprenoids, being abundantly present in plants, exist as valuable classes of chemically significant molecules such as essential oils (EOs), carotenoids, and many more. EOs and carotenoids are now being subjected to encapsulation due to the favorable pharmacological profile. The purpose is to enable the smooth release of isoprenoids at the site of action without being degraded and with high bioavailability. The process of encapsulation depends on the choice of a suitable carrier matrix with maximum encapsulation efficiency and its reusability. The most commonly used carrier matrices include polysaccharides (such as starch, inulin, pectin, cellulose, and hemicellulose) and proteins (such as lecithin, whey proteins, and soy proteins). The natural macromolecules would be discussed in detail from the standpoint of physicochemical properties and their stability issues along with the bioavailability and bioaccessibility constraints under various routes of administration.

11.2 DISTRIBUTION OF ISOPRENOIDS IN THE PLANT KINGDOM

Isoprenoids are a very important group of secondary metabolites which are made up of a variable number of isoprene units joined together in a head-to-tail fashion (Sharma et al., 2019). The isoprene units are composed of five carbon atoms, chemically known as 2-methyl butadiene. Therefore, the isoprenoids are classified on the basis of the number of isoprene units, i.e., hemiterpenes (one isoprene unit), monoterpenes (two isoprene units), sesquiterpenes (three isoprene units), diterpenes (four isoprene units), sesterterpenes (five isoprene units), triterpenes (six isoprene units), and tetraterpenes (eight isoprene units), as shown in Figure 11.1. They are also

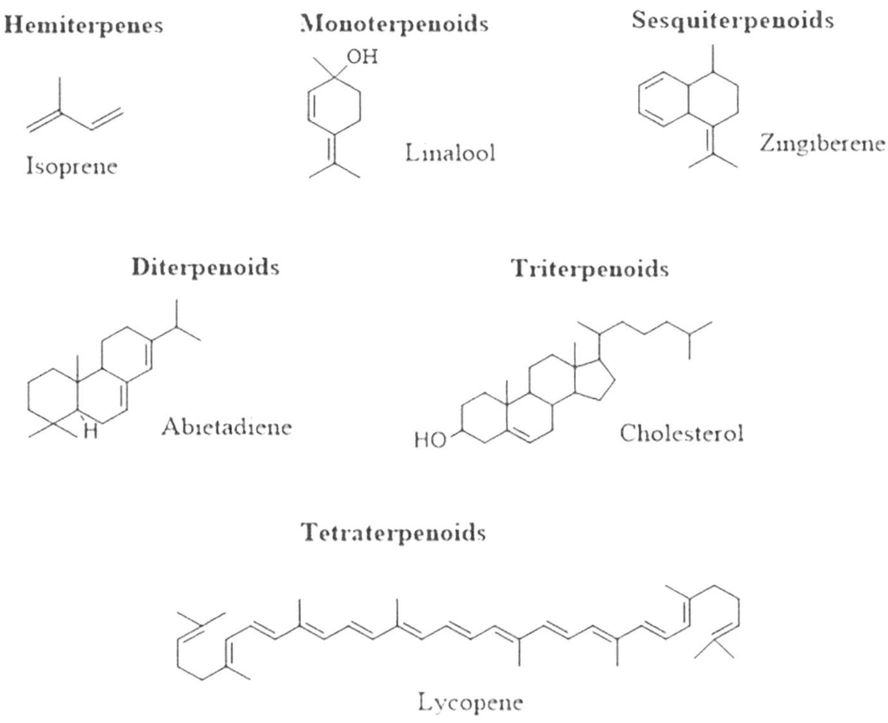

Figure 11.1 The representative isoprenoids from various chemical groups.

classified on the basis of chemical functionalities such as alcohols, phenols, ethers, carboxylic acid, aldehydes, esters, ketones, oxides, aromatic, or linear hydrocarbons (Zhou et al., 2016). The various examples of isoprenoids include geraniol, steroids, carotenoids, lycopene, and tocopherols (Leisegang, 2021).

EOs are mixtures of low-molecular-weight compounds (<300 Da) volatile in nature belonging to almost 10 chemical groups. However, carotenoids are tetraterpenoids that impart red, orange, and yellow pigmentation to their sources such as plants, algae, fungi, and bacteria. The α- and β-carotenes also known as provitamin A play an important role in ocular health with β-carotene (transform) having a higher potential to avoid night blindness (Hui-Fang et al., 2019).

Due to their production in miniscule amounts and bothersome extraction techniques, the microbial vectors such as *Saccharomyces cerevisiae* and *Escherichia coli* are utilized to keep up with the requirements of demand and supply by exploiting various synthetic biology tools and techniques (Tian et al., 2019). They require little resources and less time to generate the building blocks of isoprenoids that are isopentenyl pyrophosphate (IPP) and dimethylallyl pyrophosphate (DMAPP) naturally (as shown in Figure 11.2). Moreover, the microbes are engineered with the genes coding for plant-based IsoSs (Isoprenoid Synthases) and tailoring enzymes (Zhou et al., 2016).

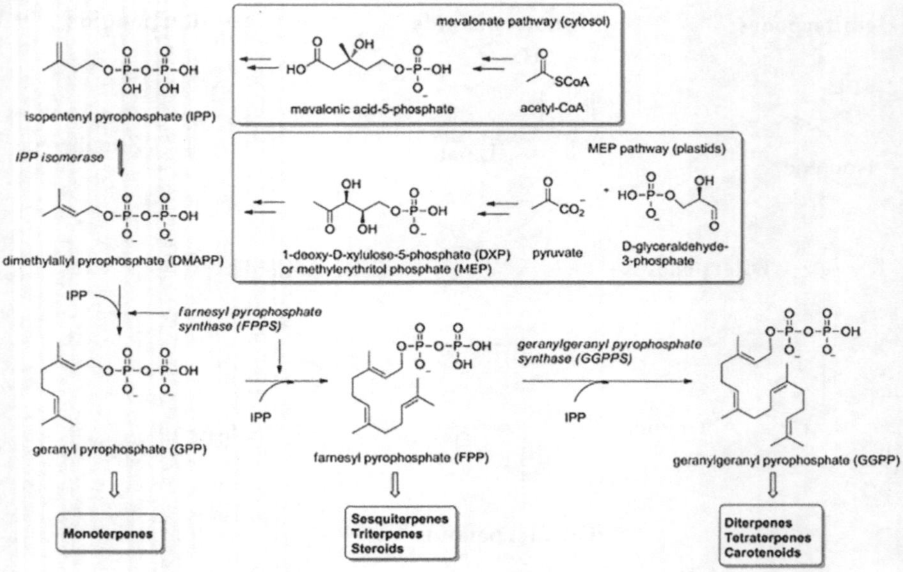

Figure 11.2 Simplistic schema involving mechanism of isoprenoids synthesis in plants via MEP and Mevalonate pathway (MVA) pathways. Both IPP and DMAPP further convert into numerous other isoprenoids with the help of plant-based isoprenoids synthases and tailoring enzymes. MEP, methylerythritol 4-phosphate. (Merino et al. (2017), the article is licensed under creative common license.)

11.3 PHYSICAL CHARACTERISTICS OF ESSENTIAL OILS

Most of the EOs are volatile at room temperature, and this is dependent on the number of carbons; the greater the number of carbons in a molecule, the lesser would be the volatility. Monoterpenes are higher in volatility, followed by sesquiterpenes, and then diterpenes, which lack volatility. The EOs comprise colorless, transparent to straw-colored liquid with the exception of a few EOs which are semisolid (guaiac wood and orris) and solid (plumeria). The EOs are insoluble in water and soluble in fats and alcohols. They have density lower than water while vetiver, cinnamon, and sassafras EOs have a density almost equal to one. Various parameters are used as quality control standards by International Organization of Standaradization (ISO)1, namely, ISO/TC 54: ISO 279:1998; ISO 592:1998; and ISO 280:1998 (Stevanovic et al., 2020). Table 11.1 describes the various standard parameters for a few EOs.

11.4 STABILITY OF ISOPRENOIDS

EOs are vulnerable to changes in their chemical composition due to fluctuations in the storage, processing, and extraction conditions in which the parameters of heat, light, and moisture are not taken care well. The polyunsaturated terpenes exhibit autoxidation by quenching one hydrogen atom leading to the formation of a resonance stabilized

Table 11.1 The Physical Characteristics of a Few Essential Oils (EOs)

Name	Organoleptic Description	Solubility	Density (g/cm^3)	Index of Refraction	Optical Rotation	Stability
Mint essential oil (EO)	**Color:** colorless to pale yellow **Odor:** peppermint-like odor **Taste:** balsamic taste accompanied by a cooling effect	70% ethanol at 20°C	0.896–0.908 at 25°C	1.460–1.464 at 20°C	−28 to −17 at 20°C	Not specified
Cinnamon bark EO	**Color:** dark yellow oily liquid **Odor:** aromatic, cinnamyl woody odor	70% ethanol at 20°C	1.010–1.030 at 25°C	1.5730–1.5910 at 25°C	−2 to 0 at 20°C	Darkened and thickened substance on exposure to air
Thyme EO	**Color:** colorless **Odor:** herbal	70% ethanol at 20°C	0.941 at 25°C	1.506 at 20°C	21 to +15	Not specified
Cumin EO	**Color:** light to dark yellow **Odor:** strong fatty	80% ethanol at 20°C; mineral oil; fixed oils; glycerin; ether; chloroform	0.908–0.958 at 20°C	1.4940–1.5160 at 20°C	+1 to +8 at 20°C	Sensitive to light, air, and moisture, alkali and metals
Nutmeg EO	**Color:** colorless to pale yellow **Odor:** nutmeg **Taste:** nutmeg	Mineral oil; fixed oil; hot alcohol, ether, chloroform	0.859–0.924 at 25°C	1.4740–1.4880 at 20°C	+10 to +30 at 20°C	Not specified

Source: Stevanovic et al. (2020), the article is an open access article.

structure, which starts the vicious cycle of oxidative degradation (Turek and Stintzing, 2012). Moreover, the exposure to oxygen may elicit degradative chain reactions involving hydroperoxide formation, which easily decomposes in the presence of acidic pH, heat, and light. Therefore, the inert gas should be filled in any leftover spaces of the containers (Turek et al., 2012). Volatile organic compounds (VOCs) being organic and volatile in nature are very susceptible to thermal degradation such as oxidative reactions via double bonds' cleavage, aromatization through dehydrogenation, epoxidation, and oxidation into alcohols, aldehydes/ketones. The exposure to oxygen causes the formation of hydroxyl or alkyl and peroxide radicals at lower and higher temperatures, respectively (Turek and Stintzing, 2012). In addition, carotenoids are sensitive toward changes in handling and storage conditions such as heat, light, and pH. They either form degradative products or undergo changes from trans-isomer to cis-isomer.

11.5 TECHNIQUES AND CARRIERS USED IN ENCAPSULATION

Encapsulation enhances the shelf life of the product by enhancing the stability against degradation via oxidation, heat, and light thus increasing its biological activity. Moreover, due to its well-designed delivery systems, the product reached the intended target with accuracy and efficiency. Isoprenoids such as EOs can be encapsulated to provide the benefits with enhanced retention abilities of volatile components along with the preservation of taste and odor characteristics. This results in prolonged shelf life by diminishing the chances of chemical degradation. For example, encapsulated peppermint oil was found to inhibit the *Listeria monocytogenes* and *Staphylococcus aureus* in a better way than its free form due to better stability and bioavailability characteristics (Liang et al., 2012). Moreover, many antibiotics when encapsulated gave a promising antibacterial spectrum (Zaman et al., 2017). Some antibiotics may produce a synergistic effect with carrier materials such as chitosan or cashew gum (Pandit et al., 2016). Figure 11.3 describes the whole process of extraction of isoprenoids (EO) leading toward the selection of a suitable carrier system and the encapsulation technique and thus the bioavailability studies according to the route of administration.

Depending upon the size of nanoparticles produced, the encapsulation processes can be broadly subdivided into two groups; the ones yielding a particle size between 1 and 5,000 μm constitute microencapsulation techniques, while the ones yielding particle sizes between 10 and 1,000 nm fall under nanoencapsulation methods, as shown in Figure 11.4.

11.5.1 Microencapsulation Methods

Various microencapsulation methods are discussed as follows.

11.5.1.1 Spray-Drying

The spray-drying method involves the atomization of the droplets containing active and carrier components either in dissolved or dispersed form followed by instant evaporation of the thus sprayed droplets by hot air at a certain temperature and pressure, for example, the encapsulation of lycopene and β-carotene (Fuenmayor et al., 2021). This yields variable results due to being sensitive to temperature and pressure parameters.

Advantages: It is characterized by a continual, fast track, cost-effective, and easy scale-up strategy. The carriers include starch, chitosan, gum arabic, and maltodextrins.

Disadvantages: In spite of the abovementioned advantages, the encapsulated material may exhibit nonuniformity in terms of particle shape and size, aggregation ability, just enough aqueous solubility of carrier material and low heat tolerance (Stevanovic et al., 2020).

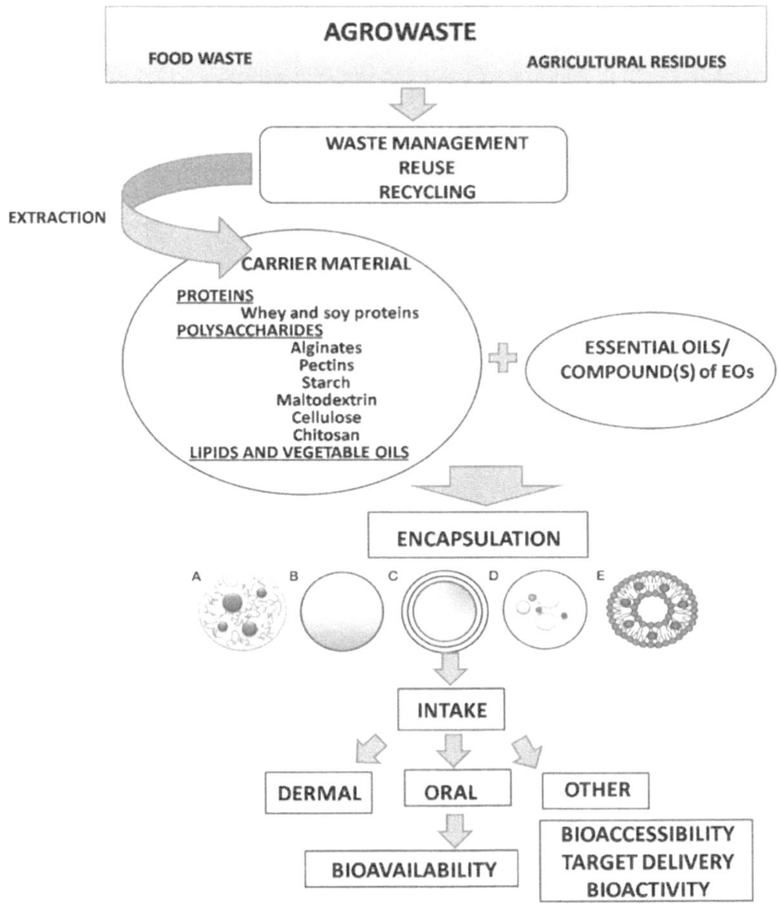

Figure 11.3 Flowchart representing the selection of carrier system and encapsulation method for the formulation of desired dosage forms for enhanced bioavailability. (A) Hydrogel bead, (B) monolayer capsule, (C) multilayer capsule, (D) composite bead, (E) liposome. (Stevanovic et al. (2020), the article is an open access article.)

11.5.1.2 Coacervation

The technique involves the formation of nanoparticles through a homogenous solution of charged macromolecules followed by the deposition of resultant coacervates at the base of the solution. The process is classified under simple coacervation if it involves the use of a single polymer, and the complex coacervation if it involves the use of multiple polymers. For example, a high content of carotenoid has been coacervated in palm oil (Fuenmayor et al., 2021).

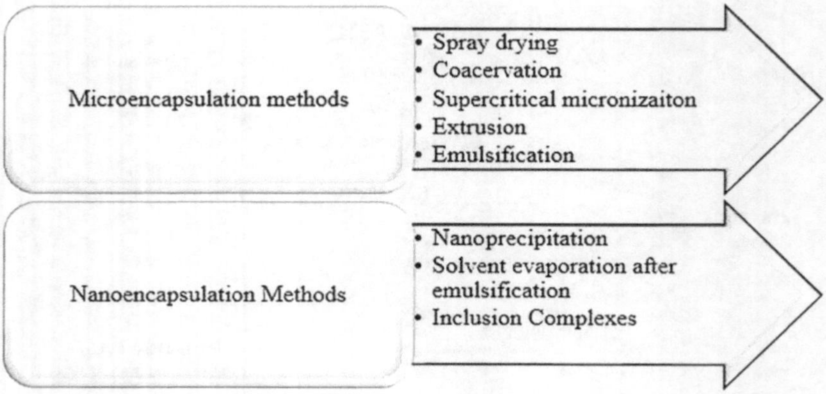

Figure 11.4 Various types of encapsulation techniques for isoprenoids.

Advantages: It imparts the characteristics of oil carrier rigidity, stability against heat, and gastric pH and solubility in intestinal fluid. In a two-layer system, the inner layer is made from an amphiphilic material (for example, proteins and/or surfactants) and the outer layer is made of polysaccharides to provide stability characteristics (Stevanovic et al., 2020).

11.5.1.3 Supercritical Micronization

This method provides better encapsulation and protection of carotenoids owing to the application of supercritical anti-solvent such as CO_2 so that a state of supersaturation is achieved, resulting in the precipitation of crystals from the solution (Fuenmayor et al., 2021). For example, lutein has been encapsulated with an efficiency of 90% using phosphatidylcholine as the carrier.

11.5.1.4 Extrusion

This technique is used in combination with/without coacervation to prepare mono- and multiphase polysaccharide hydrogels and microbeads comprising polysaccharide-protein hydrogels.

Advantages: include the stability against storage and gastric conditions and mechanical stresses and the possibility of encapsulation of substances with wider ranges of polarity. **Disadvantages**: This process is linked with a low yield and scale-up limitations (Stevanovic et al., 2020).

11.5.1.5 Emulsifications

Prior to freeze-drying or spray-drying, the bioactives are dispersed as an emulsion in a suitable carrier to give water-in-oil (w/o) or oil-in-water (o/w) emulsion. For example, improved stability has been observed with the microcapsules of β-carotene (Fuenmayor et al., 2021).

Advantages: This method facilitates the emulsification of hydrophobic materials. In contrast to the extrusion method, the emulsification process provides the benefit of smaller sized particles (0.01–1 mm) but with comparatively higher costs (Stevanovic et al., 2020).

11.5.2 Nanoencapsulation Methods

The nanoencapsulation techniques enhance the inclusion ability into liquid matrices due to the provision of larger surface area and also can easily diffuse into the epithelial walls of the lining, thus improving bioavailability to a greater extent. Various nanoencapsulation methods are discussed as follows.

11.5.2.1 Nanoprecipitation

The method involves the precipitation of a polymer solution in organic phase from the aqueous medium under constant stirring, usually followed by freeze-drying. For example, encapsulation of β-carotene in ethylcellulose polymer (particle size: 60 ± 9 nm and encapsulation efficiency: $74\% \pm 2\%$) and zein polymers (particle size: 83 ± 8 nm and encapsulation efficiency: $93\% \pm 4\%$) (Fuenmayor et al., 2021).

11.5.2.2 Solvent Evaporation after Emulsification

Emulsification of polymer first prepared by dispersing the aqueous phase followed by the addition of organic volatile solvents, such as ethyl acetate, chloroform, and dichloromethane ensured solvent evaporation by applying continuous stirring, vacuum, or heat. For example, β-carotene has been reported to be encapsulated via polymers such as polylactic acid (PLA) or polylactic co-glycolic acid (PLGA) (Fuenmayor et al., 2021).

11.5.2.3 Inclusion Complexes

A host molecule encloses a bioactive compound in the form of a cavity via van der Waals' forces, hydrogen bonding, or hydrophobic effect without significantly affecting the host's structural framework. Various volatile and small molecules can be encapsulated such as EOs and vitamins, using polymers such as β-cyclodextrin and β-lactoglobulin (Fuenmayor et al., 2021). Table 11.2 describes the various techniques to encapsulate carotenoids.

11.5.3 Carrier Systems for Encapsulation

Carriers can be classified into three major groups:

1. Polysaccharide-based carriers.
2. Protein- and polysaccharide-protein-based carriers.
3. Lipid- and lipid-protein-based carriers.

Table 11.2 Examples of Essential Oils along with the Encapsulation Techniques and Relevant Carriers

Essential Oil Source	Carrier	Encapsulation Technique
Avocado	Whey protein and maltodextrin	Spray-drying
Canola	Alginate, high methoxyl (HM) pectin-enhanced alginate	Coextrusion
Rosemary	Whey protein concentrate and maltodextrin	Freeze-drying
Sweet orange	Soybean protein isolate, gum Arabic	Complex coacervation
Holy basil	Gelatin, sodium alginate	Simple coacervation
Camphor	Gelatin, gum Arabic	Complex coacervation
Mint	Guar gum hydrolyzate, gum Arabic	Spay drying
Thyme	Alginate, soy protein isolate	Electrostatic extrusion
Linseed	Alginate, lupin protein	Extrusion
Sweet peppermint	Gelatin, gum Arabic	Complex coacervation
Pimento	Chitosan, k-carrageenan	Complex coacervation
Lippia sidoides	Maltodextrin, gum Arabic	Spray-drying
Lavandin	Soybean lecithin and cholesterol	Thin-film hydration
Blue gum	Diastearoylphosphatidylcholine, chitosan	Reverse phase evaporation
Eucalyptus camaldulensis	Soya lecithin and cholesterol	Freeze–thaw
Rosemary	Cholesterol, phosphatidyl choline	Thin-film hydration; sonication
Brazilian cherry	Hydrogenated soy lecithin	Thin-film hydration

Source: Stevanovic et al. (2020), the article is an open access article.

11.5.3.1 Polysaccharide-Based Carriers

Polysaccharide-based carriers include chitosan, alginate, and maltodextrin for the encapsulation of isoprenoids (Pasukamonset et al., 2016). Primarily obtained from brown algae, the alginate is the most commonly found naturally available linear macromolecule ranging between 32,000 and 400,000 g/mol. The aqueous soluble macromolecule contains freely distributed units of guluronic (G) and mannuronic (M) acids (Milivojevic et al., 2015). The ratio of G and M differs depending upon the season of collection, habitat, and the type of algae thus determining the physical and chemical attributes of alginate.

Also, a linear polysaccharide, the chitosan, is made up of β-(1 $\rightarrow$ 4)-linked d-glucosamine and N-acetyl-d-glucosamine units obtained after deacetylation of chitin. Alginate chains are anionic in nature while chitosan is cationic in nature and both form the gel spontaneously with a range of rheological properties due to the difference in alginate to chitosan ratios (Stevanovic et al., 2020).

Calcium alginate beads along with the covering of chitosan are used for additional strength against leakage issues (Anbinder et al., 2011). Moreover, chitosan–carrageenan carriers were used for the encapsulation of neem EO (Devi and Maji, 2009). Sarkar et al. (2013) showed that peppermint oil when encapsulated gave prolonged periods of EO components as compared to the nonencapsulated preparation.

Maltodextrin as produced by partial hydrolysis of starch contains repeated units of D-glucose as polysaccharides of variable lengths. It is used in combination with surface-active agents such proteins, gum Arabic, and modified starches. For example, the maltodextrin is co-encapsulated with gum Arabic and modified starch to formulate the cardamom EO microcapsules. The dextrose equivalent (DE) (that informs the degree of polymerization of starch sugars) dictates the oxygen barrier properties; the more the DE the more the encapsulation efficiency and lesser the oxygen permeation. The same carrier has been greatly exploited to encapsulate lemon, orange, cardamom, and ginger EO and provide resistance toward oxidation (Stevanovic et al., 2020).

The xanthan gum which is an extracellular polysaccharide is obtained by the microbe *Xanthomonas campestris* through the fermentation process. In combination with maltodextrin and chitosan, xanthum gum has been shown to provide excellent controlled release characteristics to the encapsulated material. Xanthum gum exhibited powerful emulsifying properties when combined with whey proteins and also the release of hydrophilic core material was enhanced in w/o/w emulsions (Stevanovic et al., 2020). In Table 11.2, various examples of EOs are given along with the encapsulation techniques and the relevant carrier systems.

11.5.3.2 Protein-Based Carriers

Due to exhibiting the hydrogen bonding among the molecules, proteins are used as carriers either as single or in bound form with polysaccharide hydrogels. The protein-based carriers are either heated or pH is altered in which the extent of crosslinking can be controlled. The examples include casein, gelatin, soy, and whey proteins. For example, basil EO was encapsulated using gelatin as a carrier. The main challenge is to maintain the integrity of the gelatin layer under gastric conditions (Stevanovic et al., 2020).

When proteins are used in combination with polysaccharide carriers; their chemical degradation in the stomach can be hindered, which results in a significant improvement of the release characteristics of the delivery systems. Moreover, globular proteins are shown to exhibit antioxidant behavior, which helps in the chemical stability of the EOs. The example includes the Oregano EO emulsion, which comprises of whey protein, milk powder, inulin, rice starch, and gelatin sucrose composite stabilized by Tween 80 (Stevanovic et al., 2020).

11.5.3.3 Lipid-Based Carriers

Lipid-based carriers provide remarkable encapsulation efficiency and include as much as 95% triglycerides, with monoglycerides, diglycerides, tocopherols, vitamins etc. as minor components. The other examples of lipid-based carriers are solid lipid carriers and liposomes which are very useful for the encapsulation of hydrophilic, hydrophobic, and amphiphilic materials. The later ones are costly to manufacture and have low shelf life. However, solid lipid carriers serve as the nanocarriers for the catechins (Stevanovic et al., 2020).

Proteins can be added to lipid-based carriers to increase the mechanical and chemical stabilities along with encapsulation efficiency. β-Lactoglobulin combined with the medium chain triglyceride serves as the carrier for encapsulating polyphenols (Stevanovic et al., 2020). Lipid-based carriers exhibit higher encapsulation efficiency, loading capacity, and release efficiency in small intestine as compared to the protein-polysaccharide carriers. But, being prone to the mechanical and thermal stresses, they become less useful as compared to the protein-polysaccharide carriers (Stevanovic et al., 2020). The various approaches for the selection of a suitable carrier system are given in Figure 11.5.

11.6 BIOAVAILABILITY AND BIOACCESSIBILITY CONSTRAINTS

Bioavailability defines the rate and extent to which a drug is bioavailable at the site of action to produce its desired action. It involves the processes of bioaccessibility and bioactivity. Bioaccessibility is the phenomenon which deals with the amount

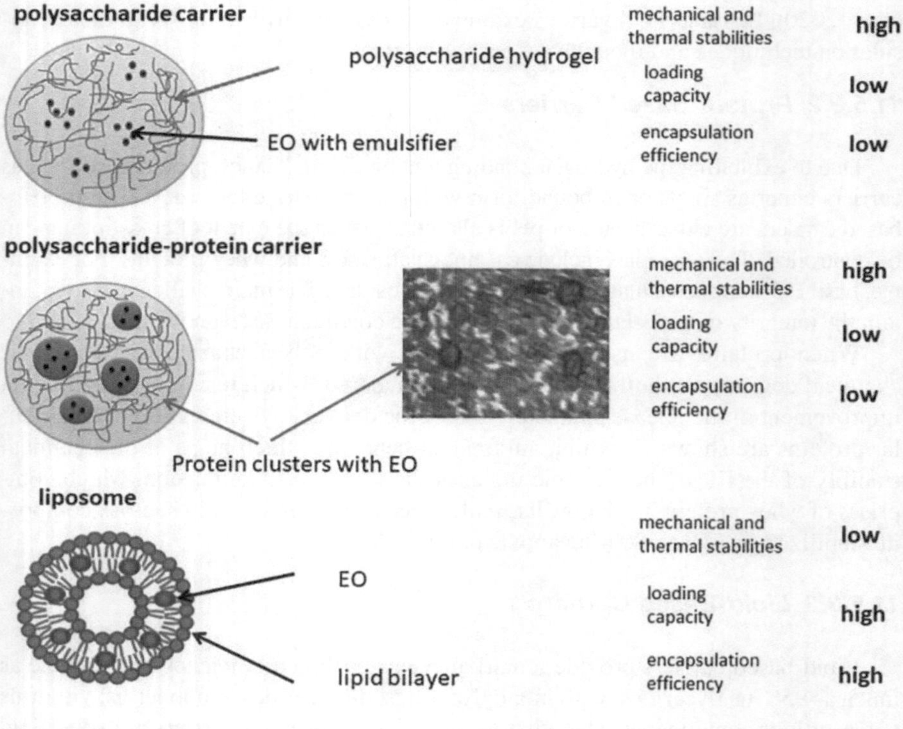

Figure 11.5 Various approaches for the selection of a suitable carrier system. (Stevanovic et al. (2020), the article is an open access article.)

of the drug that is released in the gut and is available for the absorption process. It may suffer from extensive presystemic extraction, which includes metabolism in the intestine through flora and the liver, prior to entering the bloodstream. However, bioactivity deals with the process starting from drug absorption, reaching its site of action, and interactions with the desired biomolecules to produce various physiological effects (Stevanovic et al., 2020).

Bioaccessibility of isoprenoids is studied through various in vitro and in vivo methods. The conditions of gastrointestinal system are simulated in vitro by pH adjustment and addition of several digestive enzymes such as salivary amylases, gastric lipases, pepsin, chymotrypsin, trypsin, pancreatic lipases, etc. and bile salts and integration of fermentation reactions to mimic the colon environment. The Caco-2 cell line (obtained from human colorectal adenocarcinoma) is also used as an in vitro model for the investigation of assemblage onto the basolateral side of the cell line. The bioavailability is also looked into by in vivo modeling and clinical studies.

Isoprenoids undergo pharmacokinetic and pharmacodynamic processes in order to produce their intended biological effect. A constant concentration of the isoprenoids reaches the blood stream by undergoing the processes such as absorption, metabolism, distribution, and excretion which is collectively known as pharmacokinetics. Pharmacodynamics parameters determine the effect of the isoprenoid compounds on animal and humans evident via biochemical and physiological data. Primarily, isoprenoids are taken via topical, nasal, and oral routes (Stevanovic et al., 2020).

There are a limited number of studies involving testing on humans, rather there is an ample amount of data obtained via in vitro and in vivo models. There is limited information on their findings in human as most studies are performed either in vitro or on animal models. As is known, intravenous administration of the drug makes it 100% bioavailable while bioavailability decreases upon administration through other routes. EOs are promptly absorbed via dermal, nasal, or oral routes (Stevanovic et al., 2020).

11.6.1 Routes of Administration of Isoprenoid Nanocapsules and Their Limitations

The rate and extent to which the EOs are bioavailable and distribute within the living system necessitate the assessment of safety issues related to isoprenoids, especially EOs.

11.6.1.1 Dermal Route of Administration

The constituents of EOs make their way through the stratum corneum (the outer dermis) and dermis into the blood circulation. The EOs with a high rate of percutaneous absorption should be assessed for the risk of systemic toxicity. The polar terpenes dissolve the hydrophilic drugs while the nonpolar terpenes dissolve the lipophilic compounds to a greater degree. Terpenes help to improve the efficiency of transdermal dosage forms by enhancing the percutaneous rate and lowering the cutaneous irritation. Moreover, the EOs are often encapsulated by liposomes to help prevent and heal the dermatological conditions (Stevanovic et al., 2020).

11.6.1.2 Nasal Route of Administration

The EOs produce very beneficial effects on the respiratory system. For example, the drug is inhaled via the nasal cavity, which transports the particles from trachea into the bronchi, bronchioles, and then to the lung alveoli, where they are absorbed and transported to the site of action (Stevanovic et al., 2020).

11.6.1.3 Rectal and Vaginal Route of Administration

The suppositories are designed to provide higher concentration of the drug at the site of application locally. They might cause irritation of the mucous membrane; therefore, the concentration should be carefully adjusted (Stevanovic et al., 2020).

11.6.1.4 Oral Route of Administration

The encapsulated EOs are most commonly administered as an oral dosage form. For example, menthol, thymol, limonene, chamazulene, carvacrol, and 1,8-cineole are encapsulated effectively. The administration of EOs at higher concentrations might cause Gastrointestinal (GI) irritation and inflammation. However, when given in controlled amounts, the EOs showed gastroprotective activity. For example, lemon grass reduced the gastric damage by acting on endogenous prostaglandins while orange EO and limonene produced the gastroprotective effect by inducing gastric mucous production (Stevanovic et al., 2020).

Terpenes, such as geraniol, thymol, neral, linalool, citronellol, etc., are absorbed from the stomach along with water and other substances. However, eugenol, trans-cinnamaldehyde thymol, and carvacrol were degraded to a lesser extent in the proximal gastrointestinal tract tested in vitro. Michiels et al. (2008) also observed the same results via in vivo studies.

Depending upon the molecular weight and the nature of terpenes, the rate and extent of absorption vary at different parts of the alimentary canal. The terpene, 1,8-cineole, is not well absorbed from the upper GI tract in contrary to the thymol, which is quickly absorbed. However, 1,8-cineole along with α-pinene is rapidly detected in blood plasma after oral ingestion (Zimmermann et al., 1995). Moreover, the absorption rate is enhanced from the duodenum due to the presence of bile salts and pancreatic juice, which increases the solubilization process. Also, its smaller size and lipophilic nature facilitate its absorption. The processes involved include both passive diffusion and carrier-mediated transport (Stevanovic et al., 2020).

The unionized, lipophilic molecule enters the blood stream via passive transport. Afterward, the weakly acidic terpenes are well absorbed from the gastric region (whose pH ranges between pH from 1 to 5) while the weakly basic ones are easily taken inside at the intestinal pH (which ranges from 5.5 to 7.5 and 5.7 to 6.7). The terpene alcohols that include geraniol, linalool, citral, and 1,8-cineole have amphiphilic properties; therefore, depending upon the GI microenvironment, they can act as either weakly acidic or basic, showing major absorption in the proximal GIT. The carboxylic acid functionality–containing terpenes include abietic and carnosic acids, and phenolic terpenes, such as anethole, carvacrol, and thymol, exhibit weakly acidic behavior, which makes the gastric pH as the ideal pH for their optimum absorption. Similarly, the nitrogen-containing terpenes such as amino acid derivatives with aminated terpenes and phenylalanine as an exception, behave as weakly basic molecules thus facilitating the rapid absorption via the intestine. Therefore, a variable behavior is expected out of any isoprenoid molecule due to various underlying factors such as molecular mass, stability, polarity, pH changes, etc. along with inter-individual differences in pharmacogenetics, which lead toward differences in pharmacokinetics and pharmacodynamics (Stevanovic et al., 2020).

11.6.2 Disposition Kinetics

The time period for which the isoprenoid molecules are retained in the blood stream would determine the duration of the intended biological effect. Therefore, the favorable characteristics for the active moiety would include the longer half-lives involving high absorption but low clearance rates. The isoprenoid molecules exhibit variation in metabolism and excretion processes, which is explained with the help of examples. The isoprenoids, especially EOs, are eliminated through phase II metabolism involving the glucuronidation and sulfation reactions. Many of the terpenes are cleared fast and therefore have short half-lives, for example, carvacrol, thymol, eugenol, and limonene (Stevanovic et al., 2020).

The lipophilic substances such as monoterpenes and sesquiterpenes would exhibit longer half-lives due to the redistribution to fat and muscles. Nonetheless, most fraction of EOs is cleared rapidly from the plasma, thus leaving a very little unchanged amount in the blood stream (Stevanovic et al., 2020).

11.7 APPLICATION OF ENCAPSULATED ISOPRENOIDS IN MEDICINE

The EOs (isoprenoids), due to having multifaceted pharmacological effects are heavily applied for encapsulation so as to reap the additive effects of the thus formed dosage forms.

- Microencapsulation and nanoencapsulation, particularly via lipid emulsion systems and liposomes containing lipid-based carriers, allow for the deeper penetration of the bioactives such as anticancer, anti-inflammatory, and antiaging agents. They find a wide number of applications in cosmeceuticals.
- In order for the drug delivery systems designed for the oral and respiratory routes to be well absorbed from the mucous membrane, the positively charged carrier such as chitosan is used (Kalliola et al., 2017). Chitosan has the additive advantage of being less toxic, biodegradable, and antibacterial (Kim et al., 2003). The other carriers included for the respiratory system are leucine, lactose, mannitol, and trehalose (Arpagaus et al., 2018).
- For the dosage forms requiring the deposition of drug in the deeper regions of the lungs, the particle size should be ranged between 1 and 5 μm (Arpagaus et al., 2018).

The examples include microspheres, nanoparticles, lipid vesicles, and solid lipid nanoparticles (Mehta et al., 2020). The encapsulation provides the added advantages of enhanced stability and better bioavailability.

- Although not very common, the orally administered and encapsulated EOs have a greater prospective use in the management of irritable bowel syndrome and gastrointestinal ulcers. The polymers include pectin, cellulose, alginate, maltodextrin, chitosan, sodium caseinate, etc., as they are biodegradable and nonirritant.
- There have been some reports involving the use of intravenously administered nanocapsules for the treatment of solid tumors (Arpagaus et al., 2018). Another investigation proved the promising effect of parenterally administered linalool nanocapsules and served as a novel means of therapy (Han et al., 2016).

11.7.1 Other Applications

- Apart from targeted drug delivery, encapsulation of nanocapsules has also been implemented in agriculture and textile industries such as pesticides (Kumar et al., 2019), feed additives (Stevanovic et al., 2020), and clothing as aromatic and antimicrobial agents (Bakry et al., 2016).
- Last but not the least, the encapsulation of isoprenoids has encompassed the functional foods (Dima et al., 2020), preservatives and additives for food (Stratulat et al., 2014; Bakry et al., 2016), and active food packaging (Ribeiro-Santos et al., 2017).
- Owing to the above points, it is evident that the effective utilization of isoprenoids is further enhanced due to encapsulation and results in improved bioavailability, shelf life, and bioactivity. Nevertheless, thorough studies are needed for tackling the safety and regulatory issues, developmental issues at the industrial level, and the related cost issues.

11.8 DELIVERY CONTROL OF BIOACTIVES

Bioactive compounds are being researched for their potential to reduce heart disease, inflammation, and immunological disorders. They have a variety of beneficial benefits on the body. Bioactive chemicals need a strong carrier system and effective delivery management to reach their target efficiently (Semenova, 2017). Control delivery has been highlighted as a critical stage in the administration of bioactive compounds, under which therapeutic compounds are delivered (Maghsoudi et al., 2020). The specific release of bioactive phytochemicals can be controlled by a variety of internal and external parameters such as pH, temperature, and so on (Bezerra et al., 2019).

Nanoencapsulation-based delivery techniques for bioactive components offer optimal durability, protection, and allow efficient absorption into the living system (Du et al., 2019). Due to their larger surface to mass ratio and capacity to absorb and transmit the components to the appropriate site with less damage, enhanced assimilation and bioavailability, nanoparticles are a better choice for delivering bioactive substances (Mogol et al., 2013). The outcome of bioactive components, on the other hand is determined by their physicochemical features and release site. The delivery mechanism has been classified as either solid containing lipid nanoparticles, nanocrystals, and polymeric nanoparticles or liquid mechanism including nanoemulsions, nanoliposomes, and nanopolymersomes based on the compatibility of the target bioactive component, nanoparticle features, and the desired use (Borel and Sabliov, 2014).

A delivery system is a method that entraps a bioactive chemical in a carrier and regulates the pace of bioactive release. Nanoencapsulation is useful because it protects sensitive bioactive food components from external influences like light, heat, enzyme degradation, pH, and oxygen. Discomfort caused by the ingestion of foul-tasting bioactive compounds is also reduced (Fathi et al., 2012). Based on the compliance of nanoparticle properties with bioactive substances, each of the delivery systems offers a distinctive benefit.

Nanoparticles with polymers are the most widely used nanomaterial due to their storage stability and high efficacy. To generate safer nanoparticles of nutritive quality and sustainability for food or pharmaceutical purposes, the utilization of Generally Recognized as safe (GRAS) components is required. The advantages of polymeric nanoparticles for applications include excellent stability, higher encapsulation efficacy, and sustained release of encapsulating substances (Singh et al., 2011). Furthermore, nanoparticles include a core chamber that prevents the compounds from coming into direct contact with tissues, resulting in discomfort after ingestion (Couvreur et al., 2002). As a result, nanoencapsulation is a versatile procedure used to shield, entrap, and distribute food bioactives, such as carotenoids from a variety of environmental conditions allowing for longer shelf lives and controlled release.

11.9 SAFETY ASPECTS OF ISOPRENOID NANOCAPSULES

The major safety issues with nanocapsules are whether they have significant negative effects and, generally, whether they pose a health hazard due to exposure. It may cause a health concern by overcoming biological membranes and accessing places that are normally shielded from nanoparticle contamination due to its nanoscale dimension and structural change. Although nanomaterials have been seen to be transported to secondary organs, it is still unclear if aggregation of less-toxic nanoparticles could have a toxicological impact on body parts over time (Kermanizadeh et al., 2015).

Increased formation of reactive oxygen species (ROS) and generation of free radicals are linked to the negative impacts of nanoparticles in complex biological systems, which can lead to oxidative stress and inflammation (Maurer-Jones et al., 2009). Even when using nanotechnologies, one of the fundamental limitations of administering bioactive components is that these compounds degrade in the gastrointestinal lumen prior to getting absorbed. It's time to figure out how gastrointestinal digestion affects oral bioavailability by applying useful in vitro simulation models as cost-effective methods for projecting oral bioavailability (Allen et al., 1999).

To gain an insight into the toxic effects of nanoparticles, distinct experiments with emphasis paid to interactions between nanocarriers and biological systems must be analyzed separately. This method of analyzing the existing information on nanoparticle toxicity is critical for determining the delivery systems' dependability

(Curtis et al., 2006). In fact, establishing safety profiles for nanomaterials, including immunotoxicity assessment, is critical to successfully translating them from the experimental to the clinical realm. In this context, priority should be given to the relationship between the physicochemical and functional properties of nanoparticles as well as the degree to which these features help to accomplish the functional potential intended and minimize potential health risks (Shvedova et al., 2010).

The interest of biodegradable nanoparticles and the limitations associated with their toxicity have been linked to the onset of detrimental impacts on the liver, kidney, and spleen in connection with the use of nanoparticles to transport bioactive compounds mentioned to be ingested in foods, water, or drug delivery devices. Furthermore, if the mechanisms driving the immune reaction generated by nanoparticles are unknown, there is a probability of allergic reactions (Lanone and Boczkowski, 2006). More research into the immunogenicity of nanocarriers is needed to figure out a better understanding of this issue. The immunogenic properties of nanoparticles can be used to gain benefits, as these qualities suggest that they could be useful adjuvants in the production of vaccines.

However, nanotechnology has been used to improve the bioavailability and bioactivity of bioactive compounds but they are made using physical and chemical methods which require costly and potentially dangerous chemicals thus limiting their use in vivo. These limitations are also related to the environmental impact of residues produced during nanoparticle synthesis. In this sense, using environmentally friendly and biocompatible chemicals to synthesize nanoparticles should help to reduce the negative consequences of these procedures.

11.10 CONCLUSIONS

The encapsulation of isoprenoids enables the controlled release of bioactives; optimizes the dosage; enhances the bioavailability by encountering the problems of low aqueous solubility and high volatility; and masks the features such as odor and taste (for example, EOs). Therefore, the carriers used in encapsulation should be optimized according to the requirements such as biodegradability, safety of consumption, and protection against the premature release of the active constituents. In the case of EOs, lipid-based carriers (such as liposomes and solid lipid nanocarriers) and lipid-protein and protein-polysaccharide carriers should be considered.

REFERENCES

Allen, C., Maysinger, D., & Eisenberg, A. (1999). Nano-engineering block copolymer aggregates for drug delivery. *Colloids and Surfaces B: Biointerfaces*, *16*(1–4), 3–27.

Anbinder, P. S., Deladino, L., Navarro, A. S., Amalvy, J. I., & Martino, M. N. (2011). Yerba mate extract encapsulation with alginate and chitosan systems: Interactions between active compound encapsulation polymers. *Journal of Encapsulation and Adsorption Sciences*. *1*, 80–87.

Arpagaus, C., Collenberg, A., Rütti, D., Assadpour, E., & Jafari, S. M. (2018). Nano spray drying for encapsulation of pharmaceuticals. *International Journal of Pharmaceutics, 546*(1–2), 194–214.

Bakry, A. M., Abbas, S., Ali, B., Majeed, H., Abouelwafa, M. Y., Mousa, A., & Liang, L. (2016). Microencapsulation of oils: A comprehensive review of benefits, techniques, and applications. *Comprehensive Reviews in Food Science and Food Safety, 15*(1), 143–182.

Bezerra, P. Q. M., de Matos, M. F. R., Ramos, I. G., Magalhães-Guedes, K. T., Druzian, J. I., Costa, J. A. V., & Nunes, I. L. (2019). Innovative functional nanodispersion: Combination of carotenoid from Spirulina and yellow passion fruit albedo. *Food Chemistry, 285*, 397–405.

Borel, T., & Sabliov, C. M. (2014). Nanodelivery of bioactive components for food applications: Types of delivery systems, properties, and their effect on ADME profiles and toxicity of nanoparticles. *Annual Review of Food Science and Technology, 5*, 197–213.

Couvreur, P., Barratt, G., Fattal, E., & Vauthier, C. (2002). Nanocapsule technology: A review. *Critical Reviews™ in Therapeutic Drug Carrier Systems, 19*(2), 1–21.

Curtis, J., Greenberg, M., Kester, J., Phillips, S., & Krieger, G. (2006). Nanotechnology and nanotoxicology. *Toxicological Reviews, 25*(4), 245–260.

Devi, N., & Maji, T. K. (2009). A novel microencapsulation of neem (Azadirachta Indica A. Juss.) seed oil (NSO) in polyelectrolyte complex of κ-carrageenan and chitosan. *Journal of Applied Polymer Science, 113*(3), 1576–1583.

Dima, C., Assadpour, E., Dima, S., & Jafari, S. M. (2020). Bioactive-loaded nanocarriers for functional foods: From designing to bioavailability. *Current Opinion in Food Science, 33*, 21–29.

Du, Y., Bao, C., Huang, J., Jiang, P., Jiao, L., Ren, F., & Li, Y. (2019). Improved stability, epithelial permeability and cellular antioxidant activity of β-carotene via encapsulation by self-assembled α-lactalbumin micelles. *Food Chemistry, 271*, 707–714.

Fathi, M., Mozafari, M. R., & Mohebbi, M. (2012). Nanoencapsulation of food ingredients using lipid based delivery systems. *Trends in Food Science & Technology, 23*(1), 13–27.

Fuenmayor, C. A., Baron-Cangrejo, O. G., & Salgado-Rivera, P. A. (2021). Encapsulation of carotenoids as food colorants via formation of cyclodextrin inclusion complexes: A review. *Polysaccharides, 2*(2), 454–476.

Han, H. D., Cho, Y. J., Cho, S. K., Byeon, Y., Jeon, H. N., Kim, H. S., …, Lee, J. W. (2016). Linalool-incorporated nanoparticles as a novel anticancer agent for epithelial ovarian carcinomalinalool nanoparticle as novel anticancer agent. *Molecular Cancer Therapeutics, 15*(4), 618–627.

Hui-Fang, C., You-Cheng, S., Kamesh V., & Chin-Kun, W. (2019). Food for eye health: Carotenoids and omega-3 fatty acids. In: Melton, L., Shahidi, F., & Varelis, P. (Eds.), *Encyclopedia of Food Chemistry*. Academic Press: Cambridge, MA, pp. 313–322.

Immethun, C. M., Hoynes-O'Connor, A. G., Balassy, A., & Moon, T. S. (2013). Microbial production of isoprenoids enabled by synthetic biology. *Frontiers in Microbiology, 4*, 75.

Kalliola, S., Repo, E., Srivastava, V., Heiskanen, J. P., Sirviö, J. A., Liimatainen, H., et al. (2017). The pH sensitive properties of carboxymethyl chitosan nanoparticles cross-linked with calcium ions. *Colloids and Surfaces B: Biointerfaces, 153*, 229–236.

Kermanizadeh, A., Balharry, D., Wallin, H., Loft, S., & Møller, P. (2015). Nanomaterial translocation–the biokinetics, tissue accumulation, toxicity and fate of materials in secondary organs: A review. *Critical Reviews in Toxicology, 45*(10), 837–872.

Kim, K. W., Thomas, R. L., Lee, C., & Park, H. J. (2003). Antimicrobial activity of native chitosan, degraded chitosan, and O-carboxymethylated chitosan. *Journal of Food Protection, 66*(8), 1495–1498.

Kumar, S., Nehra, M., Dilbaghi, N., Marrazza, G., Hassan, A. A., & Kim, K. H. (2019). Nano-based smart pesticide formulations: Emerging opportunities for agriculture. *Journal of Controlled Release, 294*, 131–153.

Lanone, S., & Boczkowski, J. (2006). Biomedical applications and potential health risks of nanomaterials: Molecular mechanisms. *Current Molecular Medicine*, *6*(6), 651–663.

Leisegang, K. (2021). Herbal pharmacognosy: An introduction. In *Herbal Medicine in Andrology* (pp. 17-–26). Academic Press.

Liang, R., Xu, S., Shoemaker, C. F., Li, Y., Zhong, F., & Huang, Q. (2012). Physical and antimicrobial properties of peppermint oil nanoemulsions. *Journal of Agricultural and Food Chemistry, 60*(30), 7548–7555.

Maghsoudi, S., Shahraki, B. T., Rabiee, N., Fatahi, Y., Dinarvand, R., Tavakolizadeh, M., … Tayebi, L. (2020). Burgeoning polymer nano blends for improved controlled drug release: A review. *International Journal of Nanomedicine*, *15*, 4363.

Maurer-Jones, M. A., Bantz, K. C., Love, S. A., Marquis, B. J., & Haynes, C. L. (2009). Toxicity of therapeutic nanoparticles. *Nanomedicine*, *4*(2), 219–241.

Mehta, P. P., Ghoshal, D., Pawar, A. P., Kadam, S. S., & Dhapte-Pawar, V. S. (2020). Recent advances in inhalable liposomes for treatment of pulmonary diseases: Concept to clinical stance. *Journal of Drug Delivery Science and Technology*, *56*, 101509.

Merino, P., Maiuolo, L., Delso, I., Algieri, V., De Nino, A., & Tejero, T. (2017). Chemical approaches to inhibitors of isoprenoid biosynthesis: Targeting farnesyl and geranylgeranyl pyrophosphate synthases. *RSC Advances*, *7*(18), 10947–10967.

Michiels, J., Missotten, J., Dierick, N., Fremaut, D., Maene, P., & De Smet, S. (2008). In vitro degradation and in vivo passage kinetics of carvacrol, thymol, eugenol and trans-cinnamaldehyde along the gastrointestinal tract of piglets. *Journal of the Science of Food and Agriculture, 88*(13), 2371–2381.

Milivojevic, M., Pajic-Lijakovic, I., Levic, S., Nedovic, V., &Bugarski, B. (2015). Alginic acid: Sources, modifications and main applications. In: Moore, A. (Ed.), *Alginic Acid–Chemical Structure, Uses and Health Benefits*. Nova Science Publishers, Inc.: New York, 45–88.

Mogol, B. U. R. Ç. E., Gökmen, V., & Shimoni, E. (2013). Nano-encapsulation improves thermal stability of bioactive compounds Omega fatty acids and silymarin in bread. *Agro Food Industry Hi-Tech, 24*(3).

Pandit, J., Aqil, M., & Sultana, Y. (2016). Nanoencapsulation technology to control release and enhance bioactivity of essential oils. In: Grumezescu, A. M. (Ed.), *Encapsulations*. Academic Press: Cambridge, MA, pp. 597–640.

Pasukamonset, P., Kwon, O., & Adisakwattana, S. (2016). Alginate-based encapsulation of polyphenols from Clitoriaternatea petal flower extract enhances stability and biological activity under simulated gastrointestinal conditions. *Food Hydrocolloids*, *61*, 772–779.

Ribeiro-Santos, R., Andrade, M., de Melo, N. R., & Sanches-Silva, A. (2017). Use of essential oils in active food packaging: Recent advances and future trends. *Trends in Food Science & Technology*, *61*, 132–140.

Sarkar, S., Gupta, S., Variyar, P. S., Sharma, A., & Singhal, R. S. (2013). Hydrophobic derivatives of guar gum hydrolyzate and gum Arabic as matrices for microencapsulation of mint oil. *Carbohydrate Polymers, 95*(1), 177–182.

Semenova, M. (2017). Protein–polysaccharide associative interactions in the design of tailor-made colloidal particles. *Current Opinion in Colloid & Interface Science, 28*, 15–21.

Sharma, A., Singh, R. S., Gupta, G., Ahmad, T., & Kaur, B. (2019). Metabolic engineering of enzyme-regulated bioprocesses. In: Singh, S. P. (Ed.), *Advances in Enzyme Technology.* Elsevier: Amsterdam, Netherlands, pp. 293–323.

Shvedova, A. A., Kagan, V. E., & Fadeel, B. (2010). Close encounters of the small kind: Adverse effects of man-made materials interfacing with the nano-cosmos of biological systems. *Annual Review of Pharmacology and Toxicology, 50*, 63–88.

Singh, S., Pandey, V. K., Tewari, R. P., & Agarwal, V. (2011). Nanoparticle based drug delivery system: Advantages and applications. *Indian Journal of Science and Technology, 4*, 177–180.

Stevanovic, Z. D., Sieniawska, E., Glowniak, K., Obradovic, N., & Pajic-Lijakovic, I. (2020). Natural macromolecules as carriers for essential oils: From extraction to biomedical application. *Frontiers in Bioengineering and Biotechnology*, *8*, 563.

Stratulat, I., Britten, M., Salmieri, S., Fustier, P., St-Gelais, D., Champagne, C. P., & Lacroix, M. (2014). Enrichment of cheese with bioactive lipophilic compounds. *Journal of Functional Foods, 6*, 48–59.

Tian, H., Zada, B., Singh, B. H., Wang, C., & Kim, S. W. (2019). Synthetic biology approaches for the production of isoprenoids in Escherichia coli. In: Larroche, C., Angeles Sanroman, M., Du, G., & Pandey, A. (Eds.), *Current Developments in Biotechnology and Bioengineering*. Elsevier: Amsterdam, Netherlands, pp. 311–329.

Turek, C., & Stintzing, F. C. (2012). Impact of different storage conditions on the quality of selected essential oils. *Food Research International*, *46*(1), 341–353.

Turek, C., Kirschmann, N., & Stintzing, F. C. (2012). Quality monitoring of selected essential oils upon storage at different temperature regimes. *Zeitschrift für Arznei-&Gewürzpflanzen*, *17*(2), 73–79.

Zaman, S. B., Hussain, M. A., Nye, R., Mehta, V., Mamun, K. T., & Hossain, N. (2017). A review on antibiotic resistance: Alarm bells are ringing. *Cureus, 9*(6), e1403.

Zhou, K., Edgar, S., & Stephanopoulos, G. (2016). Engineering microbes to synthesize plant isoprenoids. In: Colowick, S. P., & Kaplan, N. O (Eds.), *Methods in Enzymology*. Academic Press: Cambridge, MA, pp. 225–245.

Zimmermann, T., Seiberling, M., Thomann, P., & Karabelnik, D. (1995). The relative bioavailability and pharmacokinetics of standardized myrtol. *Arzneimittel-forschung*, *45*(11), 1198–1201.

CHAPTER 12

Nanoencapsulation and Targeted Delivery of Different Enzymes

Varun Kumar and Mohamad Anis

CONTENTS

12.1 INTRODUCTION

Encapsulation has potential applications in a variety of disciplines of science and technology, particularly in the food industry and drug delivery system. The technique of encapsulating potential active components on a nanoscale and delivering the final

DOI: 10.1201/9781003259183-12

product functionality with a regulated core dissociation is known as nanoencapsulation (Assadpour and Jafari, 2019a). Nanoencapsulation can overcome food manufacturing issues such as the effective supply of health-beneficial functional replacements and also allowing the tastes of the components to be released (Alehosseini and Jafari, 2019). The term "encapsulation" refers to encapsulated materials with nanometer sizes such as films, coatings, layers, or simply microdispersions. Studies have discovered advanced encapsulation technologies for the prevention and enhanced bioavailability of a wide range of bioactive molecules including therapeutics and food-active ingredients (nutraceuticals). Scientists previously focused on encapsulation methods, such as spray drying, extrusion, coacervation, liposomes, conventional emulsions, and other physical and chemical processes were limited. However, there has been a significant increase in the number of publications/papers on nanoencapsulation of food bioactive components using various techniques in recent years (Thiruvengadam et al., 2018; Subramani and Ganapathyswamy, 2020; Anusha Siddiqui et al., 2022). In the pharmaceutical, textile, food, feed agriculture, and chemical industries, enzymes are widely used, resulting in considerable cost savings and more energy-capable processes. Enzymes are superior than chemical catalysts for catalyzing chemical reactions with high selectivity and rapid transformation rates. Enzymes are exceptional catalysts through unrivaled chiral and positional selectivity, activating processes by unrivaled chiral and positional selectivity. Enzymes are environmentally acceptable and low-energy substitutes to conservative chemical catalysts because of their great selectivity which allows for efficient reactions with few by-products. These reactions are the foundation of all living creatures' metabolisms and they offer a significant opportunity for companies to perform beautiful, efficient, and cost-effective biocatalytic conversions (van Beilen and Li, 2002; Schmid et al., 2001; Küchler et al., 2016). Therapeutic enzymes could be used alone or in conjunction with certain medications to treat a variety of disorders in a safe manner (Kaur and Sekhon, 2012). The use of nanoparticles for the transportation of therapeutic enzymes has recently been discussed (Bosio et al., 2016). Biomaterials (materials derived from natural sources) are examined, and projected as improved resources for fabrication of nanoparticle delivery of enzymes by medicinal and ecological implications in this paper.

12.2 DESCRIPTION OF COATING MATERIALS USED FOR ENZYME ENCAPSULATION

The proper construction of nanoencapsulation by using suitable surface materials is an important aspect of the process. Food components and wall materials are crucial factors in the progression of nanocapsules. During the development of nanocapsules, various precautions need to be usually taken (Assadpour and Jafari, 2019b). The coating layer around the ingredient should be made correctly and there should be no leakage. Also, according to the established processes, the encapsulating material and technique are appropriate. Coating materials should be chosen from a diverse variety of synthetic and natural polymers with the properties desired in the final product ultimately examining the determinants. The chemical composition of the wall material has an important impact on the nanofunctional capsule capabilities. The materials should not be reactive with any ingredient throughout the storage and process period. The surface layer must be capable of holding and sealing the active component inside its structure. Under drying or other desolventization conditions, the capacity to release the solvent or other substances utilized at the time of the encapsulation process. The material should be strong enough to shield the active ingredient from extrinsic factors such as humidity, light, oxygen, and heat. In addition to that, to eliminate any relevant issues, the coating material should have these features and be affordable and food grade. Suitable enzyme transporters had to be created to shield the extremely purified enzyme from the different undesirable deposition factors and avoid any possible immunological activity in the bloodstream (Pei et al., 1994; Way, 1984). Essentially, the proper transporter structure should maintain an elevated level of enzyme activity in circulation for a long time. The optimal carrier system must be immune response robust, accessible to hazardous molecules, biodegradable, and have a high encapsulation performance with no harmful consequences as well as a consistent breakdown kinetic. Microparticles and nanoparticles can be used to create vessels that lead. An intriguing technique to enzyme encapsulation method using a biopolymeric matrix also creates an auto-degradation mechanism which is required for utilizations like solid tissue substitution as well as some medication delivery methods.

12.2.1 Poly(3-Hydroxybutyrate-co-3-Hydroxy Valerate) (PHBV) Nanocapsules

PHBV is a potential natural polymer that has lately been used to make microcapsules, microparticles, and nanoparticles (Sendil et al., 1999; Gürsel et al., 2001; Gangrade and Price, 1991; Pouton and Akhtar, 1996). Microbial and genetically altered crops both create this polymer which is completely natural. It has the potential to replace other polymers in the development of healthcare devices and systems as eco-friendly materials. For the encapsulation of model enzymes (L-asparaginase, catalase, glucose oxidase) and proteins, it is made as nanocapsules utilizing a twofold emulsion-solvent evaporation technique. To optimize the encapsulation efficiency and functionality of a contained enzyme many changes to the emulsion can be performed. To achieve molecular mass, the polymer was treated with sodium borohydride (MW). A 14-fold drop in MW (297 to 21 kDa) was also seen after 4 hours of incubation. Using low MW PHBV enhanced enzyme activity without increasing the amount of encapsulated protein, indicating that these nanocapsules have a better permeability to solutes (reactants and products). The encapsulation yields of catalase, L-asparaginase, and BSA were dramatically boosted after the second water phase was adjusted to the isoelectric precipitation of the proteins. Blanco and Alonso (1997) used the water-in-oil-in-water (W/O/W)/solvent evaporation technique to make nanocapsules from poly(lactic-co-glycolic acid) (PVA) nanospheres in an aqueous medium of protein which were then emulsified in dichloromethane containing PHBV using probe sonication for 15 seconds. A double emulsion was then mixed overnight with a magnetic stirrer in PVA in a flask and the solvent was evaporated at 4°C. Centrifugation was used to separate the nanocapsules which were then washed in Tris-HCl buffer. In a Tris-HCl buffer, the nanocapsules are resuspended (Baran et al., 2002). The "basic preparation" is the name given to this operation. Each step of the w/o/w method can be tweaked individually.

12.2.2 Dendritic Polymers (DPs)

In the mid-1980s, the use of DPs, a nanocapsule-based hyper-branched polymer with a C18 exterior modification was primary revealed (Tomalia et al., 1985). Because of its resilience to heat, solvent, and pH extremes, its application as a drug delivery carrier swiftly attracted attention (Tomalia and Esfand, 1997). The poly-hyper-branched (2-ethyl oxazoline) by CH3 (CH2)17 exterior arrangement can be synthesized (Yin et al., 1997; Xia et al., 2002). The DP can be transferred into suitable buffers and after that pooled by the enzyme's aquatic system to produce a known concentration mixture of dendrimer-enzyme with a polymer–enzyme ratio of 1:1. According to the size allocation capacity for the DP-nanodevices, adding Rh enzyme increases the relative size of the DP-particles. The most common hydrodynamic radius in the enzyme-free DP sample peak is also visible. The peaks were moved toward higher values in Rh-treated DP, showing that the enzyme stimulates the development of a complex between the dendrimer units (Petrikovics et al., 2010). Dendritic nanostructures (DP-OPAA and DP-OPH) were also used to encapsulate the OPAA and OPH enzymes. DFP and PO were used as OP substrates, and both were examined. The preventive value of nano-intercalated Rh in the occurrence of traditional sodium TS and a chemical constituent of garlic, diallyl disulfide was also tested *in vitro* and animals (Petrikovics et al., 2010). With TS alone in conjunction with SN, nanocaptured Rh dramatically improved preventive anecdotal resistance.

12.2.3 Polyethylene Glycol (PEG) Conjugation

PEG is an extensively adopted polymer for protein modification because of its low toxicity, the comparatively effortless chemistry essential for covalent protein binding and potential application. Abuchowski et al. (1977) were the first to demonstrate that PEG conjugation can extend the circulatory residency of bovine liver catalase. Since then, multiple studies have shown that PEG conjugation improves the therapeutic properties of recombinant proteins. PEG attachment to proteins is thought to prolong their circulation by interfering with protein removal processes such as kidney glomerular filtration, proteolytic breakdown, and active acquiescence through particular receptors. The key drawback of PEG protein encapsulation is that it can cause biological activity to be lost (Chapman et al., 1999; Veronese et al., 1996). In mice and monkeys, attachment of PEG chains to AChE leads to an increase in its abode duration in the circulation (Kronman et al., 2010; Cohen et al., 2001). It was also found that the duration of these enzyme forms in the circulation of Rhesus macacus correlates with their number of appended PEG moieties and is influenced by the actual location of the PEG Chains at the molecule surface by profiling the pharmacokinetic behaviour of a variety of well-defined hypolysine human mutant AChE molecules after PEGylation (Kronman et al., 2010). PEGylation's impact on OPH's biological and molecular features was investigated. The maximum enzymatic rate of PEGylated OPH was reduced, but significant efficacy was preserved against the two substrates tested: 30% with PO and 505–60% with demeton-S. In animal trials, encapsulated enzyme showed a considerable increase in pharmacokinetic parameters. In comparison to unchanged OPH, discharge from a guinea pig's bloodstream was drastically decreased leading to a rise in abode time and systemic accessibility. When compared to the unmodified enzyme, the extended PEG-OPH conjugation dramatically reduced the development of anti-OPH antibodies in the humoral immune response (Novikov et al., 2010).

12.2.4 Liposomes

The structure of "liposomes," characterized as spherical vesicles composed of lipid bilayers coincided with early research on bimolecular-based nanoparticles. These liposome bilayers can be made from a variety of phospholipids and the component attributes of the subsequent particles are reflected in the consequential molecules (Malam et al., 2009). The simultaneous encapsulation and trapping of hydrophilic and hydrophobic molecules are unusual given the well-known properties of liposomes which include a hydrophilic center and a hydrophobic bilayer. Proteins, peptides, enzymes, antibodies, hormones, antigens, and vaccinations have all been carried in liposomes to reach the target area. The development of enzyme-containing liposomes in which lipid vesicles transport water-soluble enzymes, is one of the key practical techniques. The encapsulated enzyme has two major applications: enzyme therapy in food processing and biomedical in which liposomes primarily serve as carriers (Walde and Ichikawa, 2001). The liposome-encapsulated enzymes are used to speed up the ripening process and flavor development in cheese (Piard et al., 1986) and to treat lactose intolerance (Matsuzaki et al., 1989). Other biocatalytic liposomes include those that contain butyrylcholinesterase as a toxin scavenger (Schumacher et al., 1999) and phosphotriesterase for the degradation of hazardous organophosphates (Allen, 1994).

A liposomal bilayer comprises the anthracycline drug for the treatment of tumors, AIDS, and multiple myelomas which is created by manipulating liposomal structures

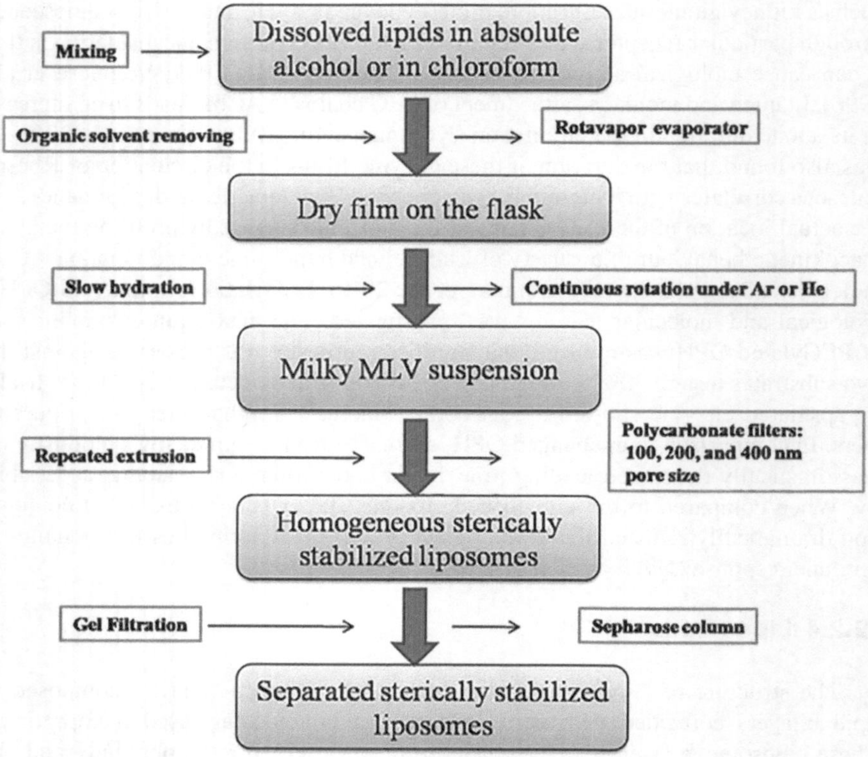

Figure 12.1 Main procedures in the manufacturing of antidotal enzyme-containing liposomes that are sterically stabilized. (Koyani et al. (2017), the article comes under the Creative commons (CC) License.)

by adding various particles in the liposome bilayer using dynamic and inert strategies (Malam et al., 2009; Ning et al., 2007). To enhance the encapsulation efficiency and stability of enzymes, optimal lipid compositions for sterically stabilized liposome can be followed as shown in Figure 12.1. The manipulated liposomes have been suggested as possible vaccines for a variety of diseases including mycobacterium disease (Smith et al., 2015). Liposomes displaying viral proteins (dubbed "virosome" by the authors) have been reported as a significant recent contribution. These formulations have been proposed for hepatitis B vaccination prophylaxis as well as to enhance immunization pandemic preparation (Smith et al., 2015; Felnerova et al., 2004; Monto et al., 2009). Other biocatalytic liposomes could undoubtedly benefit from these modifications. Liposomes have got a lot of attention since their development 50 years ago as drug delivery vehicles and their fascinating contribution to the medical sciences has got them a lot of attention. However, this technology still has a lot of potential that hasn't been fully realized. The development of customized liposomes with multifunctional features is a promising research area. Toxicology, stability, immune-promoted, tissue targeting, as well as a number of other challenges, should all be evaluated in order to create a reliable technology that the pharmaceutical industries can quickly adopt.

12.2.5 Virus-Like Particles (VLPs) Nanovehicle

VLPs are nanoparticles that are comparable to viruses without a genome. VLPs are dynamic structures that take on a variety of forms and which act on the pH and ionic strength of the environment. They are monodisperse, chemically stable, and retain a few features of the parent virus. Furthermore, the VLPs' basic structure can be involuntary in various methods to place within different sorts of molecules. VLPs can be combined through other compounds to create additional complicated nanoparticles (Destito et al., 2009; Ma et al., 2012; Strable and Finn, 2009). The application of VLPs as nanocarriers has multiple benefits: (i) It inhibits premature cargo deterioration and avoids cargo communication, (ii) it enhances cargo assimilation and/or delivery, and (iii) it allows for homeostasis of the carried compounds (Ma et al., 2012). VLPs have a lot of potentials mainly for therapeutic applications. VLPs are self-assembled, highly organized structures. (ii) They have a large surface area for ligand binding. (iii) They have a wide range of active groups for ligand attachment. (iv) They have available empty space for encapsulating cargo. (v) They have a porous structure for binding of substrate and product release which is important for reactors. Finally, some VLPs may be capable of cell internalization. It has been established that virus capsids or VLPs can be used as enzyme conjugator to produce nonreactors and as enzyme platforms (Table 12.1) (Comellas-Aragonès et al., 2007; Douglas et al., 2002; Douglas and Young, 1998; Seebeck et al., 2006; Varpness et al., 2005; Vriezema et al., 2005).

Enzymatic nanoreactors and enzyme nanocarriers are not without constraints and the need for improvement. VLPs could be coated with enzyme molecules (Cadena-Nava et al., 2011), but the chemical properties of drain capsids should undoubtedly be taken into account while developing diverse ways of producing efficient enzymatic VLPs.

VLPs can be directed to specific target tissues or cells in addition to protecting and delivering enzymes with a specified catalytic activity, which is a significant problem of fascination with nanomedicine. They might be new functionalities with particular molecules to be accepted for a receptor on targeted cell surface and subsequently internalized due to their protein environment and the occurrence of reactive molecules on the capsid (Hommersom et al., 2014). A multienzymatic VLP-based nanoparticle was recently developed. Three enzymes involved in carbohydrate metabolism, namely, CelB glucosidase, ATP-galactosidase, and ADP-glucokinase were encapsulated into VLPs from bacteriophage P22 (Patterson et al., 2014). The concentration of enzymes inside virus capsids could be in the mM range. As a result,

Table 12.1 Enzyme Encapsulation in Virus-Like Particles (VLPs)

Enzyme	VLP	$k_{cat} K_M^{-1}$ Related to Free Enzyme	Encapsulation Methods	References
Cytosine deaminase	SV40	Lower (VNR)	Protein fusion by virus capsid protein (*in vivo*)	Inoue et al. (2008)
Peptidase E	Bacteriophase Qβ	Lower 3×	RNA-based fusion protein (*in vivo*)	Fiedler et al. (2010)
Luciferase	Bacteriophase Qβ	Lower 30×	RNA-based fusion protein (*in vivo*)	Minten et al. (2011)
Pseudozyma antarctica lipase B	CCMV	Higher (k_{cat})	*Coiled-coil* motif fusion protein (*in vitro*)	Glasgow et al. (2012)
Alkaline Phosphatase	Bacteriophase MS2	Higher (k_{cat})	Charges created by electrostatic attraction. A negative peptide fusion protein (*in vitro*)	O'Neil et al. (2013)
Alcohol dehydrogenase	Bacteriophase P22	Lower 1.6×	Protein fusion with scaffold protein (*in vivo*)	Cardinale et al. (2012)
CelB glucosidase	Bacteriophase P22	Lower 1.6×	Structural protein fusion (*in vivo*)	Patterson et al. (2012b)
Phosphotriesterase	Bacteriophase P22	Lower 600×	Protein fusion with scaffolding protein (*in vivo*)	O'Neil et al. (2013)
Cytochrome P450	CCMV	Lower 10×	Complementary charges trapping (*in vitro*)	Sánchez-Sánchez et al. (2014)
Cytochrome P450	Bacteriophase P22	Lower 10×	Protein fusion with scaffold protein (*in vivo*)	Sánchez-Sánchez et al. (2015)

Abbreviations: VNR: virus neutralization reaction; CCMV: Cowpea chlorotic mottle virus

it is possible to research catalysis in crowded situations that are similar to those encountered to gain a better understanding of enzymatic catalysis within cells. Even though many encapsulated enzymes have lower catalytic activity than free enzymes, biocatalytic VLPs have been used to create nanoreactors with novel properties such as thermo-stability, proteolysis resistance, protection during the freeze-drying process (DeRoock et al., 2001; Pennington et al., 1996; Zhang et al., 2012), reduction of substrate inhibition (Patterson et al., 2012a) and improved performance under specific operational conditions (Comellas-Aragonès et al., 2007). Nonetheless, synthesizing metabolomes through enzyme encapsulation in VLPs might result in a sophisticated catalytic environment by a wide range of possible uses. The concentration of enzymes inside virus capsids could be in the mM range. As a result, it is possible to research catalysis in crowded situations that are similar to those encountered at the cellular level to gain a better understanding of enzymatic catalysis within cells. Even though many encapsulated enzymes have lower catalytic activity than free enzymes, biocatalytic VLPs have been used to create nanoreactors with novel properties (DeRoock et al., 2001; Pennington et al., 1996; Zhang et al., 2012), reduction of substrate inhibition (Patterson et al., 2012a), and improved performance under specific operational conditions (Comellas-Aragonès et al., 2007).

12.2.6 DNA Nanostructures Enzyme Carriers

Seeman suggested the idea of using DNA as structure genetic materials for the fabrication of nanostructures by molecular arrangement in 1982 (Seeman, 1982), which sparked the emerging discipline of DNA nanotechnology. The advent of the DNA origami method, which permitted DNA strands to be molded as per nanostructure (Rothemund, 2006), gave the field a substantial boost. The huge potential of DNA-based enzyme nanoreactors (Linko et al., 2016) also mentioned the uniqueness of DNA-based nanoparticles as single and various systems as well as their possible uses. Furthermore, the nontoxicity and excellent biocompatibility of DNA, combined with the desired structure, permitted researchers to investigate its use in a variety of fields, particularly biomedicine applications. Site-specific DNA nanostructures can be used as the foundation for structures such as independent walkers (Inoue et al., 2008), nanotweezers (Zhou et al., 2012), and nanocages (Douglas et al., 2012). Furthermore, DNA nanoarchitecture provides a potent foundation for the proper triggering of enzyme cascades, resulting in increased biological efficiency. Protein-DNA assemblies can coordinate cascades of enzymatic processes by controlling the relative distance and orientation of enzymatic components or by facilitating the contact between enzymes/cofactors and electrode surfaces (Teller and Willner, 2010; Erkelenz et al., 2011). The study and demonstration of multienzymatic DNA nanoreactors have been completed (Linko et al., 2015). However, in these systems, the regulation of enzyme activity remains a key concern. The utilization of programmed DNA nanostructures, as well as the modification of important factors like location, stoichiometry, and inter-enzyme distance for optimal activity, increases the possibility for the construction of multienzymatic nanoreactors (Fu et al., 2014). The notion of responsive enzymatic nanodevices (Wang et al., 2016) and the incorporation of feedback or feed-forward control loops could be used for diagnostics and the manufacturing of high-value compounds. The applications of DNA nanotechnology for enzyme-induced drug delivery and enzyme-triggered nanostructure assembly are also expanded by the rational management and regulation of DNA-based nanodevices including several enzyme systems. (Del Grosso et al., 2015).

Biocatalytic nanoreactors are predicted to be used in a variety of biological applications, including intracellular delivery of enzymes and effective cellular absorption. Noncovalent alteration of origami structures could improve cell internalization (Brglez et al., 2015; Schaffert et al., 2016). Though various techniques have been devised and attempted to achieve this aim, transport-limiting constraints such as enzyme size, charge, and stability have always hindered progress. However, using engineered DNA nanoparticles as a carrier for enzyme cargo has recently been proposed as a viable technique. Kuzuya and Komiyama (2009), Zhao et al. (2016), and Kiviaho et al. (2016) produced efficient gene carriers that preserve enzymes from biological degradation, boost catalytic performance, and may be used to control enzymatic reaction rates. Coating galactosidase with a dense shell of oligonucleotides improves the method of transporting it into cells (Brodin et al., 2015), which is also a new feature of DNA nanotechnology. Without a question, the fast advancement of DNA nanotechnology holds tremendous promise for enzyme-based nanoreactors. Nonetheless, further study is needed to achieve the objective of developing an effortless, scalable, efficient, and cost-effective technique for producing DNA-based nanostructures with desirable properties.

12.3 STABILITY AND POTENTIAL APPLICATIONS IN THE FOOD SECTOR

Nanoencapsulation is a process for making insoluble bioactive compounds more stable in aqueous media. Nanoencapsulation improves the bioactive compounds' water solubility and stability while also allowing for a regulated release to keep their biomedical activity in the body healthy. Also, optimizing the delivery improves bioavailability, lowering the possibility of re-use. This approach is a feasible alternative for redesigning functional food components that have a significant impact on product processability, bioavailability, and stability (Das and Wong, 2020; Martínez-Álvarez et al., 2020; Pateiro, et al., 2021; Jurić et al., 2022). The process of encapsulating a bioactive molecule in its liquid, solid, or gas state within a matrix or inert substance most commonly a polymer is known as encapsulation. The coated material (food or flavor molecules/ingredients) can be preserved by an enclosed wall, which increases its stability in the environment. It has a variety of uses in research and industry including cosmetics, medical products, pharmaceuticals, and food. Masking undesirable odors or flavors in food compositions is a typical application as is changing liquid phases to solid phases.

12.4 BIOAVAILABILITY CONSTRAINTS

The US FDA defines bioavailability since "the quantity and rate whereby the active ingredient is ingested and made accessible at pharmacological exploit residue pocket." The degree of bioactive molecules liberated from their environment and made accessible for intestinal absorption in the digestive system is known as bioaccessibility. Bioaccessible components may not have been bioavailable whereas inaccessible molecules are generally bioavailable. The component must be dissolved or dispersed within the gastrointestinal fluid before it can be absorbed. Low bioavailability of specific biomolecules is caused by different factors such as low solubility, limited biochemical strength in the Gastrointestinal (GI) environment, little assimilation and transportation during the GI lumen as well as fast digestion occurring in hepatocytes. Nanoparticle-based compound stability is a novel and emerging area. Nanostructured transportation methods are frequently used to improve the bioavailability and carrier particles improve their physicochemical properties (Yao et al., 2014; Chaudhry and Castle, 2015). Encapsulation helps increase the assimilation and bioavailability, resulting in unusually high plasma levels; yet a significant change in the physicochemical properties of the interactions of the particles with the human body may cause them to behave differently than the usual particles (Chaudhry and Castle, 2015).

Consumption of nanoparticles may have adverse effects not associated with conventional types due to a mixture of improved absorption, altered biokinetics/biodistribution, and modified linkages with biological systems (Chaudhry and Castle, 2015). It is essential to understand the outcome and behavior of nanoparticles in dietary components and the human system for their well-being. Also, the ability of nanostructured carriers to maintain their structure within the GI system is a key element in determining their safety. Changes in the environment may cause nanoparticles to modify their aggregation state or dimension, e.g., bile salts, strong acids, and digestive enzymes. These alterations influence the nanoparticles' potential to cross biological laminations in the human body such as the intestinal villus and epithelial cells. Nanomaterials made of biomolecules like carbohydrates, lipids, and proteins are widely believed to have a negligible safety issue to synthetic components. Smaller particles are also more reactive and hazardous than larger ones, according to research, because the increase in surface area per unit mass underlines their intrinsic properties. The acute oral toxicity evaluation on mice, which permits the determination of the median lethal dose, is one of the most commonly used approaches for validating the safety of EOs/bioactive components and their nanoencapsulated formulations. Moreover, many delivery systems still need to undergo clinical trials and *in vitro* experiments to confirm their safety (Fathi, 2018; Hosseini et al., 2021).

Nanoencapsulated active dietary ingredients and polyphenols migrate via several sections of the gastrointestinal tract, and their characteristics change as they are transported by the gastrointestinal tract and colon before their components are released. Within the gastrointestinal system, digestive enzymes such as protease, lipase, amylase, acids, ions, bile salts, and other substance are released, which will also have an impact on nanoencapsulation. Nanomaterials are designed with lipophilic micelle combination with bioactive compounds and nutraceuticals that are easily dispersed, transported, and absorbed primarily in the small intestine (Bazana et al., 2019; Li et al., 2022a,b). In the small intestine, undigested nutrients such as fiber are transported to the colon and metabolized by microbiota. Bioactive compounds and nutraceuticals in the small intestine first permeate the gastrointestinal mucosal layer via paracellular or transcellular routes before epithelial tissue. After getting through the GI tract, bioactive compounds and nutraceuticals enter the bloodstream and can then be transported throughout the system. However, the fact that nutraceuticals must pass through cell membranes to achieve efficient cellular absorption is another important consideration. Nutraceuticals are absorbed in cells in a multitude of ways some of which may have health-promoting effects.

Bioactive chemicals are delivered at the cellular target by one of these three main physical and chemical pathways in delivery systems:

1. Bioactive compounds are released through diffusion as the nanomaterial swells and hydrates.
2. Bioactive chemicals from the inner core entrapment are released via an enzymatic reaction that results in the polymer breaking, cleaving, or degrading at the delivery site.
3. Dispersion of the bioactive compounds from the encapsulated material and/or release of the swollen bioactive chemicals.

Many new products based on loading diverse combinations of preservatives, flavors, nutrients, and bioactive ingredients are predicted to develop in the future (Assadpour and Jafari, 2019c). Modifying these systems with insoluble bioactive chemicals in water is a major challenge. Controlling, triggering, and slowing the release of bioactive substances and nutritional additives into the final product that we utilize in our systems will be crucial in solubilization and bioavailability technologies. In the future, the researcher should focus on constructing or embedding features in carrier vehicles that regulate the flow of bioactive substances from the product into our blood or body system. However, it appears to be very time-consuming to develop. The pharmaceutical sector has created progress in the discipline of limiting bioactive chemical release, although these efforts are generally costly. Many of the controlled release features that are currently unachievable will they hope become a challenge in the future.

12.5 REGULATORY ASPECTS AND FUTURE PERSPECTIVES

Safety, health risks, and regulatory guidelines should be considered during the preparation, processing, packaging, and consumption of nano-processed foods. There is no official law regulating the ethical implications of nanostructures in the food and pharmaceutical sectors on an international market. The risk evaluation of encapsulated nanoproducts is still mostly unregulated in most countries. The poor reliable data interchange between nations has been identified as a concern to health and environmental risks, and it might be limiting the promotion of emerging and novel products across the globe. Furthermore, according to European Union regulations, any food item manufactured by nanotechnology applications must undergo a safety regulation before being licensed to market. The Food and Drug Administration in the United States has also established some guidelines for the use of nanotechnology in food and pharmaceutical products. Due to no specific laws for nanofoods, a database covering items that are currently available on the European market was created in 2012. Currently, the repository includes information on 3,036 items, comprising 128 nano-engineered foods accessible on the European market.

Research needs to focus to improve the process, method development, optimization, and encapsulation technique to explore the nanotechnology in food products and their functional abilities in the gastrointestinal tract and target industrial scale-up level. Also, it needs to focus on the use of nanocarriers of bioactive substances in food and biological systems with an emphasis on their effects on cell viability and their adsorption, transportation, metabolism, and excreta patterns in human systems. Effective bioactive materials to be nanoencapsulated highlight the need for global regulation to ensure their health and market. Several reports demonstrating the positive impact of nanostructured bioactive substances back up this act and point to suggesting future research.

12.6 SAFETY ASPECTS

Nanoencapsulated essential oils (EO)/bioactive components: safety and marketing are the two most significant threats associated with their use in the real food chain (Ali et al., 2020). Various regulatory authorities such as the European Commission (EC), US FDA, and other regulations have widely registered in "GRAS" group and act in the "Everything Added to Food in the US" (EAFUS) list considering the perception of "long term conventional narration of safe use" and "principle of safe-limitation" (Chaudhry and Castle, 2015; Cadena et al., 2018; Falleh et al., 2020). Food and Drug Administration recently granted GRAS status to the limited edition of EO (thyme, cinnamon, oregano, clove, and basil). However, EC has granted GRAS status to several bioactive constituents (thymol, eugenol, linalool, carvone, cinnamaldehyde, vanillin, and citral) including other EO (Maisanaba et al., 2017; Cadena et al., 2018; Chaudhari et al., 2021; Chaudhari et al., 2022).

Despite their well-documented preservation properties, there are currently just a few antifungal processes through EOs such as "DMC Base natural," made from a composite of rosemary, sage, and citrus EO; E-Rase™ made from jojoba EO; Sporan™ made from rosemary EO; Promax™ made from thyme EO; TimorexGold® made from tea tree EO; and Valero™ made from cinnamon EO; and none of the nanoencapsulated EO/bioactive constituents derived formulations (except for a few recent findings). The rationale for this is a lack of data about their unknown risks, toxicological validation on mammalian systems, and regulatory approval by regulating cell ingestion, absorption, transportation, and metabolism. Smaller size can produce cytotoxic effects which could have a detrimental impact on human health if consumed

(De Souza Simões et al., 2017). According to a study, smaller particles are more active and harmful than larger particles because their intrinsic qualities are highlighted by the augment in facade region per unit mass. The often used technique for validating the safety of EO /bioactive components and their nanoencapsulated mechanized is acute oral toxicity testing on rodents that permits for resolve of the maximum lethal concentration. The safety and acceptability for ingestion and food application will increase as the lethal dosage value increases. Several studies reported that EO and bioactive components showed acute toxicity in experimental animals and very few data were reported on nanoencapsulation-based formulations. As a result, it's essential to remember that each EO/bioactive constituent-based nanoencapsulated product should be validated for safety on a mammalian model system before being approved for use. Before they are legally allowed in real food systems, toxicological study of these nanoformulations should be done to increase their pragmatic value for broad-scale commercial application in the food sector (Marslin et al., 2020).

12.7 CONCLUSIONS

Biomaterial-based nanoparticles have indisputable promise for enzyme delivery. There are numerous diseases caused by a lack of enzymatic activity. Because of their biocompatibility, biodegradability, available functional groups, and adjustable structural properties, biodegradable polymer nanomaterials appear to be a preferable way for biological applications than synthetic preparations. Nonetheless, fundamental research needs to be conducted to address issues such as particle stability, immunogenic response in physiological fluids, and catalytic performance, among others. Based on the available evidence, it is clear that enzyme delivery through biomaterial-based nanoparticles is an important alternative in the emerging field of biomedicine.

REFERENCES

Abuchowski, A., McCoy, J. R., Palczuk, N. C., van Es, T., & Davis, F. F. (1977). Effect of covalent attachment of polyethylene glycol on immunogenicity and circulating life of bovine liver catalase. *Journal of Biological Chemistry, 252*(11), 3582–3586.

Alehosseini, E., & Jafari, S. M. (2019). Micro/nano-encapsulated phase change materials (PCMs) as emerging materials for the food industry. *Trends in Food Science & Technology, 91*, 116–128.

Ali, H., Al-Khalifa, A. R., Aouf, A., Boukhebti, H., & Farouk, A. (2020). Effect of nanoencapsulation on volatile constituents, and antioxidant and anticancer activities of Algerian Origanum glandulosum Desf. essential oil. *Scientific Reports*, *10*(1), 1–9.

Allen, T. M. (1994). Long-circulating (sterically stabilized) liposomes for targeted drug delivery. *Trends in Pharmacological Sciences, 15*(7), 215–220.

Anusha Siddiqui, S., Redha, A. A., Esmaeili, Y., & Mehdizadeh, M. (2022). Novel insights on extraction and encapsulation techniques of elderberry bioactive compounds. *Critical Reviews in Food Science and Nutrition*, *12*(1) 1–16.

Assadpour, E., & Jafari, S. M. (2019a). An overview of specialized equipment for nanoencapsulation of food ingredients. In: Jafari, S. (Ed.), *Nanoencapsulation of Food Ingredients by Specialized Equipment* (pp. 1–30). Academic Press: Cambridge, MA.

Assadpour, E., & Jafari, S. M. (2019b). Nanoencapsulation: Techniques and developments for food applications. In: Amparo, L. R., Rovira, M. J. F., Sanz, M. M., & Gomez-Mascaraque, L. G. (Eds.), *Nanomaterials for Food Applications* (pp. 35–61). Elsevier: Amsterdam, Netherlands.

Assadpour, E., & Jafari, S. M. (2019c). A systematic review on nanoencapsulation of food bioactive ingredients and nutraceuticals by various nanocarriers. *Critical reviews in food science and nutrition, 59*(19), 3129–3151.

Baran, E. T., Özer, N., & Hasirci, V. A. S. I. F. (2002). Poly (hydroxybutyrate-co-hydroxyvalerate) nanocapsules as enzyme carriers for cancer therapy: An in vitro study. *Journal of Microencapsulation*, *19*(3), 363–376.

Bazana, M. T., Codevilla, C. F., & de Menezes, C. R. (2019). Nanoencapsulation of bioactive compounds: Challenges and perspectives. *Current Opinion in Food Science, 26*, 47–56.

Blanco, M. D., & Alonso, M. J. (1997). Development and characterization of protein-loaded poly (lactide-co-glycolide) nanospheres. *European Journal of Pharmaceutics and Biopharmaceutics, 43*(3), 287–294.

Bosio, V. E., Islan, G. A., Martínez, Y. N., Durán, N., & Castro, G. R. (2016). Nanodevices for the immobilization of therapeutic enzymes. *Critical Reviews in Biotechnology, 36*(3), 447–464.

Brglez, J., Nikolov, P., Angelin, A., & Niemeyer, C. M. (2015). Designed intercalators for modification of DNA origami surface properties. *Chemistry–A European Journal, 21*(26), 9440–9446.

Brodin, J. D., Sprangers, A. J., McMillan, J. R., & Mirkin, C. A. (2015). DNA-mediated cellular delivery of functional enzymes. *Journal of the American Chemical Society, 137*(47), 14838–14841.

Cadena, M. B., Preston, G. M., Van der Hoorn, R. A., Flanagan, N. A., Townley, H. E., & Thompson, I. P. (2018). Enhancing cinnamon essential oil activity by nanoparticle encapsulation to control seed pathogens. *Industrial Crops and Products, 124*, 755–764.

Cadena-Nava, R. D., Hu, Y., Garmann, R. F., Ng, B., Zelikin, A. N., Knobler, C. M., & Gelbart, W. M. (2011). Exploiting fluorescent polymers to probe the self-assembly of virus-like particles. *The Journal of Physical Chemistry B, 115*(10), 2386–2391.

Cardinale, D., Carette, N., & Michon, T. (2012). Virus scaffolds as enzyme nano-carriers. *Trends in Biotechnology, 30*(7), 369–376.

Chapman, A. P., Antoniw, P., Spitali, M., West, S., Stephens, S., & King, D. J. (1999). Therapeutic antibody fragments with prolonged in vivo half-lives. *Nature Biotechnology, 17*(8), 780–783.

Chaudhari, A. K., Singh, V. K., Das, S., & Dubey, N. K. (2021). Nanoencapsulation of essential oils and their bioactive constituents: A novel strategy to control mycotoxin contamination in food system. *Food and Chemical Toxicology*, *149*, 112019.

Chaudhari, A. K., Singh, V. K., Das, S., & Dubey, N. K. (2022). Fabrication, characterization, and bioactivity assessment of chitosan nanoemulsion containing allspice essential oil to mitigate Aspergillus flavus contamination and aflatoxin B1 production in maize. *Food Chemistry*, *372*, 131221.

Chaudhry, Q., & Castle, L. (2015). Safety assessment of nano-and microscale delivery vehicles for bioactive ingredients. In: Sabliov, C., Chen, H., & Yada, R. (Eds.), *Nanotechnology and Functional Foods: Effective Delivery of Bioactive Ingredients*. John Wiley & Sons: Hoboken, NJ, p. 348.

Cohen, O., Kronman, C., Chitlaru, T., Ordentlich, A., Velan, B., & Shafferman, A. (2001). Effect of chemical modification of recombinant human acetylcholinesterase by polyethylene glycol on its circulatory longevity. *Biochemical Journal*, *357*(3), 795–802.

Comellas-Aragonès, M., Engelkamp, H., Claessen, V. I., Sommerdijk, N. A., Rowan, A. E., Christianen, P., … & Nolte, R. J. (2007). A virus-based single-enzyme nanoreactor. *Nature Nanotechnology, 2*(10), 635–639.

Das, S., & Wong, A. B. (2020). Stabilization of ferulic acid in topical gel formulation via nano-encapsulation and pH optimization. *Scientific Reports*, *10*(1), 1–18.

de Souza Simões, L., Madalena, D. A., Pinheiro, A. C., Teixeira, J. A., Vicente, A. A., & Ramos, O. L. (2017). Micro-and nano bio-based delivery systems for food applications: In vitro behavior. *Advances in Colloid and Interface Science, 243*, 23–45.

Del Grosso, E., Dallaire, A. M., Vallée-Bélisle, A., & Ricci, F. (2015). Enzyme-operated DNA-based nanodevices. *Nano Letters, 15*(12), 8407–8411.

DeRoock, I. B., Pennington, M. E., Sroka, T. C., Lam, K. S., Bowden, G. T., Bair, E. L., & Cress, A. E. (2001). Synthetic peptides inhibit adhesion of human tumor cells to extra-cellular matrix proteins. *Cancer Research, 61*(8), 3308–3313.

Destito, G. A. M. M., Schneemann, A., & Manchester, M. (2009). Biomedical nanotechnology using virus-based nanoparticles. *Viruses and Nanotechnology*, *327*, 95–122.

Douglas, T., & Young, M. (1998). Host–guest encapsulation of materials by assembled virus protein cages. *Nature, 393*(6681), 152–155.

Douglas, T., Strable, E., Willits, D., Aitouchen, A., Libera, M., & Young, M. (2002). Protein engineering of a viral cage for constrained nanomaterials synthesis. *Advanced Materials, 14*(6), 415–418.

Douglas, S. M., Bachelet, I., & Church, G. M. (2012). A logic-gated nanorobot for targeted transport of molecular payloads. *Science, 335*(6070), 831–834.

Erkelenz, M., Kuo, C. H., & Niemeyer, C. M. (2011). DNA-mediated assembly of cyto-chrome P450 BM3 subdomains. *Journal of the American Chemical Society, 133*(40), 16111–16118.

Falleh, H., Jemaa, M. B., Saada, M., & Ksouri, R. (2020). Essential oils: A promising eco-friendly food preservative. *Food Chemistry*, *330*, 127268.

Fathi, M., Donsi, F., & McClements, D. J. (2018). Protein-based delivery systems for the nano-encapsulation of food ingredients. *Comprehensive Reviews in Food Science and Food Safety*, *17*(4), 920–936.

Felnerova, D., Viret, J. F., Glück, R., & Moser, C. (2004). Liposomes and virosomes as delivery systems for antigens, nucleic acids and drugs. *Current Opinion in Biotechnology, 15*(6), 518–529.

Fiedler, J. D., Brown, S. D., Lau, J. L., & Finn, M. G. (2010). RNA-directed packaging of enzymes within virus-like particles. *Angewandte Chemie, 122*(50), 9842–9845.

Fu, J., Yang, Y. R., Johnson-Buck, A., Liu, M., Liu, Y., Walter, N. G., … & Yan, H. (2014). Multi-enzyme complexes on DNA scaffolds capable of substrate channelling with an artificial swinging arm. *Nature Nanotechnology, 9*(7), 531–536.

Gangrade, N., & Price, J. C. (1991). Poly (hydroxybutyrate-hydroxyvalerate) microspheres containing progesterone: Preparation, morphology and release properties. *Journal of Microencapsulation, 8*(2), 185–202.

Glasgow, J. E., Capehart, S. L., Francis, M. B., & Tullman-Ercek, D. (2012). Osmolyte-mediated encapsulation of proteins inside MS2 viral capsids. *ACS Nano, 6*(10), 8658–8664.

Gürsel, İ., Korkusuz, F., Türesin, F., Alaeddinoğlu, N. G., & Hasırcı, V. (2001). In vivo application of biodegradable controlled antibiotic release systems for the treatment of implant-related osteomyelitis. *Biomaterials, 22*(1), 73–80.

Hommersom, C. A., Matt, B., van der Ham, A., Cornelissen, J. J. L. M., & Katsonis, N. (2014). Versatile post-functionalization of the external shell of cowpea chlorotic mottle virus by using click chemistry. *Organic & Biomolecular Chemistry, 12*(24), 4065–4069.

Hosseini, S. F., Ramezanzade, L., & McClements, D. J. (2021). Recent advances in nano-encapsulation of hydrophobic marine bioactives: Bioavailability, safety, and sensory attributes of nano-fortified functional foods. *Trends in Food Science & Technology, 109*, 322–339.

Inoue, T., Kawano, M. A., Takahashi, R. U., Tsukamoto, H., Enomoto, T., Imai, T.,... & Handa, H. (2008). Engineering of SV40-based nano-capsules for delivery of heterologous proteins as fusions with the minor capsid proteins VP2/3. *Journal of biotechnology, 134* (1–2), 181–192.

Jurić, S., Jurić, M., Siddique, M. A. B., & Fathi, M. (2022). Vegetable oils rich in polyunsaturated fatty acids: Nanoencapsulation methods and stability enhancement. *Food Reviews International, 38*(1), 32–69.

Kaur, R., & Sekhon, B. S. (2012). Enzymes as drugs: An overview. *Journal of Pharmaceutical Education & Research, 3*(2), 29–41.

Kiviaho, J. K., Linko, V., Ora, A., Tiainen, T., Järvihaavisto, E., Mikkilä, J., … Kostiainen, M. A. (2016). Cationic polymers for DNA origami coating–examining their binding efficiency and tuning the enzymatic reaction rates. *Nanoscale, 8*(22), 11674–11680.

Koyani, R., Pérez-Robles, J., Cadena-Nava, R. D., & Vazquez-Duhalt, R. (2017). Biomaterial-based nanoreactors, an alternative for enzyme delivery. *Nanotechnology Reviews, 6*(5), 405–419.

Kronman, C., Cohen, O., Mazor, O., Ordentlich, A., Raveh, L., Velan, B., & Shafferman, A. (2010). Next generation OP-bioscavengers: A circulatory long-lived 4-PEG hypolysine mutant of F338A-HuAChE with optimal pharmacokinetics and pseudo-catalytic characteristics. *Chemico-Biological Interactions, 187*(1–3), 253–258.

Küchler, A., Yoshimoto, M., Luginbühl, S., Mavelli, F., & Walde, P. (2016). Enzymatic reactions in confined environments. *Nature Nanotechnology, 11*(5), 409–420.

Kuzuya, A., & Komiyama, M. (2009). Design and construction of a box-shaped 3D-DNA origami. *Chemical Communications, 28*, 4182–4184.

Li, H., Zhang, X., Zhao, C., Zhang, H., Chi, Y., Wang, L., … Zhang, X. (2022a). Entrapment of curcumin in soy protein isolate using the pH-driven method: Nanoencapsulation and formation mechanism. *LWT, 153*, 112480.

Li, X., Zhou, P., Luo, Z., Feng, R., & Wang, L. (2022b). Hohenbuehelia serotina polysaccharides self-assembled nanoparticles for delivery of quercetin and their anti-proliferative activities during gastrointestinal digestion in vitro. *International Journal of Biological Macromolecules, 203*, 244–255.

Linko, V., Eerikäinen, M., & Kostiainen, M. A. (2015). A modular DNA origami-based enzyme cascade nanoreactor. *Chemical Communications, 51*(25), 5351–5354.

Linko, V., Nummelin, S., Aarnos, L., Tapio, K., Toppari, J. J., & Kostiainen, M. A. (2016). DNA-based enzyme reactors and systems. *Nanomaterials, 6*(8), 139.

Ma, Y., Nolte, R. J., & Cornelissen, J. J. (2012). Virus-based nanocarriers for drug delivery. *Advanced Drug Delivery Reviews, 64*(9), 811–825.

Maisanaba, S., Llana-Ruiz-Cabello, M., Gutiérrez-Praena, D., Pichardo, S., Puerto, M., Prieto, A. I., … & Cameán, A. M. (2017). New advances in active packaging incorporated with essential oils or their main components for food preservation. *Food Reviews International, 33*(5), 447–515.

Malam, Y., Loizidou, M., & Seifalian, A. M. (2009). Liposomes and nanoparticles: Nanosized vehicles for drug delivery in cancer. *Trends in Pharmacological Sciences, 30*(11), 592–599.

Marslin, G., Khandelwal, V., & Franklin, G. (2020). Cordycepin nanoencapsulated in poly (Lactic-Co-Glycolic acid) exhibits better cytotoxicity and lower hemotoxicity than free drug. *Nanotechnology, Science and Applications, 13*, 37.

Martínez-Álvarez, Ó., Calvo, M. M., & Gómez-Estaca, J. (2020). Recent advances in astaxanthin micro/nanoencapsulation to improve its stability and functionality as a food ingredient. *Marine Drugs, 18*(8), 406.

Matsuzaki, M., McCafferty, F., & Karel, M. (1989). The effect of cholesterol content of phospholipid vesicles on the encapsulation and acid resistance of β-galactosidase from E. coli. *International Journal of Food Science & Technology, 24*(4), 451–460.

Minten, I. J., Claessen, V. I., Blank, K., Rowan, A. E., Nolte, R. J., & Cornelissen, J. J. (2011). Catalytic capsids: The art of confinement. *Chemical Science, 2*(2), 358–362.

Monto, A. S., Ansaldi, F., Aspinall, R., McElhaney, J. E., Montaño, L. F., Nichol, K. L., … & Stephenson, I. (2009). Influenza control in the 21st century: Optimizing protection of older adults. *Vaccine, 27*(37), 5043–5053.

Ning, Y. M., He, K., Dagher, R., Sridhara, R., Farrell, A. T., Justice, R., & Pazdur, R. (2007). Liposomal doxorubicin in combination with bortezomib for relapsed or refractory multiple myeloma. *Oncology (Williston Park, NY), 21*(12), 1503–1508.

Novikov, B. N., Grimsley, J. K., Kern, R. J., Wild, J. R., & Wales, M. E. (2010). Improved pharmacokinetics and immunogenicity profile of organophosphorus hydrolase by chemical modification with polyethylene glycol. *Journal of Controlled Release, 146*(3), 318–325.

O'Neil, A., Prevelige, P. E., & Douglas, T. (2013). Stabilizing viral nano-reactors for nerve-agent degradation. *Biomaterials Science, 1*(8), 881–886.

Pateiro, M., Gómez, B., Munekata, P. E., Barba, F. J., Putnik, P., Kovačević, D. B., & Lorenzo, J. M. (2021). Nanoencapsulation of promising bioactive compounds to improve their absorption, stability, functionality and the appearance of the final food products. *Molecules, 26*(6), 1547.

Patterson, D. P., Prevelige, P. E., & Douglas, T. (2012a). Nanoreactors by programmed enzyme encapsulation inside the capsid of the bacteriophage P22. *ACS Nano, 6*(6), 5000–5009.

Patterson, D. P., Schwarz, B., El-Boubbou, K., van der Oost, J., Prevelige, P. E., & Douglas, T. (2012b). Virus-like particle nanoreactors: Programmed encapsulation of the thermostable CelB glycosidase inside the P22 capsid. *Soft Matter, 8*(39), 10158–10166.

Patterson, D. P., Schwarz, B., Waters, R. S., Gedeon, T., & Douglas, T. (2014). Encapsulation of an enzyme cascade within the bacteriophage P22 virus-like particle. *ACS Chemical Biology, 9*(2), 359–365.

Pei, L., Omburo, G., McGuinn, W. D., Petrikovics, I., Dave, K., Raushel, F. M., ... & Way, J. L. (1994). Encapsulation of phosphotriesterase within murine erythrocytes. *Toxicology and Applied Pharmacology, 124*(2), 296–301.

Pennington, M. E., Lam, K. S., & Cress, A. E. (1996). The use of a combinatorial library method to isolate human tumor cell adhesion peptides. *Molecular iversity, 2*(1), 19–28.

Petrikovics, I., Baskin, S. I., Beigel, K. M., Schapiro, B. J., Rockwood, G. A., Manage, A. B., ... & Szilasi, M. (2010). Nano-intercalated rhodanese in cyanide antagonism. *Nanotoxicology, 4*(2), 247–254.

Piard, J. C., El Soda, M., Alkhalaf, W., Rousseau, M., Desmazeaud, M., Vassal, L., & Gripon, J. C. (1986). Acceleration of cheese ripening with liposome-entrapped proteinase. *Biotechnology Letters, 8*(4), 241–246.

Pouton, C. W., & Akhtar, S. (1996). Biosynthetic polyhydroxyalkanoates and their potential in drug delivery. *Advanced Drug Delivery Reviews, 18*(2), 133–162.

Rothemund, P. W. (2006). Folding DNA to create nanoscale shapes and patterns. *Nature, 440*(7082), 297–302.

Sánchez-Sánchez, L., Cadena-Nava, R. D., Palomares, L. A., Ruiz-Garcia, J., Koay, M. S., Cornelissen, J. J., & Vazquez-Duhalt, R. (2014). Chemotherapy pro-drug activation by biocatalytic virus-like nanoparticles containing cytochrome P450. *Enzyme and Microbial Technology, 60*, 24–31.

Sánchez-Sánchez, L., Tapia-Moreno, A., Juarez-Moreno, K., Patterson, D. P., Cadena-Nava, R. D., Douglas, T., & Vazquez-Duhalt, R. (2015). Design of a VLP-nanovehicle for CYP450 enzymatic activity delivery. *Journal of Nanobiotechnology, 13*(1), 1–10.

Schaffert, D. H., Okholm, A. H., Sørensen, R. S., Nielsen, J. S., Tørring, T., Rosen, C. B., ... & Kjems, J. (2016). Intracellular delivery of a planar DNA origami structure by the transferrin-receptor internalization pathway. *Small, 12*(19), 2634–2640.

Schmid, A., Dordick, J. S., Hauer, B., Kiener, A., Wubbolts, M., & Witholt, B. (2001). Industrial biocatalysis today and tomorrow. *Nature, 409*(6817), 258–268.

Schumacher, I., Arad, A., & Margalit, R. (1999). Butyrylcholinesterase formulated in liposomes. *Biotechnology and Applied Biochemistry, 30*(3), 225–230.

Seebeck, F. P., Woycechowsky, K. J., Zhuang, W., Rabe, J. P., & Hilvert, D. (2006). A simple tagging system for protein encapsulation. *Journal of the American Chemical Society, 128*(14), 4516–4517.

Seeman, N. C. (1982). Nucleic acid junctions and lattices. *Journal of Theoretical Biology, 99*(2), 237–247.

Sendil, D., Gürsel, I., Wise, D. L., & Hasırcı, V. (1999). Antibiotic release from biodegradable PHBV microparticles. *Journal of Controlled Release, 59*(2), 207–217.

Smith, J. D., Morton, L. D., & Ulery, B. D. (2015). Nanoparticles as synthetic vaccines. *Current Opinion in Biotechnology, 34*, 217–224.

Strable, E., & Finn, M. G. (2009). Chemical modification of viruses and virus-like particles. In: Manchester, M., & Steinmetz, N. F. (Eds.), *Viruses and Nanotechnology*. Springer Science & Business Media: Berlin/Heidelberg, Germany, pp. 1–21.

Subramani, T., & Ganapathyswamy, H. (2020). An overview of liposomal nano-encapsulation techniques and its applications in food and nutraceutical. *Journal of Food Science and Technology*, *57*(10), 3545–3555.

Teller, C., & Willner, I. (2010). Organizing protein–DNA hybrids as nanostructures with programmed functionalities. *Trends in Biotechnology, 28*(12), 619–628.

Thiruvengadam, M., Rajakumar, G., & Chung, I. M. (2018). Nanotechnology: Current uses and future applications in the food industry. *3 Biotech*, *8*(1), 1–13.

Tomalia, D. & Esfand, R. (1997). Dendrons, dendrimers and dendrigrafts. *Chemistry and Industry*, *11*, 416–20.

Tomalia, D. A., Baker, H., Dewald, J., Hall, M., Kallos, G., Martin, S., … & Smith, P. (1985). A new class of polymers: Starburst-dendritic macromolecules. *Polymer Journal, 17*(1), 117–132.

van Beilen, J. B., & Li, Z. (2002). Enzyme technology: An overview. *Current Opinion in Biotechnology, 13*(4), 338–344.

Varpness, Z., Peters, J. W., Young, M., & Douglas, T. (2005). Biomimetic synthesis of a H2 catalyst using a protein cage architecture. *Nano Letters, 5*(11), 2306–2309.

Veronese, F. M., Monfardini, C., Caliceti, P., Schiavon, O., Scrawen, M. D., & Beer, D. (1996). Improvement of pharmacokinetic, immunological and stability properties of asparaginase by conjugation to linear and branched monomethoxy poly (ethylene glycol). *Journal of Controlled Release, 40*(3), 199–209.

Vriezema, D. M., Comellas Aragonès, M., Elemans, J. A., Cornelissen, J. J., Rowan, A. E., & Nolte, R. J. (2005). Self-assembled nanoreactors. *Chemical Reviews, 105*(4), 1445–1490.

Walde, P., & Ichikawa, S. (2001). Enzymes inside lipid vesicles: Preparation, reactivity and applications. *Biomolecular Engineering, 18*(4), 143–177.

Wang, D., Zhang, Y., Wang, M., Dong, Y., Zhou, C., Isbell, M. A., … & Liu, D. (2016). A switchable DNA origami nanochannel for regulating molecular transport at the nanometer scale. *Nanoscale, 8*(7), 3944–3948.

Way, J. L. (1984). Cyanide intoxication and its mechanism of antagonism. *Annual Review of Pharmacology and Toxicology, 24*(1), 451–481.

Xia, X., Hu, Z., Gao, J., Qin, D., Durst, H. D., & Yin, R. (2002). Light scattering study of the self-association behavior of long chain branched poly (2-ethyloxazoline) in solvents. *Langmuir, 18*(22), 8302–8308.

Yao, M., Xiao, H., & McClements, D. J. (2014). Delivery of lipophilic bioactives: Assembly, disassembly, and reassembly of lipid nanoparticles. *Annual Review of Food Science and Technology, 5*, 53–81.

Yin, R., Qin, D., Tomalia, D. A., KukowskaLatallo, J., & Baker, J. R. (1997). From dendrimers to dendrigrafts: Synthesis, characterization and applications. In: *Abstracts of Papers of the American Chemical Society*, vol. 214. AMER CHEMICAL SOC: Washington, DC, pp. 228-PMSE).

Zhang, J., Jia, G., Wanbin, Z., Minghao, J., Wei, Y., Hao, J., … & Sun, A. (2021). Nanoencapsulation of zeaxanthin extracted from *Lyciumbarbarum L.* by complex coacervation with gelatin and CMC. *Food Hydrocolloids, 112*, 106280.

Zhao, Z., Fu, J., Dhakal, S., Johnson-Buck, A., Liu, M., Zhang, T., … & Yan, H. (2016). Nanocaged enzymes with enhanced catalytic activity and increased stability against protease digestion. *Nature Communications, 7*(1), 1–9.

Zhou, C., Yang, Z., & Liu, D. (2012). Reversible regulation of protein binding affinity by a DNA machine. *Journal of the American Chemical Society, 134*(3), 1416–1418.

CHAPTER 13

Nanoencapsulated Probiotics and Prebiotics

Lisa F. M. Lee Nen That, Emmanuel Kyereh, Francisca Aba Ansah, and Jessica Pandohee

CONTENTS

DOI:10.1201/9781003259183-13

13.1 INTRODUCTION

Healthcare and well-being are a priority for young adults and the general population. There has been increasing popular trends in changing our current lifestyle through healthy eating, exercises, meditation/mindfulness, and reduction or avoidance of pollutants or contaminants among others. The adoption of those practices has been associated with increased immunity, cognition, physical, and mental health and mood. At a molecular level, there is evidence that a change in lifestyle affects the microbiota. The latter which is the ensemble of bacteria, viruses, archaea, and unicellular eukaryotes living in the human gastrointestinal tract is interconnected to several metabolic networks (Peterson et al. 2009; Bäckhed et al. 2012). The diversity of the microbiota and its particular population changes our health status through the production of metabolites that regulate pathways in the neural, inflammatory, or metabolic diseases (Bäckhed et al. 2012).

Nutrition and consumption of specific compounds is one of the lifestyle factors that is believed to have the greatest effect on the microbiota and therefore the fastest acting in evolving the microbiota into a healthy one (Kau et al. 2011; Salazar et al. 2017). Two main approaches have been studied in the last couple of decades; first is the introduction of probiotics followed by the introduction of prebiotics. Probiotics are live microorganisms that are administered in adequate amounts with the aim that they will in turn populate the host's gut and confer health benefits (FAO/WHO 2002). Prebiotics on the other hand are nondigestible compounds that are metabolized by microorganisms in the gut, produced by metabolites, absorbed by the host, and confer beneficial physiological effects (Slavin 2013). For these reasons, the consumption of prebiotics and probiotics has been encouraged in the society. These include vegetables, legumes, fruits, cereals, nuts, and seeds as prebiotics and fermented food such as yogurt, sauerkraut, kimchi, miso, or kefir as probiotics (Fontana et al. 2013; Davani-Davari et al. 2019).

The use of prebiotics and probiotics in medicine and nutritional intervention studies requires adequate and controlled release of the entities being studied in order to ensure that the desired effect is achieved or any change in health is due to the supplementation administrated (Razavi et al. 2021). For example, it has been recommended that food consisting of probiotic bacteria should contain at least 10^6 of live microorganism per gram or milliliter at the time of consumption in order to have a therapeutic effect (FAO/WHO 2002; Chotiko and Sathivel 2014; Šipailienė and Petraitytė 2018; Charalampopoulos and Pandiella 2010; Kyereh and Sathivel 2021).

Nanoencapsulation is a technique that can advance the use of prebiotics and probiotics immensely in medicine and in the treatment of noncommunicable diseases. It is a technology used to surround active ingredients such as prebiotics or probiotics with a functional barrier resulting in a nano-sized particle that is protected from the surrounding environment (Pateiro et al. 2021). In the case of probiotics, nanoencapsulation prevents probiotic damage, improves survival of the microorganisms, and ensures site delivery. In order for nanoencapsulation to be used for the delivery of prebiotics and probiotics in medicine, extensive research is underway to understand the safety, efficacy, and viability of various nanoencapsulation techniques.

This chapter discusses the most recent development in nanoencapsulated prebiotics and probiotics research. In particular, we discuss the viability of encapsulated entities, the choice of carriers used, the biocompatibility of nanoparticles, and the applications of nanoencapsulated prebiotics and probiotics in human health.

13.2 VIABILITY OF PREBIOTICS AND PROBIOTICS FOR ENCAPSULATION

As a specific amount of probiotic bacteria (10^6 cfu/g) incorporated in food is required to have a beneficial health effect, the encapsulation method is crucial in delivering an adequate amount of these live bacteria cells in the gastrointestinal tract (Gao et al. 2021). Probiotics strain selection is also a critical step in ensuring viability of the bacteria during encapsulation. Each probiotic strain has different physical and environmental needs that require tailor-made encapsulating method. For example, *Lactobacillus plantarum* is a facultative anaerobic bacterium that grows at a temperature of 15°C–30°C and at a pH of 5–7. Several factors have to be considered when choosing the probiotic strain for encapsulation and they consist of tolerance to heat, oxygen, and acidic conditions; ability to attach and colonize the epithelium; stimulation of immune response by the host; antagonistic behavior toward pathogens; and improvement of host digestion (Šipailienė and Petraitytė 2018).

13.2.1 Factors Affecting Viability during Nanoencapsulation

The therapeutic effect of probiotics depends on a number of viable and active cells in the food product at the moment of consumption. In stress conditions, bacteria may be in a state where they are viable but not culturable (Davis 2014). Therefore, it is also important to ensure that from the process of encapsulation to the delivery, the viable bacteria can further multiply as they adapt to the environment of the host. There are many factors that may affect the viability of probiotics during nanoencapsulation.

13.2.2 Encapsulating Material

The health effect conferred by probiotic bacteria may be positively affected by the material used for encapsulation which makes the interaction of the bacteria with the coating material an important factor to consider for optimal production of nanoencapsulated probiotics. Other important uses of the coating material include protecting the probiotics in extreme conditions such as in stomach and releasing their content at their destination (Atraki and Azizkhani 2021; Razavi et al. 2021). In a study by Sathyabama et al. (2014), *Staphylococcus succinus* (MAbB4) and *Enterococcus fecium* (FIdM3) were co-encapsulated with prebiotics on an alginate matrix. It was reported that the encapsulation enhanced the survival and viability (about 88.75%–98.75%) of both strains in a simulated gastric environment (Sathyabama et al. 2014). The encapsulating material matrix protected the bacteria and reduced loss of cell viability as compared to the free cells. The encapsulation material namely sugarbeet and chicory were suitable as prebiotics with their fermentation rate and effective in co-encapsulation with probiotics in the alginate matrix (Sathyabama et al. 2014). The addition of carrageenan in the coating material used to encapsulate *Bifidobacterium animalis* subsp. *lactis* enhanced both viability and survival after a storage period of 30 days (Frakolaki et al. 2021). Inulin which is another prebiotic used in encapsulating probiotics also increased the rate of survival of *B. animalis* subsp. *lactis*. Other studies have demonstrated that *Lactobacillus acidophilus* La-05 was encapsulated using flaxseed mucilage (FM) and flaxseed soluble protein (FSP) by spray dry technology. Bacteria cell viability for the encapsulated cells with FM and FSP after incubation in simulated gastric juice showed a stable cell count after 4 hours and reaching relative viability of almost 1.20, whereas the free cells experienced a sharp decline (0.27) after 4 hours (Bustamante et al. 2015). Dosage formulation is also an important factor in probiotics cell viability. An increase in the concentration of gum Arabic led to a rise in moisture content, encapsulation efficiency, porosity, and reduced water activity (Arepally and Goswami 2019).

13.2.3 Temperature

Temperature is a crucial factor that needs to be considered to ensure viability of probiotics for encapsulation. Probiotic strains may be sensitive to temperature at varying degrees (Šipailienė and Petraitytė 2018). Moreover, the methods used to encapsulate probiotics may require exposure to extreme temperatures such as in freeze drying, spray drying, and air-suspension coating (Vidhyalakshmi, Bhakyaraj, and Subhasree 2009). This could lead to disruption of the cell membrane and leakage of its content (Sipailienė and Petraitytė 2018; Smelt and Brul 2014). Proteins such as α- and β-subunits of Ribonucleic acid (RNA) polymerase which are also heat-sensitive unfold leading to the death of the microorganism (Šipailienė and Petraitytė 2018). A study looking at the encapsulation of *Lactobacillus reuteri* in whey revealed that the temperatures (55°C and 65°C) used for spray drying resulted in a loss of viability from 1.6×10^9 to 2.5×10^7 cfu/g in the bacteria (Jantzen, Göpel, and Beermann 2013). Arepally and Goswami (2019) reported that the drying process, especially inlet temperature affected the viability of probiotics. Indeed as inlet temperature increased, there was a decline in the encapsulation efficiency. Initially, the viable count ranged from 10.81 to 11.36 log cfu/g and a gradual decrease was observed from 9.97 to 7.3 log cfu/g. The membrane rupture and denaturation of proteins may have caused this loss in viability. The optimal inlet temperature was determined to be 150°C.

13.2.4 Oxygen and pH

Oxygen levels have to be considered when determining bacterial viability especially for anaerobic bacteria (Šipailienė and Petraitytė 2018). Indeed, oxygen toxicity can become an issue in probiotics viability (Talwalkar and Kailasapathy 2004). To solve this problem, anaerobic bacteria such as *Bifidobacterium* spp. may be co-cultured with a strain which consumes high levels of oxygen such as *Streptococcus thermophilus* (Lourens-Hattingh and Viljoen 2001). Thus, encapsulated anaerobic bacteria are protected from lyophilization and rehydration. Moreover, alginate which is used as encapsulating material inhibits oxygen diffusion and creates an anoxic environment for the bacteria (Talwalkar and Kailasapathy 2004). Studies have shown that *Bifidobacteria* have lower tolerance to oxygen than *Lactobacilli* while oxygen level has no impact on other facultative anaerobes such as *Bacillus coagulans* (Smelt and Brul 2014; Avila-Reyes et al. 2014; Sarkar 2010). *L. acidophilus*, a microaerophile, and two strains of *Bifidobacterium* spp. were encapsulated in calcium alginate and grown in aerobic conditions in skim milk for 24 hours (Talwalkar and Kailasapathy 2003). The encapsulated bacteria had higher cell viability than free bacteria suggesting that encapsulation protected them from oxygen toxicity.

pH is also of critical importance in bacterial viability. To preserve the functionality and integrity of cells, bacterial cytoplasm has to be kept at a stable pH during the process of encapsulation (Šipailienė and Petraitytė 2018). Once ingested, probiotic bacteria have to travel through the alimentary system where pH levels vary (Barajas-Álvarez, González-Ávila, and Espinosa-Andrews 2021; Yoha et al. 2022). The pH in the mouth, small and large intestines is 6–7 while the stomach has a pH of 1–3, which could be detrimental to the viability of bacteria. The shift from alkaline to acidic environments may be stressful and harmful for probiotic bacteria (Padan et al. 2005). Several studies have looked at the effect of encapsulated bacteria in simulated gastrointestinal conditions (Barajas-Álvarez, González-Ávila, and Espinosa-Andrews 2021; Zhang et al. 2021). Fayed et al. (2018) demonstrated that the encapsulation of inulin nanoparticles and *Bifidobacterium* spp. with gum Arabic and alginate protected the bacteria and 88.29% of the initial population survived the simulated gastric conditions.

13.2.5 Moisture Content and Water Activity

Moisture content and water activity may also have an impact on the viability of bacteria and are used as a measure quality of food product, especially during long periods of storage (Wang et al. 2020). The water activity for bacteria has a lower limit of 0.6 and a higher value could be detrimental to the viability of probiotic bacteria (Huang et al. 2017). The mixture of dried whole milk powder and *Lactobacillus paracasei* was stored at 25°C at stable water activity (a_w=0.11, 0.33 and 0.52) for long periods (105 days) and it was determined that the lowest water activity resulted in minimal loss of cell viability (Poddar et al. 2014). The hygroscopic properties of the coating material may also affect bacterial viability. It is dependent upon the presence of hydrophilic molecules in the material, which will bind to water molecules in the air (Šipailienė and Petraitytė 2018). Y. Wang et al. (2020) demonstrated that a combination of octenyl-succinic anhydride, chitosan, and inulin as encapsulating material had high hygroscopicity and remained stable although the humidity levels were high and no morphological changes in the particles were observed. Inulin was suitable as a drying aid agent and the combination of coating material had high threshold of X_m value which is defined as the limit for essential moisture and where dried ingredients have better stability.

13.3 CHOICE OF CARRIERS

The purpose of nanoencapsulating prebiotics and probiotics is to preserve their viability during processing, handling, storage, to enhance their nutritive value in food products, and to facilitate their passage, targeted release, and uptake in the human gut. This can be achieved using nanocarrier materials. However, probiotics have limited survival rates, which is attributable to environmental stress (Terpou et al. 2019; Yoha et al. 2022). As a result, the choice of nanocarrier materials should depend on their ability to withstand environmental stress such as dehydration in dry food products; temperature and oxidation during processing; moisture, heat, and oxygen during packaging and storage; and low survival rate in the stomach due to low pH (Šipailienė and Petraitytė 2018; Yoha et al. 2022).

Nanocarriers for encapsulation are usually made from several natural polymers, such as celluloses, proteins, lipids, and gums or synthetic materials. Recent studies on nanocarriers have shown their successful use in encapsulating probiotics against harsh environmental conditions due to their specific chemical and physical properties (Razavi et al. 2021). It is critical to also note that nanocarriers with poor encapsulation stability can cause untimely release of substances like probiotics, limiting their use at targeted locations (Palanikumar et al. 2018). A variety of nanocarriers and their uses are summarized in Figure 13.1 and reviewed below.

13.3.1 Polymeric Nanoparticles and Nanofibers

13.3.1.1 Nanocellulose

These are cellulose nanostructures which come from various natural sources such as plants, sea invertebrate animals, and microorganisms like bacteria and algae (Osorio et al. 2018; Shak , Pang, and Mah 2018). The types of nanocellulose carriers available include hairy nanocellulose, bacterial nanocellulose, cellulose nanocrystals, and cellulose nanofibrils (Sheikhi et al. 2019). The use of nanocellulose as carriers for encapsulating probiotics has been appreciated in recent years due to its ability to improve

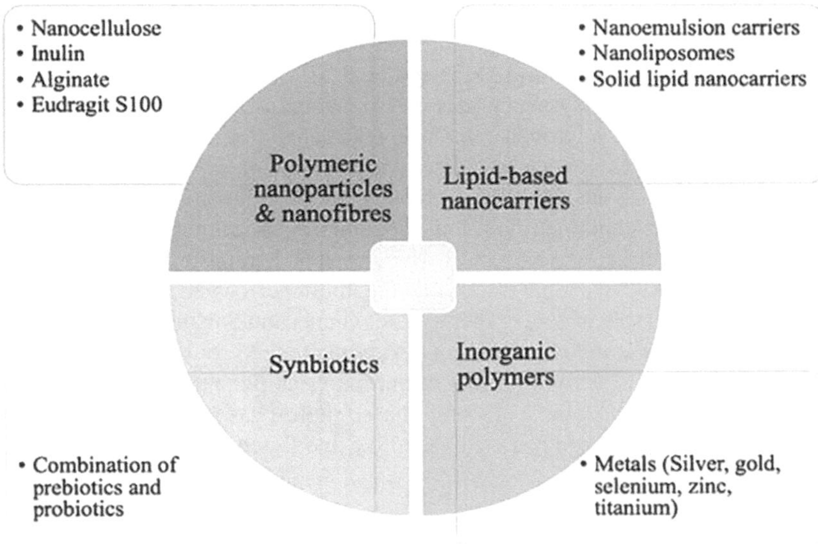

Figure 13.1 Choice of carriers used for encapsulation of prebiotics and probiotics.

the viability of probiotics during processing, storage, and gastrointestinal digestion. S. Razavi et al. (2021) has reported that nanocellulose is effective in cryoprotection, as it can be absorbed and viscous layer can be created on bacteria. This protects them against the development of ice crystals during freeze drying. Similarly, bacterial nanocellulose as a carrier was able to maintain the viability of the starter probiotic, *L. plantarum* during storage for 9 days at 4°C (Yordshahi et al. 2020). Furthermore, bacterial nanocellulose, which was impregnated with *L. plantarum* and used to wrap ground meat caused a significant reduction in *Listeria monocytogenes* (~5 log cycles). The use of pectin-bacterial nanocellulose as encapsulation material has also enhanced the viability of probiotic bacteria for long periods of storage at different temperatures (Khorasani and Shojaosadati 2016). *L. plantarum* was encapsulated with sodium alginate and cellulose nanofiber, and in acidic conditions, such as the gastric fluid, nanocellulose forms strong, stable bonds with sodium alginate, which demonstrates the pH responsive nature of nanocellulose and ensures delivery at its intended destination (S. Razavi et al. 2021). Despite the exceptional characteristics and the extensive research carried out on nanocellulose, it has some drawbacks that limit its commercial use. This includes expensive nanomaterials, lack of standards and safety regulations, and limited access to characterization capacities and skills (Balea et al. 2021).

13.3.1.2 Inulin Nanoparticles

This is another group of nanocarriers which has proven to be effective for the delivery of probiotics in the gastrointestinal tract. Not only can it deliver probiotics, but inulin is also a prebiotic (mostly dietary fiber) providing double health benefits when ingested. Inulin is a carbohydrate commonly used for industrial purposes because it can be produced in large quantities from food sources such as Jerusalem artichokes, garlic, chicory, banana, and grains (Mensink et al. 2015). It has superior carrier properties over other natural polysaccharides due to its organ-targeting ability, controlled release of biomolecules, improved cellular uptake capabilities, stabilization, and protective effects as well as flexible structure (Afinjuomo et al. 2021). Inulin plays a primary role in enhancing the ability of probiotics to inhibit infections (Cui et al. 2018). When curcumin was combined with inulin nanoparticles (curcumin-inulin nanoparticles), a combined effect was observed whereby 90% curcumin was released at the duodenum of the gastrointestinal tract while the inulin performed its prebiotic activity providing a potential role in cancer therapeutics (Fares and Salem 2015). Curcumin is an anti-inflammatory compound that can also promote the growth of probiotic bacteria (Yazdi et al. 2019). Due to the ability of inulin to stabilize probiotics, whey protein concentrate – inulin nanocomplex (protein-based inulin) has also been used as a probiotic delivery system in food products (Krithika and Preetha 2019).

13.3.1.3 Alginate

Alginate is a well-known polymer in nanomaterial studies. It is versatile, nontoxic, mucoadhesive, biocompatible, and economical. Owing to these characteristics, it is commonly used in the encapsulation of probiotics (Martau et al. 2019). Alginates are naturally occurring compounds derived from algae. Application of alginate as a carrier in encapsulation shows that although it may not provide optimum protection for probiotics during oral delivery, it may be effective in maintaining the viability of cells to some extent (Martoni et al. 2007). However, alginate in combination with other compounds has proven to be more effective. Typically, alginate chitosan was used to encapsulate *Escherichia coli* strain Nissle 1917 probiotic, and this led to a decrease in the growth *Campylobacter jejuni* in humans and poultry (Mawad et al. 2018).

13.3.1.4 Eudragit (Eu) S100

It is an anionic (negatively charged) copolymer, made from methyl methacrylate and metacrylic acid in the ratio of 2:1. It is a nontoxic, food-grade polymer but pH dependent; it is soluble under neutral to alkaline conditions but not soluble in acid (Gibson et al. 2006; Ansari et al. 2017). It has been used to encapsulate probiotics, thus leading to delivery and targeted release of probiotics in the gastrointestinal tract. Eu S100, chitosan, and calcium alginate were used to create a double coating to encapsulate Lactobacilli probiotics. This protected the probiotics in acidic conditions and allowed their release in the colon zone which is the optimal place for probiotics to function (Pourjafar et al. 2020).

13.3.2 Lipid-Based Nanocarriers

Prebiotic and probiotic substances that are water sensitive are usually encapsulated using lipid-based nanocarriers. These nanocarriers exhibit superior characteristics by providing barriers against acid, moisture, and oxygen (Reque and Brandelli

2021). Lipid-based nanocarriers have been found to be most applicable for use in the food industry as they do not dissolve and are derived from food products such as fatty acids, triglycerides, phospholipids, waxes, and cholesterol, which are also biodegradable and compatible with living tissues (Severino et al. 2012; Dabholkar et al. 2021). Lipid nanocarriers comprise nanoemulsions, nanoliposomes, solid lipid nanoparticles, nanostructured lipid carriers, and many others (Yousefi, Ehsani, and Jafari 2019).

13.3.2.1 Nanoemulsion Carriers

Nanoemulsion carriers have been in existence since the 1950s. They are nano-sized emulsions (droplet size 20–200 nm), composed of thermodynamically stable isotropic systems, in which emulsifying agents such as surfactants and co-surfactants are used to combine two immiscible liquids into a single phase (Jaiswal, Dudhe, and Sharma 2015). In the food industry, food proteins like milk protein and soy protein isolates are used as emulsifiers (Vaishanavi and Preetha 2021). Nanoemulsion carriers have several benefits over common emulsions, such as high surface area, relatively high physical stability, better bioavailability, and transparent appearance (Jin et al. 2016; Kumar et al. 2019). The structure and composition of nanoemulsions can be optimized for the encapsulation of lipophilic bioactive compounds and allow a successful delivery of these compounds (Aswathanarayan and Vittal 2019). Nanoemulsion carriers protect probiotics during storage until consumption. Vaishanavi and Preetha (2021) reported that they can stabilize probiotics during storage. *Lactobacillus delbrueckii* subsp. *bulgaricus* encapsulated in soy protein nanoemulsion was stable and viable for a storage period of 40 days. Currently, the use of nanoemulsions in the food industry is very limited due to a number of regulatory aspects that need to be addressed (Wilson 2020).

13.3.2.2 Nanoliposomes

Nano-sized liposomes are submicrometric double layer vesicles composed of phospholipid molecules and are used for the encapsulation of a wide range of substances, including flavors, vitamins, essential oils, amino acids, minerals, probiotics, antioxidants, redox agents, antimicrobials, and colorants (Mohammadi et al. 2014; Danaei et al. 2018). These nanocarriers have also been widely studied in probiotic drug delivery systems for stimulating the immune system (Ghaffar et al. 2014). This is due to their distinct ability to encapsulate both hydrophilic and hydrophobic substances (Siler-Marinkovic et al. 1997). In addition, encapsulation by liposomes has proven to be effective through its ability to accumulate at target sites, minimize toxicity, and secure outer surfaces of encapsulated substances from degradation (Ferreira et al. 2021). Amnuaikit et al. (2019) reported that the lyophilized cell-free supernatant of the bacteria *L. paracasei* SD1 and *Lactobacillus rhamnosus* SD11 is considered toxic. When they were encapsulated by liposomes, they became nontoxic.

Moreover, a study by Taylor et al. (2007) demonstrated that when nisin was encapsulated in nanoliposomes made with distearoyl phosphatidylcholine and distearoyl phosphatidylglycerol, the nanoliposomes maintained high stability with about 70%–90% of the enclosed nisin regardless of its exposure to high temperatures of 25°C–75°C and varying pH suggesting that nanoliposomes are highly stable. Although small interfering RNA (siRNA) have been considered as a potential drug in gene therapy, they have poor cellular uptake and degrade quickly (Morrissey et al. 2005; Lorenzer et al. 2015). Liposomes were used to protect them and facilitate their targeted delivery in tissues. Mokhtarieh et al. (2018) also showed that the encapsulated siRNAs were protected from degradation by RNase under refrigerated conditions for more than a month. Despite the benefits of nanoliposomes for encapsulation, it is still a difficult task to use liposomes in the encapsulation of probiotics due to the large size of probiotics (Sarao and Arora 2015). Also, it still requires human clinical trials, extensive toxicity risk analysis, more cost-effective production techniques as well as regulations for accountable management, storage, and waste disposal after production to ensure its scalability and transfer from laboratory into a commercial product (Aguilar-Pérez et al. 2020).

13.3.2.3 Lipid Nanoparticles

These lipid nanocarriers were created in the 1990s with the objective of having the combined benefits of polymer nanocarriers like controlled drug delivery, effective drug loading capacity, as well as biocompatibility of lipid emulsions and good bioavailability, and as a substitute carrier system for liposomes (Müller, Lippacher, and Gohla 2000a; Pardeike, Hommoss, and Müller 2009). They address some challenges in using polymeric nanocarriers, in terms of their availability for regulatory approval. Solid lipid nanoparticles (SLNs) are submicron colloidal carriers made up of physiological lipid, distributed in water and in an aqueous surfactant solution (Mukherjee, Ray, and Thakur 2009). These nanocarriers have the ability to keep particle structures and allow the controlled release of desired properties in encapsulated compounds, which may be impossible for emulsions with liquid droplets (Zhong and Zhang 2019). In comparison to nanoliposomes, SLNs have superior qualities including large-scale production potential, release of bioactive substances after a long period of time, better encapsulation efficiency of lipophilic compounds, and lesser degradation rate (Müller, Mäder, and Gohla 2000b; Fathi, Mozafari, and Mohebbi 2012).

Lipid-based nanoparticles can be categorized into solid lipid nanoparticles (SLNs) and nanostructured lipid carriers (NLCs) (Naseri, Valizadeh, and Zakeri-Milani 2015). They are produced through high or cold pressure homogenization methods and double emulsions by ultrasonication (Pozo-Rodríguez, Solinís, and Rodríguez-Gascón 2016; Mehnert and Mäder 2012). Encapsulation of drugs including probiotics into SLNs and NLCs for different aspects of drug targeting has good potential in the cosmetic, gene transfer, and the food industry (Shastri 2017). Applications of SLNs in the food industry are promising. In a test by Jo, Choi, and Kwon (2015), the viability of encapsulated *L. plantarum* in palm or coconut SLNs was greater in ultra-high temperature milk than in distilled water. The authors suggested that the production of SLNs containing palm or coconut can be used in *Lactobacilli*-fortified beverages. NLCs can be used as a nanoencapsulating material where SLNs may not be possible. The less ordered solid lipid matrix resulting in liquid lipid inner phase of NLCs enhances loading capacity, lowers drug leakage during storage, makes it possible to capture active ingredients which are better solubilized in liquid lipid, and improves release properties (Pyo, Müller, and Keck 2017). NLCs by themselves have also shown bactericidal effects without altering important microbiota (*Lactobacillus*, *E. coli*, *S. epidermidis*, and *S. aureus*) in the gut. Indeed, the NLCs had a bactericidal effect on *Helicobacter pylori*, a harmful bacterium which can cause gastric cancer at low concentrations (Seabra et al. 2018).

13.3.3 Inorganic Nanoparticles

In recent years, nanotechnology has been used in different fields such as food, biomedicine, materials, and chemistry due to the advantages that nanoparticles provide such as nanosize, high surface-to-volume ratio, and optical, catalytic, and magnetic characteristics (Ladj et al. 2013; Wang et al. 2021). In the biomedical sector, they are used in drug delivery as they are nontoxic, robust, low-cost to produce, and have high-loading capacity (Baeza, Ruiz-Molina, and Vallet-Regí 2016). Some examples of inorganic nanoparticles include silver, gold, platinum, copper, and zinc oxide (Chandrakala, Aruna, and Angajala 2022). These nanoparticles act as carriers and transport their content to their targeted delivery destination with reduced risk of side effects. Food-grade nanoparticles have also been developed as food additives, food packaging material, and antimicrobial agents to improve shelf life and add nutritive value (McClements and Xiao 2017). Silver, iron oxide, titanium dioxide, silicon dioxide, and zinc oxide are the most common inorganic nanoparticles used.

Food-grade nanoparticles have been tested for their effect on probiotics (Mikiciuk et al. 2016; Khan et al. 2019; Wang et al. 2021). They have had antimicrobial activity against certain probiotic strains such as *L. plantarum* and *Lactobacillus fermentum* as the nanoparticles inhibited growth by disrupting the cell membrane (Wang et al. 2021). However, probiotics have also demonstrated their ability to synthesize their own inorganic nanoparticles. As the different conditions and processes involved in the synthesis of inorganic nanoparticles have had a detrimental effect on the environment, there is a need to find other environment-friendly alternatives (Saravanan et al. 2021). Silver and gold nanoparticles are the main nanoparticles that have been synthesized by probiotic bacteria. Mousavi et al. (2020) reported the synthesis of silver nanoparticles from *L. rhamnosus*, which was also tested for its anticancer activities. The nanoparticles inhibited the growth of colorectal cancer cells. Gold nanoparticles possessed high antioxidant activity when they were synthesized from *Lactobacillus kimchicus* highlighting their beneficial health effect (Markus et al. 2016).

13.3.4 Synbiotics

Probiotics along with prebiotics are termed "synbiotics" that improve the survivability of probiotics, while stimulating the growth of specific native strains present in the Gastrointestinal (GI) tract through a synergistic approach for the benefit of the host (Swanson et al. 2020; Yoha et al. 2022). Encapsulation of probiotics has been adopted as an approach to improve the survival of probiotics in food and the gut (Rashidinejad et al. 2020). Co-encapsulation of prebiotics and probiotics by spray drying is on the rise and has improved stability and viability of probiotics immensely (Pinto et al. 2019; Colin-Cruz et al. 2019; Pinto et al. 2019; Tao et al. 2019). Atia et al. (2016) have reported on the development of an oral-synbiotic supplement whereby inulin was added to the alginate beads and used to encapsulate and protect three probiotic strains, namely, *Pediococcus acidilactici*, *L. salivarius*, and *L. reuteri*. The addition of inulin did not affect the antimicrobial activity of the probiotics and offered more protection than without inulin.

13.4 BIOCOMPATIBILITY OF NANOPARTICLES

Biocompatibility refers to the interactions (whether beneficial or harmful) of a foreign entity in the human body. The concept was introduced in the field of medicine; in particular, when a medical device was implanted in a patient's body, the term biocompatibility conveyed feedback of the host response to the new implants. Ideally, a biocompatible material or object will have the following characteristics. Firstly, to be described as biocompatible, it means that the foreign entity has the ability to perform its function once it has reached its target site. Secondly, biocompatible materials will cause an effect that is related to its purpose in the body. Finally, the nature of the host's reaction to the foreign object and its suitability might differ for other patients. In the field of nanoencapsulation, all these criteria are expected to be pertinent. Thus, for nanoencapsulated prebiotics and probiotics to be considered biocompatible, they will perform their intended function in the hosts without causing side effects such as toxic, immunogenic, thrombogenic, and carcinogenic reactions.

Copper nanoparticles have been shown to be highly compatible with probiotic strains of *Lactobacillus, Enterococcus*, and *Enterobacterium*. While reports of copper nanoparticles biotoxicity were as low as 30 to 15 µg/mL for *Lactobacillus* and 30 µg/mL for *Bifidobacterium*, the behavior of the nanoparticles in humans was not investigated (Sizentsov et al. 2018). Inorganic nanoparticles, on the other hand, act as antibacterial agents, while they are useful in preventing food spoilage by fighting off harmful bacteria, inorganic nanoparticles in the host body would destroy bacterial populations in the gut microbiota unselectively. M. Wang et al. (2021) showed that inorganic nanoparticles, such as SiO_2, Ag, and ZnO nanoparticles, bind to the surfaces of *L. plantarum* and *L. fermentum* and inhibit the reproduction of two *Lactobacilli* in a dose-dependent manner. In other words, careful considerations are needed before inorganic nanoparticles are used widely at high concentration in food products. Moreover, this emphasizes the need for further research to identify safe levels for the use of inorganic nanoparticles in the food sector (Fröhlich and Fröhlich 2016).

Polymer-based nanospheres such as poly(lactic-co-glycolic acid) (PLGA) induce subjectively mild tissue reaction. PLGA nanoparticles that have a rapid clearance from the body do not usually cause peritoneal adhesion, whereas PLFA microparticles that stay longer in peritoneal cavity do cause peritoneal adhesion. Lastly biocompatibility is a relative matter that depends on the risk–benefit ratio, and relies on a subjective declaration since, in general, inflammation would totally vanish over time, and the neighboring tissues do not exemplify a good proof of damage.

13.5 FOOD APPLICATIONS OF NANOENCAPSULATED PREBIOTICS AND PROBIOTICS

The beneficial effects of encapsulated prebiotics and probiotics make it a good supplement to food products. There is a growing market for functional foods as they have improved nutritive value. In fact, many studies have focused on developing novel foods where nanoencapsulated prebiotics and probiotics have been incorporated. Furthermore, studies have demonstrated the direct health effect that they have on humans.

13.5.1 Dairy Products

Nanoencapsulated prebiotics and probiotics have been widely incorporated in dairy products. The International Dairy Federation recommends a minimum concentration of probiotic bacteria of 10^7 cfu/g or mL of product or $\geq 10^8$–10^9 cfu per daily portion of the final product to experience health benefits (Frakolaki et al. 2021).

Frakolaki et al. (2021) reported that *B. animalis* subsp. *lactis* was encapsulated by extrusion and emulsification methods using formulations containing sodium alginate, glycerol, carrageenan, and inulin. Encapsulated probiotic bacteria experienced a lower drop in viability from 9.5 to ~6 log cfu/g compared to free probiotic bacteria (from 9.5 to 4.6 log cfu/g) in yogurt manufactured with whole fat (3.5% w/v). Moreover, probiotic bacteria encapsulated by the extrusion method provided significantly better protection than the emulsification method as the former method allowed for slower diffusion of acids and hydrogen peroxide due to bigger bead size. Probiotic cells encapsulated with the addition of carrageenan (5.9–6.9 log cfu/g after 30 days of storage) had better viability and survival rate in both encapsulation methods after 30 days of storage. Inulin also improved the survival rate of probiotics. Encapsulated *B. animalis* subsp. *lactis* also led to a reduction in the fermentation time compared to conventional methods. Firmness, adhesiveness, elasticity, and gumminess of yogurt were significantly improved with the addition of the encapsulated bacteria, which could be due to the possible interaction of hydrocolloids, namely, alginate and carrageenan with milk proteins.

Nanoencapsulation of *L. paracasei* KS-199 within alginate-based electrospun nanofiber mats was successfully produced and incorporated in kefir (Yilmaz et al. 2020). The high melting temperature of the electrospun nanofiber mats could provide a possible application in foods that require stability at high temperatures (more than 220°C) such as heat-processed food and bakery foods. When free and encapsulated *L. paracasei* KS-199 were exposed to simulated gastrointestinal conditions, the viability of the bacteria decreased by 3.51 and 2.49 log cfu/g, suggesting that the alginate polymer had a protective effect and increased survival of the bacteria as it passes through the stomach to the intestine for colonization. The viability of free and encapsulated *L. paracasei* KS-199 in kefir samples was also tested and it was determined that encapsulated bacteria with alginate polymer had an increase in viability (~0.7 log).

Zinc nanoparticles (ZnO-NPs) were synthesized from Lactobacillus strains and incorporated in yogurt as fortified food (El-Sayed et al., 2021). Among the *Lactobacteria* strains tested for biosynthesis of ZnO-NPs (*L. plantarum*, *L. gasseri*, *L. rhamnosus*, *L. salivarius*), *L. gasseri* had the highest quantity of ZnO-NPs (164 mg/g).

The mean size of ZnO-NPs was 282 nm. Antimicrobial activity of ZnO-NPs (2–3 mg) had an inhibition zone ranging from 25 to 37 nm. The size and positive charge of ZnO-NPs allow them to enter bacteria creating damage and leading to cell death. The bacteria tested were *Staphylococcus aureus*, *Bacillus cereus*, *L. monocytogenes*, *Candida albicans*, *E. coli*, *Salmonella typhimurium*, *Yersinia enterocolitica*, and *Aspergillus niger*. Counts for mold and yeast were below detectable limits for yogurt supplemented with 40 and 60 mg ZnO-NPs during the first 2 weeks of storage. At the third and fourth weeks of storage, the count for mold and yeast ranged from 1.9 to 2.83 log cfu/g. There was also a significant decrease in the counts of mesophilic bacteria, *L. acidophilus*, *S. thermophiles*, and *Lactobacillus bulgaricus* in yogurt fortified with ZnO-NPs and untreated yogurt over a 4-week storage. The population count was lowest in the highest concentration of ZnO-NPs in yogurt after 4 weeks of storage for all bacteria. Sensory tests including color, odor, and appearance reveal that the scores for overall acceptability for control yogurt dropped (from 4.5 to 3.5) after 2 weeks while fortified yogurt with 40 and 60 mg of ZnO-NPs remained constant at approximately 4–4.5 for 3 weeks and decreased to 3 at the fourth week. Addition of ZnO-NPs to yogurt can be a source of zinc for people who suffer from zinc deficiency.

Malheiros et al. (2016) described the effect of encapsulated bacteriocins produced by *Lactobacillus sakei* subsp. *sakei* 2a in goat milk against *L. monocytogenes*. The bacteriocin was encapsulated in nanovesicles from soybean phosphatidylcholine and 1,2 dioleoyloxy-3-trimethylammonium-propane. After 24 hours, encapsulated bacteriocins were significantly more effective in reducing the growth of *L. monocytogenes* than free bacteriocins in Brain heart infusion (BHI) medium. In Ultra-high temperature (UHT) goat milk, antimicrobial activity of both free and encapsulated bacteriocins was effective during 5 days of storage at 7°C as growth of *L. monocytogenes* was suppressed with a 5-log difference from control. The 5 days of storage usually correspond to the recommended period of consumption for the UHT goat milk. Beyond the 5 days of storage, growth of *L. monocytogenes* increased indicating the possible adaptation of bacteria to the bacteriocin. Goat milk is usually considered an alternative for people allergic to cow milk.

In a study by Jiao et al. (2020), CAMT2 produced by probiotics *Bacillus amyloliquefaciens* was encapsulated in nanovesicles prepared from soybean phosphatidylcholine (Woldemariam Yohannes et al. 2020). Antimicrobial activity was tested in skim and whole milk at 4°C and it was reported that in skim milk free CAMT2 and encapsulated CAMT2 inhibited the growth of *L. monocytogenes* at day 2 and 6, respectively. However, in whole milk complete inhibition was observed for encapsulated CAMT2 at day 8, while treatment with free CAMT2 had the lowest growth inhibition at ~5×10^2 CFU/mL after 2 days before increasing to the original bacterial concentration. Free CAMT2 molecules could be sticking to the lipid surface, leading to reduced concentration of CAMT2 being utilized for their antimicrobial activity.

13.5.2 Meat Products

Nanoencapsulated prebiotics and probiotics have been incorporated in fish and meat and their effect has been evaluated. Nisin was encapsulated in alginate chitosan nanoparticles and its antimicrobial activity was tested against *L. monocytogenes* which was present in lean beef (Zimet et al. 2018). The encapsulation efficiency was at 36.1%±0.6% and the size of the nanoparticles was 60–70 nm. Nanoencapsulated nisin (60%±9%) had an enhanced antimicrobial effect on *L. monocytogenes* compared to free nisin (43%± 2%) at a concentration of 500 IU/mL after 48 hours at 37°C. Although *L. monocytogenes* is able to grow at a low temperature of 4°C in lean meat, both encapsulated and free nisin inhibited bacterial growth at a concentration of 125 IU/mL. The antibacterial activity of nisin was more potent at 4°C. The nanoencapsulation of nisin provided a controlled delivery of nisin while maintaining the release of nisin for 21 days. *L. monocytogenes* and either encapsulated or free nisin were added to lean beef during storage at 4°C, and it was demonstrated that nisin in both free and encapsulated forms had a prompt antimicrobial effect. The encapsulated nisin at the highest concentration (800 IU/g) over a period of 24 days led to the highest inhibitory effect on *L. monocytogenes*. This demonstrates the anti-listerial effect of nisin and the improved activity of encapsulated nisin suggesting its application as a preservative in meat.

Several studies on encapsulation of both prebiotics and probiotics have been carried out on fish fillets. *L. rhamnosus* was encapsulated with polyvinyl alcohol and sodium alginate–based nanofibers and used as a coating on rainbow trout fish fillets for food preservation (Ceylan et al. 2018a,b). The small size of nanofibers allows it to come into contact with a larger surface area of the fish than bigger size molecules, thus providing better protection of essential fatty acids present in fish (Ceylan et al. 2018a). Sensory parameters including odor, color, and texture deteriorated more rapidly in uncoated fish fillets (5 days of storage) than coated samples (9 days of storage), suggesting that the coating of the nanofibers reduced the release of volatile ingredients. Linoleic acid, eicosapentaenoic acid, and docosahexaenoic acid, which were the major essential fatty acids present in fish fillets were preserved and more stable in coated samples. Ceylan et al. (2018b) also tested the effect of encapsulated *L. rhamnosus* on antimicrobial growth in fish fillets. The size of the nanofibers was <580 nm and the zeta potential was −7.74 mV. Several microbiological tests were carried out against total mesophilic aerobic bacteria, psychrophilic bacteria, yeast, and mold. The coating by nanoencapsulated *L. rhamnosus* had an antimicrobial effect on both total mesophilic, aerobic, and psychrophilic bacteria by 38%. However, the coating had a very limited effect on the growth of yeast and mold. This study suggests that encapsulated *L. rhamnosus* could be used as a coating against bacterial growth.

Nanoencapsulation of *L. reuteri* by electrospinning was produced as a coating for fish fillet to enhance the nutritional value of the fish fillets (Ceylan et al. 2019). *L. reuteri* is an important microorganism that has previously demonstrated inhibitory activity against *H. pylori* infection. Viability of encapsulated *L. reuteri* was at 78% and over 7 days of storage, there was only a slight decrease in viability from 4.33±0.03 to 3.39±0.03 log 10 cfu/g which suggests the potential role of encapsulated *L. reuteri* to obtain functional fish fillets. The encapsulated *L. reuteri* was also tested in rainbow trout and mackerel fish fillets. Bacterial growth was observed in both fish fillets. In rainbow trout fillets, the bacterial count remained constant over the 7 days (2.92 log cfu/g) period while there was a slight decrease from 3.27 to 2.66 log cfu/g in the bacterial count in mackerel fish fillets. Moreover, the antioxidant assays showed that coated fish fillet had higher antioxidant activity as compared to the uncoated samples. This demonstrates that coated fish fillet has the potential to be used in marinating in the fish industry, cold-smoked fish, or lightly processed fish products.

Polyvinyl alcohol was used to produce nanomats containing electrospun nisin and curcumin and used as a coating on rainbow trout fish fillets (Meral et al. 2019). The effect of the coating on antimicrobial growth and sensory parameters was evaluated. The nanomats had a diameter of 172 nm and were stable up to a temperature of 200°C. There was a significant delay in the growth of total mesophilic aerobic bacteria in coated fish samples and the nanomats also reduced the growth of unwanted lactic acid bacteria. Hardness of the fish is an important factor determining the freshness and coating of nanomats, which reduced the decline in the hardness of the fish as compared to the control samples over a period of 12 days. The evaluation of sensory parameters which included odor, color, and texture was acceptable for control and coated samples by the 4th and 10th days of storage, respectively. This demonstrates that coating with nisin and curcumin had increased the shelf life of the fish fillets.

Another study showed that nisin was loaded into nanofibers made of polyvinyl alcohol and was used to coat rainbow trout fillets (Oner, Meral, and Ceylan 2021). A range of 70.35–213.5 nm was observed for the diameter of the nanofiber and the zeta potential was -3.38 ± 1.51 mV. The addition of nanoencapsulated nisin (336.10^{-4} g) on fish fillets (100 g) delayed the growth of total mesophilic aerobic bacteria by 0.15–0.75 log cfu/g. Although the sensory parameters which included odor, color, and texture exhibited a declining trend over the storage period, the decline was slower in coated fish samples. The sensory quality was stable over the 7-day storage period.

13.5.3 Other Food Applications

Nanoencapsulated prebiotics and probiotics have been used in other food products. The use of prebiotics as a coating material has enhanced the quality of food products. Algal oil, which is a good source of omega-3 fatty acids was encapsulated with prebiotic carbohydrates such as inulin and maltodextrin (Wang et al. 2020). Nanocapsules of algal oil were produced using five emulsion formulations, namely, octenyl-succinic anhydride (OSA), OSA+inulin, OSA+maltodextrin, OSA+chitosan, OSA+maltodextrin+inulin, and OSA+chitosan+inulin. The combination of OSA+chitosan+inulin provided the most satisfactory results. Its encapsulation efficiency was at 98.57% and did not present any cracks or dents on the surface of the particles. This emulsion formulation had reduced the moisture content making it stable at high humidity levels and had the highest oxidative stability. The surface oil content, which indicates the quantity of unencapsulated oil was the lowest for the combination of OSA+chitosan+inulin. This also suggests high encapsulation efficiency and the coating material protected the algal oil from high temperature. These findings indicate that a prebiotic system as the coating material can improve the value of algal oil.

Clove essential oil has antifungal, antibacterial, and antioxidant properties making it a suitable food preservative. Nanoencapsulation of clove essential oil is necessary to protect against oxidation and evaporation (Hasheminejad et al., 2019). In this study, chitosan nanoparticles loaded with clove essential oil were tested for their antifungal efficacy against *A. niger*. The encapsulated essential oil had higher antifungal activity than free clove essential oil against *A. niger* in spoiled pomegranate. This was possible due to the controlled delivery of the nanocapsules to provide an enhanced inhibitory effect. The nanoencapsulated clove essential oil using chitosan nanoparticles may also be used for extending the shelf life of fruits and vegetables.

Bacteriocins, namely, Bac4463 and Bac22 produced by *Lactobacillus* strains were enclosed in silver nanoparticles to provide antimicrobial properties against food pathogens with no adverse effects (Sidhu and Nehra 2020). The antimicrobial activity was higher for encapsulated bacteriocins against different food pathogens compared to free bacteriocins. The minimum inhibitory concentration (MIC) observed for encapsulated Bac4463 against *S. aureus* was 8 μg/mL and concentration of encapsulated Bac22 was 2 μg/mL against *Shigella flexneri*. The addition of nanoencapsulated bacteriocins had a negative effect on the integrity of cell membranes of the food pathogens. This highlights the beneficial role of bacteriocins as food preservatives against food pathogens.

Thymol, cardamom essential oil, and *L. plantarum* were nanoencapsulated and incorporated into candies and all three with their nanoparticle candies were tested for their antibacterial effect against *Streptococcus mutans*, which causes formation of dental caries (Karimi et al. 2020). They all had an inhibitory effect on the bacterial activity of *S. mutans*. The MIC ranged from 0.01 to 0.06 mg/mL with thymol having the highest inhibitory effect. Encapsulated *L. plantarum* (0.01 mg/mL) had a more pronounced effect than free *L. plantarum* (0.06 mg/mL). However, the sensory analysis showed that cardamom essential oil candy had the highest acceptance score among participants.

13.5.4 Health Applications

The consumption of nanoencapsulated prebiotics and probiotics produces functional foods, which have an effect on health. These probiotic bacteria directly influence the gut microbiota. Hong et al. (2021) demonstrated that the combination of *L. plantarum* and phthalyl pullulan nanoparticles was used to improve its antimicrobial activity and to assess the effect of the synbiotics in mice suffering from dysbiosis. Gut dysbiosis, which is characterized by a microbial imbalance in the gut microbiota, has been closely linked to many health disorders such as obesity, cancer, cardiovascular disease, and inflammatory bowel disease (Carding et al. 2015). Mice, which were administered these symbiotic formulations, had their gut microbiota restored as changes in bacterial diversity, richness, and composition were observed. The synbiotics also suppressed the growth of pathogenic *E. coli* and *L. monocytogenes*. Side effects such as loss of appetite, inflammation, and weight loss caused by pathogenic bacteria were reduced. This highlights the importance of using synbiotics to alleviate dysbiosis of the gut microbiota.

Probiotics could also be considered as an alternative to antibiotics due to the health benefits they confer to humans (Kim et al. 2019). Antimicrobial molecules (lactic acid and bacteriocin) are produced by probiotic bacteria for inhibition of pathogen growth, modulation of immune system, improved and absorption of nutrients. *P. acidilactici* internalized with phthalyl dextran nanoparticles was co-cultured with Gram-negative pathogens, *Salmonella gallinarum*, *E. coli* K88, *E. coli* O157:H7, and the Gram-positive pathogen *L. monocytogenes*. Antimicrobial activity was higher in *P. acidilactici* internalized with phthalyl dextran nanoparticles compared to untreated *P. acidilactici* or *P. acidilactici* treated with dextran only. Production of the antimicrobial pediocin and its activity were also higher in *P. acidilactici* internalized with phthalyl dextran nanoparticles, thus indicating that the phthalyl dextran nanoparticles improved the antimicrobial activity of *P. acidilactici* and the production of antimicrobial molecules. Moreover, mice fed with *P. acidilactici* internalized with phthalyl dextran nanoparticles had enhanced antimicrobial activity as infection by pathogenic *E. coli* O157:H7 was reduced. *E. coli* O157:H7 is known to cause hemolytic uremic syndrome. The analysis of the gut microbiota showed that the antimicrobial activity caused by *P. acidilactici* internalized with PDN prevents a decrease in diversity in the gut microbiota.

Synthesis of antimicrobials by probiotic bacteria could be an effective way to combat multidrug-resistant bacteria in wounds (Gomaa 2019). In fact, bacteriocins and silver nanoparticles produced by *L. paracasei* demonstrated higher antibacterial activity against multidrug-resistant bacteria in wounds than treatment with either bacteriocin or silver nanoparticles alone. The combination of bacteriocin and silver nanoparticles led to the damage of cell membrane, release of proteins and DNA from the cells, and production of reactive oxygen species.

Inorganic nanoparticles produced by probiotic bacteria also have a potential role as safe nanodrug carriers. Indeed, Markus et al. (2016) showed that gold nanoparticles synthesized by *L. kimchicus* DCY51^T did not cause any cytotoxicity in murine macrophage cells (RAW264.7) and human colon cancer cell lines (HT29), which indicate that they are biocompatible. Furthermore, free radical scavenging activity by gold nanoparticles increased in a dose-dependent manner, i.e., at increasing concentration of gold nanoparticles and could be due to the antioxidant activity of *L. kimchicus* DCY51^T. The IC_{50} value was 233.75 µg/mL.

The protective layer of proteins that encapsulate the nanoparticles could play a positive role in antioxidant activity and low cytotoxicity and be beneficial in medical applications such as drug delivery.

13.6 CONCLUSIONS

There has been enormous progress achieved in the field of nanoencapsulation of prebiotics and probiotics such as the different choice of carriers available for encapsulation and the importance of understanding the different factors that affect the viability of probiotics. This has also highlighted the limitations and difficulties in the development of functional foods with the incorporation of encapsulated prebiotics and probiotics. Research on microencapsulation has been very extensive and has been ongoing for many years while the study of nanoencapsulation in the food industry is quite recent (Durazzo et al. 2020). Additional work is important to determine whether nanoencapsulated prebiotics and probiotics may be applied in other food products. Safety and regulations are equally important with the increasing use of nanotechnology in the food sector. Few studies have also looked at the biosynthesis of inorganic nanoparticles from bacteria, which could be used in other fields such as the medical sector.

REFERENCES

Afinjuomo, F., S. Abdella, S. H. Youssef, Y. Song, and S. Garg. (2021). "Inulin and its application in drug delivery." *Pharmaceuticals* 14: 855.

Aguilar-Pérez, K. M., J. I. Avilés-Castrillo, D. I. Medina, R. Parra-Saldivar, and H. Iqbal. (2020). "Insight into nanoliposomes as smart nanocarriers for greening the twenty-first century biomedical settings." *Frontiers in Bioengineering and Biotechnology* 8: 579536.

Amnuaikit, T., M. M. Htwe, R. Teanpaison, and P. Khongkow. (2019). "Liposomes of probiotic's lyophilized cell free supernatant: A potential cosmeceutical product." *Pharmazie* 74: 462–466.

Ansari, F., H. Pourjafar, V. Jodat, J. Sahebi, and A. Ataei. (2017). "Efect of Eudragit S100 nanoparticles and alginate chitosan encapsulation on the viability of *Lactobacillus acidophilus* and *Lactobacillus rhamnosus*." *AMB Express* 7: 144.

Arepally, D., and T. K. Goswami. (2019). "Effect of inlet air temperature and gum Arabic concentration on encapsulation of probiotics by spray drying." *LWT - Food Science and Technology* 99: 583–593.

Aswathanarayan, J. B., and R. R. Vittal. (2019). "Nanoemulsions and their potential applications in food industry." *Frontiers in Sustainable Food Systems* 3: 95.

Atia, A., A. Gomaa, I. Fliss, E. Beyssac, G. Garrait, and M. Subirade. (2016). "A prebiotic matrix for encapsulation of probiotics: Physicochemical and microbiological study." *Journal of Microencapsulation* 33(1): 89–101.

Atraki, R., and M. Azizkhani. (2021). "Survival of probiotic bacteria nanoencapsulated within biopolymers in a simulated gastrointestinal model." *Innovative Food Science and Emerging Technologies* 72: 102750.

Avila-Reyes, S. V., Garcia-Suarez, F. J., Jiménez, M. T., San Martín-Gonzalez, M. F., and Bello-Perez, L. A. (2014). Protection of L. rhamnosus by spray-drying using two prebiotics colloids to enhance the viability. *Carbohydrate Polymers*, 102: 423–430.

Bäckhed, F., C. M. Fraser, Y. Ringel, M. E. Sanders, R. Balfour Sartor, P. M. Sherman, J. Versalovic, V. Young, and B. Brett Finlay. (2012). "Defining a healthy human gut microbiome: Current concepts, future directions, and clinical applications." *Cell Host & Microbe* 12(5): 611–622.

Baeza, A., D. Ruiz-Molina, and M. Vallet-Regí. (2016). "Recent advances in porous nanoparticles for drug delivery in antitumoral applications: Inorganic nanoparticles and nanoscale metal-organic frameworks." *Expert Opinion on Drug Delivery* 14(6): 783–796.

Balea, A., A. Blanco, M. Delgado-Aguilar, M. C. Monte, Q. Tarrés, E. Fuente, P. Mutjé, and C. Negro. (2021). "Nanocellulose characterization challenges." *Bioresources* 16(2): 4382–4410.

Barajas-Álvarez, P., M. González-Ávila, and H. Espinosa-Andrews. (2021). "Recent advances in probiotic encapsulation to improve viability under storage and gastrointestinal conditions and their impact on functional food formulation." *Food Reviews International* 1–22.

Bustamante, M., Villarroel, M., Rubilar, M., & Shene, C. (2015). Lactobacillus acidophilus La-05 encapsulated by spray drying: Effect of mucilage and protein from flaxseed (Linum usitatissimum L.). *LWT-Food Science and Technology*, 62(2): 1162–1168.

Carding, S., K. Verbeke, D. T. Vipond, B. M. Corfe, and L. J. Owen. (2015). "Dysbiosis of the gut microbiota in disease." *Microbial Ecology in Health and Disease* 26(1): 26191.

Ceylan, Z., R. Meral, I. Cavidoglu, C. Y. Karakas, and M. T. Yilmaz. (2018a). "A new application on fatty acid stability of fish fillets: Coating with probiotic bacteria-loaded polymer-based characterized nanofibers." *Journal of Food Safety* 38: e12547.

Ceylan, Z., R. Meral, C. Y. Karakaş, E. Dertli, and M. T. Yilmaz. (2018b). "A novel strategy for probiotic bacteria: Ensuring microbial stability of fish fillets using characterized probiotic bacteria-loaded nanofibers." *Innovative Food Science and Emerging Technologies* 48: 212–218.

Ceylan, Z., E. Uslu, H. İspirli, R. Meral, M. Gavgalı, M. T. Yilmaz, and E. Dertli. (2019). "A novel perspective for Lactobacillus reuteri: Nanoencapsulation to obtain functional fish fillets." *LWT - Food Science and Technology* 115: 108427.

Charalampopoulos, D., and Pandiella, S. S. (2010). Survival of human derived Lactobacillus plantarum in fermented cereal extracts during refrigerated storage. *LWT-Food Science and Technology* 43(3): 431–435.

Chandrakala, V., V. Aruna, and G. Angajala. (2022). "Review on metal nanoparticles as nanocarriers: Current challenges and perspectives in drug delivery systems." *Emergent Materials* 5: 1593–1615.

Chotiko, A., and Sathivel, S. (2016). Development of a combined low-methoxyl-pectin and rice-bran-extract delivery system to improve the viability of Lactobacillus plantarum under acid and bile conditions. *LWT-Food Science and Technology* 66: 420–427.

Colín-Cruz, M. A., Pimentel-González, D. J., Carrillo-Navas, H., Alvarez-Ramírez, J., and Guadarrama-Lezama, A. Y. (2019). Co-encapsulation of bioactive compounds from blackberry juice and probiotic bacteria in biopolymeric matrices. *Lwt* 110: 94–101.

Cui, L.-H., C.-G. Yan, H.-S. Li, W.-S. Kim, L. Hong, S.-K. Kang, Y.-J. Choi, and C.-S. Cho. (2018). "A new method of producing a natural antibacterial peptide by encapsulated probiotics internalized with inulin nanoparticles as prebiotics." *Journal of Microbiology and Biotechnology* 28(4): 510–519.

Dabholkar, N., T. Waghule, V. K. Rapalli, S. Gorantla, A. Alexander, R. N. Saha, and G. Singhvi. (2021). "Lipid shell lipid nanocapsules as smart generation lipid nanocarriers." *Journal of Molecular Lipids* 339: 117145.

Danaei, M., M. Kalantari, M. Raji, H. Samareh Fekri, R. Saber, G. P. Asnani, S. M. Mortazavi, M. R. Mozafari, B. Rasti, and A. Taheriazam. (2018). "Probing nanoliposomes using single particle analytical techniques: Effect of excipients, solvents, phase transition and zeta potential." *Heliyon* 4(12): e01088.

Davani-Davari, D., M. Negahdaripour, I. Karimzadeh, M. Seifan, M. Mohkam, S. J. Masoumi, A. Berenjian, and Y. Ghasemi. (2019). "Prebiotics: Definition, types, sources, mechanisms, and clinical applications." *Foods* 8(3): 92.

Davis, C. (2014). "Enumeration of probiotic strains: Review of culture-dependent and alternative techniques to quantify viable bacteria." *Journal of Microbiological Methods* 103: 9–17.

Durazzo, A., A. Nazhand, M. Lucarini, A. G. Atanasov, E. B. Souto, E. Novellino, R. Capasso, and A. Santini. (2020). "An updated overview on nanonutraceuticals: Focus on nanoprebiotics and nanoprobiotics." *International Journal of Molecular Sciences* 21: 2285.

El-Sayed, H. S., S. M. El-Sayed, and A. M. Youssef. (2021). "Novel approach for biosynthesizing of zinc oxide nanoparticles using *Lactobacillus gasseri* and their influence on microbiological, chemical, sensory properties of integrated yogurt." *Food Chemistry* 365: 130513.

Fares, M. M., and M. S. Salem. (2015). "Dissolution enhancement of curcumin via curcumin-prebiotic inulin nanoparticles." *Drug Development and Industrial Pharmacy* 41(11): 1785–1792.

Fathi, M., M. R. Mozafari, and M. Mohebbi. (2012). "Nanoencapsulation of food ingredients using lipid based delivery systems." *Trends in Food Science & Technology* 23: 13–27.

Fayed, B., A. Abood, H. S. El-Sayed, A. M. Hashem, and N. S. H. Mehanna. (2018). "A synbiotic multiparticulate microcapsule for enhancing inulin intestinal release and Bifidobacterium gastro-intestinal survivability." *Carbohydrate Polymers* 193: 137–143.

Ferreira, M., M. Ogren, J. N. R. Dias, M. Silva, S. Gil, L. Tavares, F. Aires-da-Silva, M. M. Gaspar, and S. I. Aguiar. (2021). "Liposomes as antibiotic delivery systems: A promising nanotechnological strategy against antimicrobial resistance." *Molecules* 26(7): 2047.

Food and Agriculture Organization (FAO). *Guidelines for the Evaluation of Probiotics in Food*. FAO; London, ON, Canada: Report of a Joint FAO/WHO Working Group on Drafting Guidelines for the Evaluation of Probiotics in Food. 30 April–1 May 2002.

Fontana, L., M. Bermudez-Brito, J. Plaza-Diaz, S. Muñoz-Quezada, and A. Gil. (2013). "Sources, isolation, characterisation and evaluation of probiotics." *British Journal of Nutrition* 109(S2): S35–S50.

Frakolaki, G., T. Kekes, F. Lympaki, V. Giannou, and C. Tzia. (2021). "Use of encapsulated Bifidobacterium animalis subsp. lactis through extrusion or emulsification for the production of probiotic yogurt." *Journal of Food Processing Engineering* 47(7): e13792.

Fröhlich, E. E., and E. Fröhlich. (2016). "Cytotoxicity of nanoparticles contained in food on intestinal cells and the gut microbiota." *International Journal of Molecular Sciences* 17(4): 509.

Gao, Y., X. Wang, C. Xue, and Z. Wei. (2021). "Latest developments in food-grade delivery systems for probiotics: A systematic review." *Critical Reviews in Food Science and Nutrition* 21: 1–18.

Ghaffar, K. A., A. K. Giddam, M. Zaman, M. Skwarczynski, and I. Toth. (2014). "Liposomes as nanovaccine delivery systems." *Current Topics in Medicinal Chemistry* 14(9): 1194–1208.

Gibson, P. R., B. Fixa, B. Pekárková, M. Bátovský, G. Radford-Smith, J. Tibitanzl, L. Gabalec, T. H. J. Florin, and R. Greinwald. (2006). "Comparison of the efficacy and safety of Eudragit-L-coated mesalazine tablets with ethylcellulose-coated mesalazine tablets in patients with mild to moderately active ulcerative colitis." *Alimentary Pharmacology & Therapeutics* 23(7): 1017–1026.

Gomaa, E. K. (2019). "Synergistic antibacterial efficiency of bacteriocin and silver nanoparticles produced by probiotic *Lactobacillus paracasei* against multidrug resistant bacteria." *International Journal of Peptide Research and Therapeutics* 25: 1113–1125.

Hasheminejad, N., F. Khodaiyan, and M. Safari. (2019). "Improving the antifungal activity of clove essential oil encapsulated by chitosan nanoparticles." *Food Chemistry* 275: 113–122.

Hong, L., S.-M. Lee, W.-S. Kim, Y.-J. Choi, S.-H. Oh, Y.-L. Li, S.-H. Choi, D. H. Chung, E. Jung, S.-K. Kang, and C.-S. Cho. (2021). "Synbiotics containing prebiotics: A novel therapeutic strategy to restore gut dysbiosis." *Frontiers in Microbiology* 12: 715241.

Huang, S., M. Vignolles, X. D. Chen, Y. L. Loir, G. Jan, P. Schuck, and R. Jeantet. (2017). "Spray drying of probiotics and other food-grade bacteria: A review." *Trends in Food Science & Technology* 63: 1–17.

Jaiswal, M., R. Dudhe, and P. K. Sharma. (2015). "Nanoemulsion: An advanced mode of drug delivery system." *3 Biotech* 5(2): 123–127.

Jantzen, M., A. Göpel, and C. Beermann. (2013). "Direct spray drying and microencapsulation of probiotic Lactobacillus reuteri from slurry fermentation with whey." *Journal of Applied Microbiology* 115: 1029–1036.

Jiao, D., Y. Liu, Y. Liu, R. Zeng, X. Hou, G. Nie, L. Sun, and Z. Fang. (2020). "Preparation of phosphatidylcholine nanovesicles containing bacteriocin CAMT2 and their anti-listerial activity." *Food Chemistry* 314: 126244.

Jin, W., H. Liang, Y. Li, Liu. S., and B. Li. (2016). "Nanoemulsions for food: Properties, production, characterization, and applications." *Emulsions* 3: 1–36.

Jo, S. J., M. -J. Choi, and Y. -J. Kwon. (2015). "Effect of palm or coconut solid lipid nanoparticles (SLNs) on growth of Lactobacillus plantarum in milk." *Food Science of Animal Resources* 35(2): 197–204.

Karimi, N., V. Jabbari, A. Nazemi, K. Ganbarov, N. Karimi, A. Tanomand, S. Karimi, A. Abbasi, B. Yousefi, E. Khodadadi, and H. S. Kafil. (2020). "Thymol, cardamom and *Lactobacillus plantarum* nanoparticles as a functional candy with high protection against *Streptococcus mutans* and tooth decay." *Microbial Pathogenesis* 148: 104481.

Kau, A. L., P. P. Ahern, N. W. Griffin, A. L. Goodman, and J. I. Gordon. (2011). "Human nutrition, the gut microbiome and the immune system." *Nature* 474(7351): 327–336.

Khan, S. T., S. Saleem, M. Ahamed, and J. Ahmad. (2019). "Survival of probiotic bacteria in the presence of food grade nanoparticles from chocolates: An in vitro and in vivo study." *Applied Microbiology and Biotechnology* 103: 6689–6700.

Khorasani, A. C., and S. A. Shojaosadati. (2016). "Bacterial nanocellulose-pectin bionanocomposites as prebiotics against drying and gastrointestinal condition." *International Journal of Biological Macromolecules* 83: 9–18.

Kim, W., G. G. Han, L. Hong, S. Kang, M. Shokouhimehr, Y. Choi, and C. Cho. (2019). "Novel production of natural bacteriocin via internalization of dextran nanoparticles into probiotics." *Biomaterials* 218: 119360.

Krithika, B., and R. Preetha. (2019). "Formulation of protein based inulin incorporated synbiotic nanoemulsion for enhanced stability of probiotic." *Material Research Express* 6: 114003.

Kumar, M., R. S. Bishnoi, A. K. Shukla, and C. P. Jain. (2019). "Techniques for formulation of nanoemulsion drug delivery system: A review." *Preventive Nutrition and Food Science* 24(3): 225–234.

Kyereh, E., and Sathivel, S. (2021). Viability of Lactobacillus plantarum NCIMB 8826 immobilized in a cereal-legume complementary food "weanimix" with simulated gastrointestinal conditions. *Food Bioscience*, 40: 100848.

Ladj, R., A. Bitar, M. M. Eissa, H. Fessi, Y. Mugnier, R. Le Dantec, and A. Elaissari. (2013). "Polymer encapsulation of inorganic nanoparticles for biomedical applications." *International Journal of Pharmaceutics* 458: 230–241.

Lorenzer, C., M. Dirin, A. M. Winkler, V. Baumann, and J. Winkler. (2015). "Going beyond the liver: Progress and challenges of targeted delivery of siRNA therapeutics." *Journal of Controlled Release* 203: 1–15.

Lourens-Hattingh, A., and B. C. Viljoen. (2001). "Yogurt as probiotic carrier food." *International Dairy Journal* 11: 1–17.

Malheiros, P. S., I. M. Cuccovia, and B. D. G. M. Franco. (2016). "Inhibition of *Listeria monocytogenesin vitro* and in goat milk by liposomal nanovesicles containing bacteriocins produced by *Lactobacillus sakei* subsp. *sakei* 2a." *Food Control* 63: 158–164.

Markus, J., R. Mathiyalagan, Y. Kim, R. Abbai, P. Singh, S. Ahn, Z. E. J. Perez, J. Hurh, and D. C. Yang. (2016). "Intracellular synthesis of gold nanoparticles with antioxidant activity by probiotic *Lactobacillus kimchicus* DCY51T isolated from Korean kimchi." *Enzyme and Microbial Technology* 95: 85–93.

Martău, G. A., M. Mihai, and D. C. Vodnar. (2019). "The use of chitosan, alginate, and pectin in the biomedical and food sector: Biocompatibility, bioadhesiveness, and biodegradability." *Polymers* 11: 1837.

Martoni, C., J. Bhatena, M. L. Jones, A. M. Urbanska, H. Chen, and S. Prakash. (2007). "Investigation of microencapsulated BSH active Lactobacillus in the simulated human GI tract." *Journal of Biomedicine and Biotechnology* 2007: 13684.

Mawad, A., Y. A. Helmy, A. -G. Shalkami, D. Kathayat, and G. Rajashekara. (2018). "*E. coli* Nissle microencapsulation in alginate-chitosan nanoparticles and its effect on Campylobacter jejuni in vitro." *Applied Microbiology and Biotechnology* 102: 10675–10690.

McClements, D. J., and H. Xiao. (2017). "Is nano safe in foods? Establishing the factors impacting the gastrointestinal fate and toxicity of organic and inorganic food-grade nanoparticles." *NPJ Science of Food* 1: 6.

Mehnert, W., and K. Mäder. (2012). "Solid lipid nanoparticles." *Advanced Drug Delivery Reviews* 64: 83–101.

Mensink, M. A., H. W. Frijlink, K. V. D. V. Maarschalk, and W. L. J. Hinrichs. (2015). "Inulin, a flexible oligosaccharide I: Review of its physicochemical characteristics." *Carbohydrate Polymers* 130: 405–419.

Meral, R., A. Alav, C. Y. Karakas, E. Dertli, M. T. Yilmaz, and Z. Ceylan. (2019). "Effect of electrospun nisin and curcumin loaded nanomats on the microbial quality, hardness and sensory characteristics of rainbow trout fillet." *LWT - Food Science and Technology* 113: 108292.

Mikiciuk, J., E. Mikiciuk, A. Wrońska, and A. Szterk. (2016). "Antimicrobial potential of commercial silver nanoparticles and the characterization of their physical properties toward probiotic bacteria isolated from fermented milk products." *Journal of Environmental Science and Health, Part B* 51(4): 222–229.

Mohammadi, R., M. Mahmoudzadeh, M. Atefi, K. Khosravi-Darani, and M. R. Mozafari. (2014). "Applications of nanoliposomes in cheese technology." *International Journal of Dairy Technology* 68(1): 11–23.

Mokhtarieh, A. A., J. Lee, S. Kim, and M. K. Lee. (2018). "Preparation of siRNA encapsulated nanoliposomes suitable for siRNA delivery by simply discontinuous mixing." *Biochimica et Biophysica Acta (BBA) - Biomembranes* 1860(6): 1318–1325.

Morrissey, D. V., K. Blanchard, L. Shaw, K. Jensen, J. A. Lockridge, B. Dickinson, J. A. McSwiggen, C. Vargeese, K. Bowman, C. S. Shaffer, and B. A. Polisky. (2005). "Activity of stabilized short interfering RNA in a mouse model of Hepatitis B virus replication." *Hepatology* 41(6): 1349–1356.

Mousavi, S. M. A. A., S. A. Mirhosseini, M. R. S. Panahi, and H. M. Hosseini. (2020). "Characterization of biosynthesized silver nanoparticles using *Lactobacillus rhamnosus* GG and its in vitro assessment against colorectal cancer cells." *Probiotics and Antimicrobial Proteins* 12: 740–746.

Mukherjee, S., S. Ray, and R. S. Thakur. (2009). "Solid lipid nanoparticles: A modern formulation approach in drug delivery system." *Indian Journal of Pharmaceutical Sciences* 71(4): 349–358.

Müller, R. H., A. Lippacher, and S. Gohla. (2000a). "Solid lipid nanoparticles (Sln) as a carrier system for the controlled release of drugs." In: *Handbook of Pharmaceutical Controlled Release Technology* (edited by D. L. Wise). Boca Raton, FL: CRC Press, p. 377.

Müller, R. H., K. Mäder, and S. Gohla. (2000b). "Solid lipid nanoparticles (SLN) for controlled drug delivery: A review of the state of the art." *European Journal of Pharmaceutics and Biopharmaceutics* 50: 161–177.

Naseri, N., H. Valizadeh, and P. Zakeri-Milani. (2015). "Solid lipid nanoparticles and nanostructured lipid carriers: Structure, preparation and application." *Advanced Pharmaceutical Bulletin* 5(3): 305–313.

Oner, B., R. Meral, and Z. Ceylan. (2021). "Determination of some quality indices of rainbow trout fillets treated with nisin-loaded polyvinylalcohol-based nanofiber and packed with polyethylene package." *LWT - Food Science and Technology* 149: 111854.

Osorio, M., A. Cañas, R. Zuluaga, P. Gañán, I. Ortiz, and C. Castro. (2018). "Nanocellulose-based composites in biomedical applications." In: *Advanced Green Composites*. Hoboken, NJ: Wiley, pp. 369–401.

Padan, E., E. Bibi, M. Ito, and T. A. Krulwich. (2005). "Alkaline pH homeostasis in bacteria: New insights." *Biochimica et Biophysica Acta* 1717: 67–88.

Palanikumar, L., E. S. Choi, J. Y. Oh, S. A. Park, H. Choi, K. Kim, C. Kim, and J. Ryu. (2018). "Importance of encapsulation stability of nanocarriers with high drug loading capacity for increasing in vivo therapeutic efficacy." *Biomacromolecules* 19: 3030–3039.

Pardeike, J., A. Hommoss, and R. H. Müller. (2009). "Lipid nanoparticles (SLN, NLC) in cosmetic and pharmaceutical dermal products." *International Journal of Pharmaceutics* 366: 170–184.

Pateiro, M., B. Gómez, P. E. S. Munekata, F. J. Barba, P. Putnik, D. B. Kovačević, and J. M. Lorenzo. (2021). "Nanoencapsulation of promising bioactive compounds to improve their absorption, stability, functionality and the appearance of the final food products." *Molecules* 26(6): 1547.

Peterson, J., S. Garges, M. Giovanni, P. McInnes, L. Wang, J. A. Schloss, V. Bonazzi, J. E. McEwen, K. A. Wetterstrand, C. Deal, C. C. Baker, V. Di Francesco, T. K. Howcroft, R. W. Karp, R. D. Lunsford, C. R. Wellington, T. Belachew, M. Wright, C. Giblin, H. David, M. Mills, R. Salomon, C. Mullins, B. Akolkar, L. Begg, C. Davis, L. Grandison, M. Humble, J. Khalsa, A. R. Little, H. Peavy, C. Pontzer, M. Portnoy, M. H. Sayre, P. Starke-Reed, S. Zakhari, J. Read, B. Watson, and M. Guyer. (2009). "The NIH human microbiome project." *Genome Res* 19(12): 2317–2323.

Pinto, S. S., C. B. Fritzen-Freire, C. O. Dias, and R. D. Amboni. (2019). "A potential technological application of probiotic microcapsules in lactose-free Greek-style yoghurt." *International Dairy Journal* 97: 131–138.

Poddar, D., S. Das, G. Jones, J. Palmer, G. B. Jameson, R. G. Haverkamp, and H. Singh. (2014). "Stability of probiotic *Lactobacillus paracasei* during storage as affected by the drying method." *International Dairy Journal* 39: 1–7.

Pourjafar, H., N. Noori, H. Gandomi, A. A. Basti, and F. Ansari. (2020). "Viability of microencapsulated and non-microencapsulated Lactobacilli in a commercial beverage." *Biotechnology Reports* 25: e00432.

Pozo-Rodríguez, A. D., M. A. Solinís, and A. Rodríguez-Gascón. (2020). "Applications of lipid nanoparticles in gene therapy." *European Journal of Pharmaceutics and Biopharmaceutics* 109: 184–193.

Pyo, S. -M., R. H. Müller, and C. M. Keck. (2017). "Encapsulation by nanostructured lipid carriers." In: *Nanoencapsulation Technologies for the Food and Nutraceutical Industries* (edited by S. M. Jafari). Cambridge, MA: Academic Press, pp. 114–137.

Rashidinejad, A., A. Bahrami, A. Rehman, A. Rezaei, A. Babazadeh, H. Singh, and S. M. Jafari. (2020). "Co-encapsulation of probiotics with prebiotics and their application in functional/synbiotic dairy products." *Critical Reviews in Food Science and Nutrition 62*(9), 2470–2494.

Razavi, S., S. Janfaza, N. Tasnim, D. L. Gibson, and M. Hoorfar. (2021). "Nanomaterial-based encapsulation for controlled gastrointestinal delivery of viable probiotic bacteria." *Nanoscale Advances* 3(10): 2699–2709.

Reque, P. M., and A. Brandelli. (2021). "Encapsulation of probiotics and nutraceuticals: Applications in functional food industry." *Trends in Food Science & Technology* 114: 1–10.

Salazar, N., L. Valdés-Varela, S. González, M. Gueimonde, and C. G. de los Reyes-Gavilán. (2017). "Nutrition and the gut microbiome in the elderly." *Gut Microbes* 8(2): 82–97.

Sarao, L. K., and M. Arora. (2015). "Probiotics, prebiotics and microencapsulation: A review." *Critical Reviews in Food Science and Nutrition* 57: 344–371.

Saravanan, A., P. S. Kumar, S. Karishma, D. N. Vo, S. Jeevanantham, P. R. Yaashikaa, and C. S. George. (2021). "A review on biosynthesis of metal nanoparticles and its environmental applications." *Chemosphere* 264: 128580.

Sarkar, S. (2010). Approaches for enhancing the viability of probiotics: a review. *British Food Journal* 112(4): 329–349.

Sathyabama, S., M. R. Kumar, P. B. Devi, R. Vijayabharathi, and V. B. Priyadharisini. (2014). "Co-encapsulation of probiotics with prebiotics on alginate matrix and its effect on viability in simulated gastric environment." *LWT - Food Science and Technology* 57: 419–425.

Seabra, C. L., C. Nunes, M. Brás, M. Gomez-Lazaro, C. A. Reis, I. C. Gonçalves, S. Reis, and M. C. L. Martins. (2018). "Lipid nanoparticles to counteract gastric infection without affecting gut microbiota." *European Journal of Pharmaceutics and Biopharmaceutics* 127: 378–386.

Severino, P., T. Andreani, A. S. Macedo, J. F. Fangueiro, M. H. A. Santana, A. M. Silva, and E. B. Souto. (2012). "Current state-of-art and new trends on lipid nanoparticles (SLN and NLC) for oral drug delivery." *Journal of Drug Delivery* 2012: 750891.

Shak, K. P. Y., Y. L. Pang, and S. K. Mah. (2018). "Nanocellulose: Recent advances and its prospects in environmental remediation." *Beilstein Journal of Nanotechnology* 9: 2479–2498.

Shastri, D. H. (2017). "Effective delivery routes and strategies for solid lipid nanoparticles (Sln) and nanostructured lipid carriers." *Current Pharmaceutical Design* 23: 6592–6601.

Sheikhi, A., J. Hayashi, J. Eichenbaum, M. Gutin, N. Kuntjoro, D. Khorsandi, and A. Khademhosseini. (2019). "Recent advances in nanoengineering cellulose for cargo delivery." *Journal of Controlled Release* 294: 53–76.

Sidhu, P. K., and K. Nehra. (2020). "Bacteriocin-capped silver nanoparticles for enhanced antimicrobial efficacy against food pathogens." *IET Nanobiotechnology* 14(3): 245–252.

Siler-Marinkovic, S., L. Mojovic, V. Davinic, and B. Bugarski. (1997). "Liposomes as carriers of antimicrobial drugs." *Drug Development and Industrial Pharmacy* 23: 483–488.

Šipailienė, A., and S. Petraitytė. (2018). "Encapsulation of probiotics: Proper selection of the probiotic strain and the influence of encapsulation technology and materials on the viability of encapsulated microorganisms." *Probiotics and Antimicrobial Proteins* 10: 1–10.

Sizentsov, A. N., O. V. Kvan, E. P. Miroshnikova, I. A. Gavrish, V. A. Serdaeva, and A. V. Bykov. (2018). "Assessment of biotoxicity of Cu nanoparticles with respect to probiotic strains of microorganisms and representatives of the normal flora of the intestine of broiler chickens." *Environmental Science and Pollution Research* 25(16): 15765–15773.

Slavin, J. (2013). Fiber and prebiotics: mechanisms and health benefits. *Nutrients*, 5(4): 1417–1435.

Smelt, J. P. P. M., and Brul, S. (2014). Thermal inactivation of microorganisms. *Critical Reviews in Food Science and Nutrition*, 54(10): 1371–1385.

Swanson, K. S., G. R. Gibson, R. Hutkins, R. A. Reimer, G. Reid, K. Verbeke, K. P. Scott, H. D. Holscher, M. B. Azad, N. M. Delzenne, and M. E. Sanders. (2020). "The International Scientific Association for Probiotics and Prebiotics (ISAPP) consensus statement on the definition and scope of synbiotics." *Nature Reviews Gastroenterology & Hepatology* 17: 687–701.

Talwalkar, A., and K. Kailasapathy. (2003). "Effect of microencapsulation on oxygen toxicity in probiotic bacteria." *The Australian Journal of Dairy Technology* 58(1): 36–39.

Talwalkar, A., and K. Kailasapathy. (2004). "A review of oxygen toxicity in probiotic yogurts: Influence on the survival of probiotic bacteria and protective techniques." *Comprehensive Reviews in Food Science and Food Safety* 3: 117–124.

Tao, T., Ding, Z., Hou, D., Prakash, S., Zhao, Y., Fan, Z., ... and Han, J. (2019). Influence of polysaccharide as co-encapsulant on powder characteristics, survival and viability of microencapsulated Lactobacillus paracasei Lpc-37 by spray drying. *Journal of Food Engineering* 252: 10–17.

Taylor, T. M., S. Gayinsky, P. M. Davidson, B. D. Bruce, and J. Weiss. (2007). "Characterization of antimicrobial-bearing liposomes by zeta-potential, vesicle size, and encapsulation efficiency." *Food Biophysics* 2: 1–2.

Terpou, A., A. Papadaki, I. K. Lappa, V. Kachrimanidou, L. A. Bosnea, and N. Kopsahelis. (2019). "Probiotics in food systems: Significance and emerging strategies towards improved viability and delivery of enhanced beneficial value." *Nutrients* 11: 1591.

Vaishanavi, S., and R. Preetha. (2021). "Soy protein incorporated nanoemulsion for enhanced stability of probiotic *(Lactobacillus delbrueckii subsp. bulgaricus)* and its characterization." *Materials Today: Proceedings* 40: S148–S153.

Vidhyalakshmi, R., R. Bhakyaraj, and R. S. Subhasree. (2009). "Encapsulation "The future of probiotics": A review." *Advances in Biological Research* 3: 96–103.

Wang, Y., Z. Zheng, K. Wang, C. Tang, Y. Liu, and J. Li. (2020). "Prebiotic carbohydrates: Effect on physicochemical stability and solubility of algal oil nanoparticles." *Carbohydrate Polymers* 228: 115372.

Wang, M., Y. Li, J. Yang, R. Shi, L. Xiong, and Q. Sun. (2021). "Effects of food-grade inorganic nanoparticles on the probiotic properties of *Lactobacillus plantarum* and *Lactobacillus fermentum*." *LWT - Food Science and Technology* 139: 110540.

Wilson, S. (2020). "Nano-emulsions and its uses in food industry." *Journal of Food Processing and Technology* 11: 842.

Woldemariam Yohannes, K., Z. Wan, Q. Yu, H. Li, X. Wei, Y. Liu, J. Wang, and B. Sun. (2020). "Prebiotic, probiotic, antimicrobial, and functional food applications of *Bacillus amyloliquefaciens*." *Journal of Agricultural and Food Chemistry* 68: 14709–14727.

Yazdi, F. G., S. Soleimanian-Zad, E. V. D. Worm, and G. Folkerts. (2019). "Turmeric extrac: Potential use as prebiotic and anti-inflammatory compound?" *Plant Foods for Human Nutrition* 74: 293–299.

Yilmaz, M. T., O. Taylan, C. Y. Karakas, and E. Dertli. (2020). "An alternative way to encapsulate probiotics within electrospun alginate nanofibers as monitored under simulated gastrointestinal conditions and in kefir." *Carbohydrate Polymers* 244: 116447.

Yoha, K. S., S. Nida, S. Dutta, J. A. Moses, and C. Anandharamakrishna. (2022). "Targeted delivery of probiotics: Perspectives on research and commercialization." *Probiotics and Antimicrobial Proteins* 14: 15–48.

Yordshahi, A. S., M. Moradi, H. Tajik, and R. Molaei. (2020). "Design and preparation of antimicrobial meat wrapping nanopaper with bacterial cellulose and postbiotics of lactic acid bacteria." *International Journal of Food Microbiology* 321: 108561.

Yousefi, M., A. Ehsani, and S. M. Jafari. (2019). "Lipid-based nano delivery of antimicrobials to control food-borne bacteria." *Advances in Colloid and Interface Science* 270: 263–277.

Zhang, Z., M. Gu, X. You, D. A. Sela, H. Xiao, and D. J. McClements. (2021). "Encapsulation of bifidobacterium in alginate microgels improves viability and targeted gut release." *Food Hydrocolloids* 116: 106634.

Zhong, Q., and L. Zhang. (2019). "Nanoparticles fabricated from bulk solid lipids: Preparation, properties, and potential food applications." *Advances in Colloid and Interface Science* 273: 102033.

Zimet, P., Á. W. Mombrú, R. Faccio, G. Brugnini, I. Miraballes, C. Rufo, and H. Pardo. (2018). "Optimization and characterization of nisin-loaded alginate-chitosan nanoparticles with antimicrobial activity in lean beef." *LWT - Food Science and Technology* 91: 107–116.

CHAPTER 14

Trends and Future Perspectives in Nanoencapsulation of Plant-Based Polyphenolics (Flavonoids, Anthocyanins, and Tannins)

Sakshi Thakur, Garima Bhardwaj, Vishal Mutreja, and Ajay Sharma

CONTENTS

DOI:10.1201/9781003259183-14

14.1 INTRODUCTION

Public health is completely influenced by food science and pharmaceutical science (Ann et al., 2009). It has become important to assure the production of healthier food, which ultimately facilitates a healthier lifestyle (Carrión-Granda et al., 2016). Various secondary metabolites (SMs) (mainly natural polyphenolics, PPHs) obtained from nature exhibit a wide range of pharmaceutical properties (Pinelo et al., 2005; Chem, 2010). Nowadays, these SMs are extensively utilized in nutraceuticals, food ingredients, food preservatives, natural medications, and in a variety of other traditional formulations. As a conclusion, these SMs have become an important aspect of current health care and preservation technology. Among these SMs, natural PPHs are the major ones, which are easily procured from different crude extracts obtained from every part of the plants like leaves, bark, rhizome, stems, roots, and flowers. These PPHs are widely known for their diverse array of biological properties. These natural PPHs are known to be the biggest class of natural antioxidants that occur naturally in the different plant parts.

These natural PPHs are also vital for the growth of the plant from which these are isolated. These natural PPHs help to protect these plants from a range of pathogens, predators, other diseases, and ultraviolet radiation exposure (Beckman, 2000; Klejdus et al., 2010). These natural PPHs also aid in human health promotion and bring down the risk of persistent disorders and illnesses in human beings (Liu, 2003). Several studies have found a link between the consumption of natural PPHs and the decreased risk of prevalent diseases such as osteoporosis, cancer, diabetes, and cardiovascular and neurological diseases (Pandey & Rizvi, 2009). The health promotion potential of natural PPHs may be attributed to their property to interact efficiently with proteins, membrane receptors, and enzymes, thereby altering their activity in a specific manner (Mcmanus et al., 1985).

Nowadays, nanotechnology plays an imperative role in enhancing the bioavailability and effectiveness of natural bioactive compounds (BACs). The efficiency of PPHs in nurturing health and lowering the probability of persistent ailments is mainly correlated to their bioavailability and metabolic rate (Scalbert & Williamson, 2018). Nanotechnology ensures the production, application, and design of materials on the nanometer scale within the size range of 10^{-9}m with the quality assurance of materials (Mondal, 2020). Its advancements have transformed several realms of food and pharmaceutical sciences, specifically those that involve the manufacturing, processing, packing, storage, delivery, and other safety measurements (Bajpai et al., 2018). BACs, especially PPHs, are the compounds that are found in minute amounts in different plant parts and are mostly hydrophobic, lipophilic, and have intensive potential in curing a wide range of ailments (Kris-Etherton et al., 2002). Presently, nanoencapsulation is the emerging technique for enhancing the bioavailability of PPHs.

Food nanotechnology offers applications for designing different types of noble methods of delivery systems, i.e., encapsulation for BACs (Pateiro et al., 2021). In order to address the problems related to food production, packaging, consumption, and nutritional value of BACs presently, the researchers and scientific community from all around the world are focusing on essential edible systems that are capable of protecting, encapsulating, and releasing of the functional bioactive elements at the target site (Silva et al., 2012; Shukla et al., 2014). PPHs, flavoring agents, essential oils, peptides, carotenoids, essential fatty acids, water-insoluble vitamins, and aroma components are all the examples of BACs (Bazana et al., 2019), which are extensively studied for food preservation against oxidants and microbial spoilage (Tabatabaei et al., 2018). These BACs generally have poor solubility, stability, bioavailability; have high rate of degradation under the biological environment; usually have undesirable taste; are prone to oxidation (upon exposure to heat, light oxygen, and humid conditions), and also have low solubility (Ne et al., 2008), (Figure 14.1). Owing to all of these issues, extensive work on the nanoencapsulation of these BACs has been going on all around the globe.

14.2 NANOENCAPSULATION OF NATURAL BIOACTIVE COMPOUNDS

Targeted delivery and encapsulation are considered as the major concerns in food and pharmaceutical sciences. The objective of BACs encapsulation is to minimize the harm and unwanted degradation through processing stages in addition to digestion conditions; therefore, developing a proper carrier system is pretty vital. Before proceeding to the nanoencapsulation methods, it is worthwhile to have knowledge about the microencapsulation process and its evolution (Jafari, 2017a). Encapsulation refers to the delivering system for BACs. The raw material for the encapsulation of BACs and nutraceuticals can be obtained from biocompatible phytochemicals, which in turn help to improve the solubility status, pH,

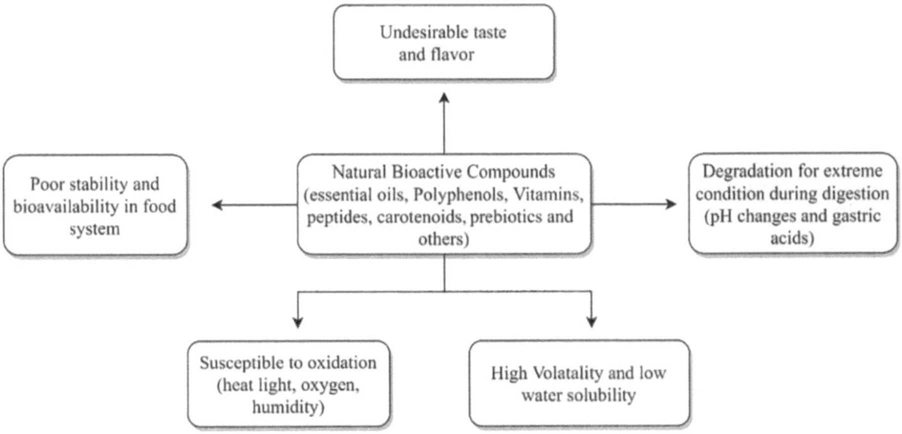

Figure 14.1 General drawbacks of using bioactive compounds in the original form.

protection against oxidants, and enable the prolonged absorption of various BACs and nutraceuticals (Pisoschi et al., 2018). Further improvements in the encapsulation process can be achieved by reducing the dimensions of encapsulating material at a nanoscale, thus increasing the area per unit volume. Nanoencapsulation necessitates the substance's encapsulation in nanocarriers by the methods of absorption, chemical interaction, incorporation, or dispersion (Anandharamakrishnan, 2014). Nanoencapsulation is one of the adopted and favorable alternative branches of nanotechnology with hydrophilic carriers to maintain and protect BACs and nutraceuticals in unfavorable environmental conditions. The technique is based on encapsulating a BAC in solid/liquid/gaseous states within the matrix of nonreactive material for protecting the coated ingredient (food/flavor ingredients) (Pateiro et al., 2021), and thus increasing its bioavailability, solubility, and stability (shelf-life) (Figure 14.2), as a consequence favoring their applications in food industries and pharmaceuticals (Assadpour et al., 2016; Rezaei et al., 2019).

14.2.1 Polyphenols and the Factors Affecting Their Bioavailability

PPHs are the ubiquitous class of antioxidants which are naturally present in different parts of vascular plants (Soobrattee et al., 2005). Presently, around 8,000 naturally occurring PPHs were identified form different plants species. These PPHs are biologically active and have more than one aromatic ring in their structure, which possess at least one Hydrozyl (OH) group, and they have been extensively used in various traditional and herbal medications all around the world for primary health care. Furthermore, according to associated meta-analysis and epidemiological studies, continuing consumption of diets, rich in plants PPHs, contribute toward the protection against various chronic diseases and, therefore, promote the speedy recovery of health (Figure 14.3) (Pandey & Rizvi, 2009).

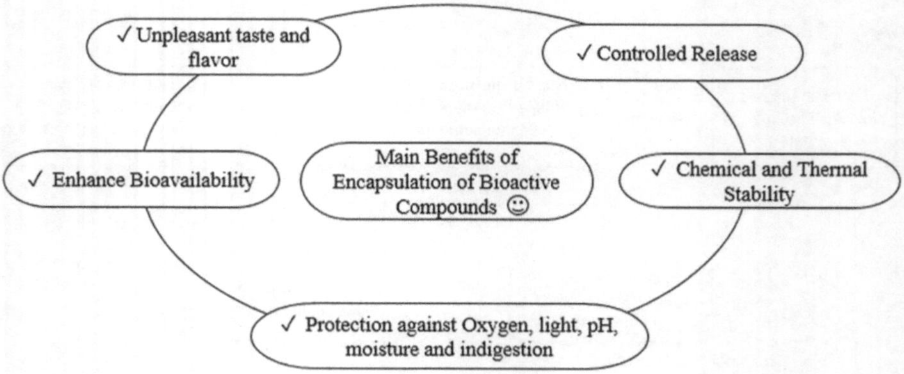

Figure 14.2 Main benefits of BACs nanoencapsulation.

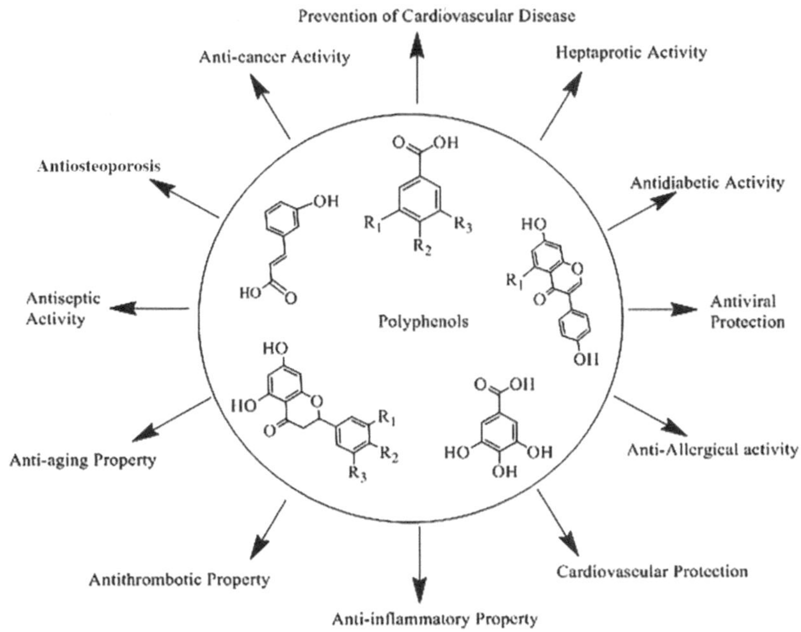

Figure 14.3 Some of the therapeutic properties shown by natural PPHs.

14.3 DISTRIBUTION AND CLASSIFICATION OF NATURAL POLYPHENOLICS

Natural PPHs compounds are one of the biggest class of natural SMs which are extensively disseminated across the realm of plants (King & Young, 1999). The composition of these natural PPHs varies a lot across the realm of plants both qualitatively and quantitatively. Some of the PPHs are omnipresent in different plant species/families, while the presence of others is restricted to only certain species/families (e.g., isoflavones in legumes) (Cheynier, 2018). Furthermore, these natural PPHs are rare in bacteria, algae, and fungi. These natural PPHs (both simple and complex ones) are primarily synthesized biogenetically from the phenylpropanoid or shikimic acid or from the acetate/malonate path (also recognized as the "polyketide" path) (Figures 14.4 and 14.5) (Arts & Hollman, 2005).

Natural PPHs are mainly categorized into two categories – flavonoids and nonflavonoids, which are further subdivided into different subclasses, The flavonoids are further subdivided into seven classes, viz., chalcones, flavones, flavonols, flavanols, flavanones, anthocyanins, and isoflavones, while stilbenes, coumarins, lignins, lignans, phenolic acids (hydroxycinnamic and hydroxybenzoic acids), and tannins (hydrolysable and condensed tannins) are the subclasses of nonflavonoids (Munin & Edwards-lévy, 2011; Manach et al., 2004) (Figure 14.6).

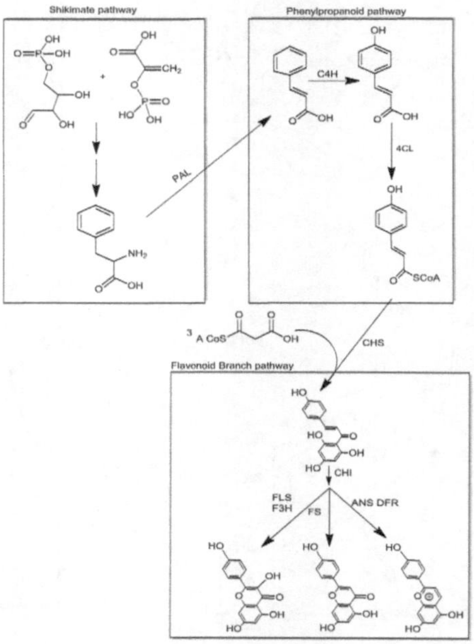

Figure 14.4 Pathways for the biosynthesis of natural PPHs. C4H, cinnamate-4-hydroxylase; PAL, phenylalanine ammonia-lyase; 4CL, 4-CoumaroylCoA-ligase; CHI, chalcone isomerase; CHS, chalcone synthase; F3H, flavanone 3-hydroxylase; FS, flavone synthase; ANS, anthocyanidin synthase; DFR, dihydroflavonol reductase; FLS, flavonol synthase.

14.3.1 Flavonoid Compounds

Flavonoids consist of a benzo-γ-pyrone structure that is found throughout the plants kingdom (Kumar & Pandey, 2013). Flavonoids are well acknowledged for their health-promoting and antifungal, antioxidant, antiviral, anticancer, antiaging, anti-inflammatory, antidiabetic properties (Shashank Kumar & Pandey, 2013; Cook & Samman, 2006). Flavonoids also play an active role in the antioxidant defense system of plant tissues that have been subjected to various abiotic and biotic stresses. These are mainly found in the nuclei of mesophyll cells and inside ROS-generating points (Agati et al., 2012).

Flavonoids were first isolated in 1930 from oranges. Initially, these were known as vitamin P and were believed to be a novel category of vitamins at the time. It was later discovered that these chemicals were flavonoids. Till now around 4,000 flavonoids including aglycones (fundamental structure) (Figure 14.6), glycosides, and methylated derivatives have been isolated (Middleton, 1998). This group of chemicals has a basic C_6-C_3-C_6 structure with15-C skeleton (Marais et al., 2006), which consists of two benzene rings (A and B in Figure 14.6) joined by a heterocyclic pyrane ring (C in Figure 14.6) (Sharma et al., 2020).

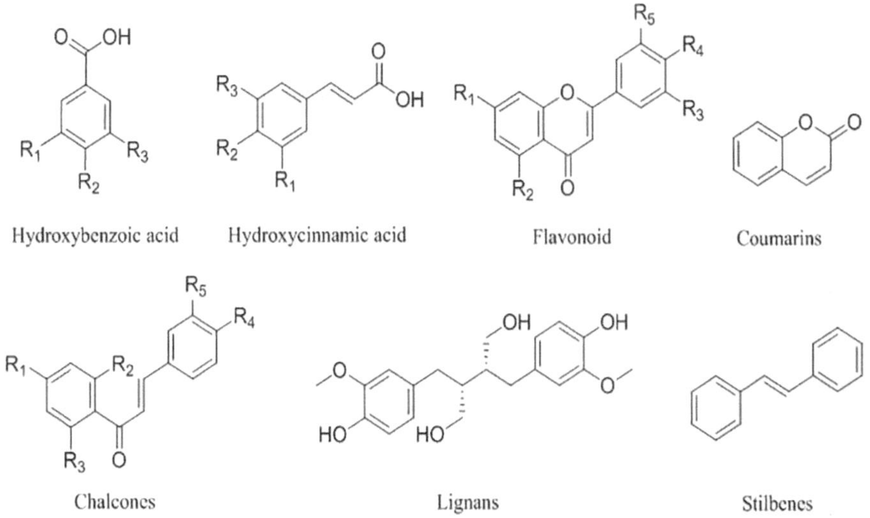

Figure 14.5 Basic chemical structures of some representative PPHs (different R may correspond to H, OH, OCH_3, or any alkyl group mainly).

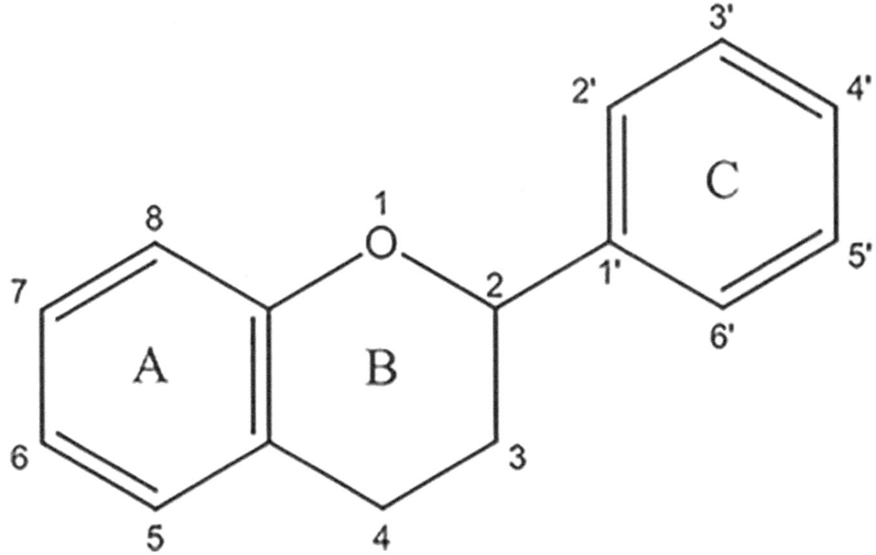

Figure 14.6 Basic benzo-γ-pyrone structure of flavonoids.

14.3.1.1 Medicinal Plants and Food Rich in Flavonoids

Humans beings have been consuming flavonoids (for around 4 million years) since the beginning of human life on the planet owing to their diverse array of therapeutic potential (Belink, 2018). Several epidemiological investigations have proposed that consuming black/green tea rich in flavonoids may reduce cholesterol level and blood pressure, thereby minimizing the probability of cardiovascular disease (Craig, 2018; Kumar et al., 2012). Higher plants are the key source of natural flavonoids and these are present in every part of the plant, but their distribution in different parts of the plant is not uniform. Table 14.1 presents some medicinal plants that are high in flavonoids.

Fruits and vegetables were also known to contain a variety of flavonoids in higher concentration. These fruits and vegetables contain different complex polyphenols in combinations (Cheynier, 2018); for instance, apples and grapes contain anthocyanins and hydroxycinnamic acids (primarily found as quinic esters such as chlorogenic acid), flavanols (procyanidins and catechins), monomers or oligomers, and dihydrochalcones (Manach et al., 2004). Anthocyanin 3-glucosides, acylated anthocyanins,

Table 14.1 Name of Some Medicinal Plants Which Are Rich Sources of Flavonoids (Self-Creation)

Plant	Family	Flavonoid	References
Acalypha indica	Euphorbiaceae	Kaempferol glycosides	Lopez (2001)
Aloe vera	Asphodelaceae	Luteolin	Lopez (2001)
Andrographis paniculate	Acanthaceae	5-hydroxy-7,8-dimethoxyflavone	Gupta et al. (1983)
Azadirachta indica	Meliaceae	Quercetin	Tripoli et al. (2007)
Bacopa moneirra	Scrophulariaceae	Luteolin	López-lázaro (2009)
Brysonimacrassa	Malpighiaceae	(+)-catechin	Aderogba et al. (2006)
Betula pendula	Betulaceae	Quercetrin	Gupta et al. (1983)
Clerodendrum phlomidis	Verbenaceae	Pectolinarigenin	Tripoli et al. (2007)
Citrus medica	Rutaceae	Hesperidin	López-lázaro (2009)
Calendula officinalis	Compositae	Isorhamnetin	Gupta et al. (1983)
Clitoria ternatea	Fabaceae	Kaempferol-3-neohesperidoside	Sankaranarayanan et al. (2010)
Cannabis sativa	Compositae	Quercetin	Gupta et al. (1983)
Limnophila indica	Scrophulariaceae	3,4-methlenedioxyflavone	Sannomiya et al. (2005)
Glycyrrhiza glabra	Leguminosae	Liquiritin	Gupta et al. (1983)
Mimosa pudica	Mimosoideae	Isoquercetin	Sannomiya et al. (2005)
Mentha longifolia	Lamiaceae	Luteolin-7-O-glycoside	Kogawa et al. (2007)
Momordica charantia	Curcurbitaceae	Luteolin	Ghoulami et al. (2001)
Oroxylum indicum	Bignoniaceae	Chrysin	Sannomiya et al. (2005)
Pongamia pinnata	Fabaceae	Ponga flavonol	Agarwal and Kamal (2007)
Passiflora incarnata	Passifloraceae	Vitexin	Gupta et al. (1983)
Tephrosia purpurea	Fabaceae	Purpurin	Sannomiya et al. (2005)
Tilia cordata	Tiliaceae	Hyperoside	Gupta et al. (1983)

flavanol, flavonols, anthocyanins, proanthocyanidins, hydroxycinnamates, phenolic acids, and stilbenes were the key PPHs present in the wine prepared from the grape berries (Koes et al., 2005).

Flavonols: Flavonols are 3-hydroxyflavones (Herrmann, 1976) found extensively in fruits and vegetables. These flavonoids are biologically most active. As a result, the intake of a fruit and vegetable diet having flavonol content plays a crucial role against several diseases. Therefore, flavonoids are sold as nutritional/dietary supplements, either in pure form or as a blend of different flavonoids or extracts. Morin, rhamnetin, myricetin, quercetin, azaleatin, fisetin, and kaempferol occur naturally, while astragalin, rutin, azalein, robinin, and myricetin are the glycosides derivatives of flavonols (Ignat et al., 2011).

Flavones: Flavones are oxygen heterocycles with a backbone of 2-phenylchromen-4-one or 2-phenyl-1-benzopyran-4-one (Marais et al., 2006). Fruits, cereals, vegetables, and herbs are the most common sources of naturally occurring flavones. Flavones have an unrivaled ability to control a wide range of enzyme systems, and thus they perform tremendous biological actions including antioxidant, antiviral, antibacterial, anticancer, antiallergic, antiosteoporotic, and so on. Some naturally occurring flavones include luteolin, chrysin, wogonin, tangeritin, acacetin, apigenin, and zapotin (Singh et al., 2014).

Isoflavones: In contrast to flavones that contain an aromatic ring at C_2, isoflavones contain an aromatic ring at C_3. Isoflavones have –OH groups at C_7 and C_4 that are similar to those found in estrogens (Klejdus et al., 2010). Isoflavones occur naturally in around 300 plant species, which belongs to 59 different families. Isoflavones are primarily found in species of Leguminosae family and Papilionoideae subfamily. Some bacteria and fungi also manufacture isoflavones naturally. Isoflavones are known to show anti-inflammatory, antioxidant, antiestrogenic, antiproliferation, osteoporosis prevention, and immunomodulatory properties that promote health. Biochanin A, genistein, wighteone, luteone, daidzein, glycitein, and formononetin are some of the isoflavone aglycones that occur naturally. The most important physiologically active isoflavones in humans are daidzein and genistein (Reinli and Block, 1996; Herrmann, 1976; Sharma et al., 2020).

Flavanones: Flavanones contain saturated three-carbon chain and one with ketonic functional group at C_4 (Marais et al., 2006). These are different from flavones by a single C_2–C_3 bond. These are mainly found in higher plants families, such as Compositae, Rutaceae, Leguminosae, and Lamiaceae. At position C_7, flavanones are typically glycosylated by many different disaccharides. These are mainly present in higher concentrations in the citrus fruit, particularly in the peel. Other than citrus plants, aromatic plants are also a source of natural flavanones. Hesperidin, sakuranin naringin, and poncirin are key flavanone glycosides, whereas naringin, sakuranetin, eriodictyol, and sterubin are major flavanone aglycones obtained from various higher plant species (Khan & Dangles, 2013).

Flavanols: Monomeric flavanols or catechins are also known as flavan-3-ols and they are the most common kind of monomeric C_6-C_3-C_6 flavanol. Catechin and epicatechin are the most widely occurring flavanols. Plant-derived food typically contains epicatechin (EC), epigallocatechin (EGC), catechin (C), gallocatechin (GC), and their galloyl substituted derivatives. Some of the flavanols occur in polymerized form and are frequently found as oligomers or polymers in plant-based food products (de Pascual-teresa et al., 2010). Other examples of some other naturally occurring flavanols are mesquitol, fisetinidol, and guiboustinidol.

Anthocyanins: Anthocyanins are water-soluble vacuolar pigments found usually in fruits and vegetables (Blando, 2004). Anthocyanins are most often found in nature as heterosides. Anthocyanidin (or aglycons) is structurally based on the 2-phenylbenzopyrilium or flaviliumcation, with the substituents (polyhydroxy and polymethoxy groups) present at different positions in the basic structure (de Pascual-teresa et al., 2010). The anthocyanidin core is made up of an aromatic ring A fused to a heterocyclic ring C that contains oxygen as the heteroatom, which is then joined to a third aromatic ring B by a C–C linkage (Sharma et al., 2020; Blando, 2004).

Chalcones: Chalcones occur naturally with a 1,3-diphenylprop-2-en-1-one backbone and the two rings (aromatic) are joint together by α, β-unsaturated 3-carbon carbonyl linkage. Chalcones are abundant in eatable plants and play the role of metabolic forerunners for synthesizing several flavonoids and isoflavonoids. Vegetables (shallots, tomatoes, potatoes, etc.), beer, hops, spices (licorice, cardamom), citrus fruits, and bean sprouts are all rich sources of natural chalcones. Chalcone-rich plants are also used in a variety of traditional herbal medicine systems (Gaonkar & Vignesh, 2017).

14.3.2 Nonflavonoid Compound

Phenolic acids: Phenolic acids (PA) are the aromatic SMs obtained from plants subcategorized into two main classes, namely, benzoic acid (hydroxybenzoic) derivatives and cinnamic acid (hydroxycinnamic acid) derivatives (Robbins, 2003; Herrmann, 1989; Croteau et al., 2000). Except for certain red fruits, onion, potato peel, etc., the content of hydroxybenzoic acid in fruits and vegetables is normally low (Herrmann, 1989; Törrönen et al., 1997; Lewis et al., 1998). 4-Hydroxybenzoic, gallic, protocatechuic, and ellagic acids are the most common phenolic acids of the hydroxybenzoic acids subclass, while sinapic, p-coumaric, caffeic, and ferulic acid are the most common phenolic acids of cinnamic acid subclass. Caffeic acid is the most prevalent (70%) among all the hydroxycinnamic acid derivatives found in fruits (Van Dam & Hu, 2005). These phenolic acids are powerful natural antioxidants and are also known to show antibacterial, cytotoxic, chemopreventive, anticancer, neuroprotective, antifungal, and antiviral properties (Saibabu et al., 2015).

Coumarins: Coumarins (Figure 14.7) have a 2H-1-benzopyran-2-one framework and are one of the most important nonflavonoid polyphenolic groups. Coumarins are also called benzopyrones because the α-pyrone ring is fused with benzene. Coumarins are separated into distinct families from simple to polycyclic coumarins, such as pyranocoumarins and furocoumarins. These are naturally present in stems, fruits, leaves, roots, flowers, and seeds of different plants. Flowers and fruits are known to have the highest concentration of coumarins. Over 1,300 distinct coumarins have been isolated and identified till date. Coumarins are naturally found (in higher concentration) in wine, tea, fruits, seeds, vegetables, nuts, and coffee. Coumarins are also known for their antiadipogenic, antioxidant, antihyperglycemic, anti-inflammatory, antihypertensive, anticonvulsant, anticancer, antitubercular, antibacterial, antifungal, antiviral, and anticoagulant properties. Esculetin, osthole, umbelliferone, fraxidin, scopoletin, ostruthin, novobiocin, and ammoresinol are the few naturally and usually occurring coumarins (Venugopala et al., 2013).

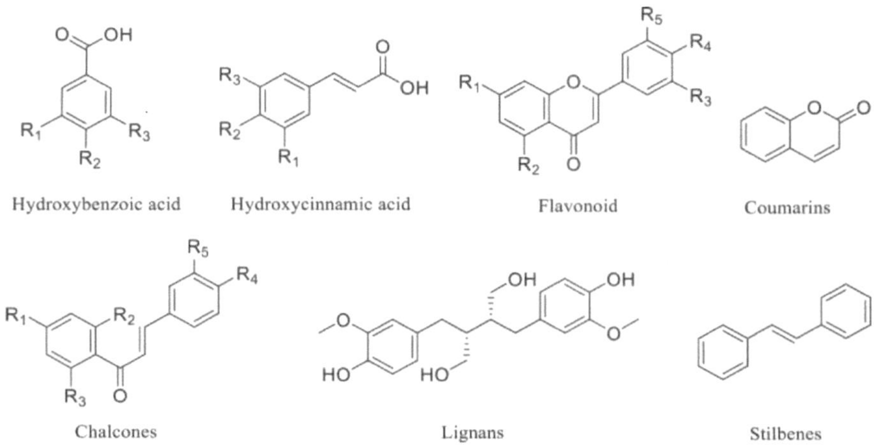

Stilbenes: Stilbenes are made up of 2-phenyl rings linked by a 2-Carbon methylene bridge. Stilbenes exist in stereoisomer form cis and trans, and natural-source of stilbenes mainly have trans-form. Dimeric, trimeric, and polymeric stilbenes are naturally found in both free and glycosylated forms. Stilbenes are found in extremely small amounts in the human diet. Plants synthesize stilbenes in response to stressful conditions and pathogen infestations. These have been extracted from more than 70 species (peanuts, grapes, berries, etc.). The naturally occurring stilbenes are resveratrol, pterostilbene, and piceatannol (Figure 14.8). Stilbenes are also known to exhibit anti-inflammatory, antioxidant, antifungal, and anticancer properties (Ignat et al., 2011).

Lignins: Lignin is an organic polymer (complex) made up of various phenylpropane units that are bonded together. Coniferyl, sinapyl, and p-coumaryl alcohols are three aromatic precursors (monolignols) that help to generate these phenylpropane units. Lignin is a major component found in the structural elements of a vascular plant and algal support tissues (bark, wood, and cell walls). Lignin offers mechanical pillars for holding plant fibers together and provides structural strength to cell walls, resistance to biochemical stress, and aids in the passage of plant liquid (water) and nutrients. Lignins and their derivatives have immunomodulatory, antiviral, antidiabetic, antiobesity, and anticancer properties (Laurichesse & Avérous, 2014; Vinardell, 2017).

Tannins: Tannins have high molecular weight ranging from 500 Da (Dalton) to more than 3,000 Da and are mainly present in wood, roots, bark, leaves, and fruit (Hagerman & Butler, 1981). These are water soluble (20°C–35°C) except for some higher molecular weight derivatives. These can be present in both flowering and nonflowering plants. Tannins are categorized into two main classes based on their chemical structure and properties: condensed tannins obtained from a combination of flavonoids and hydrolyzable tannins mainly synthesized or obtained from self-polymerization of phenolic acids or with sugar. Gallic acid is the primary source of hydrolyzable tannins, esterified to a core polyol, followed by the esterification of galloyl groups, which are then oxidatively cross-attached to form the more complicated hydrolyzable tannins. Proanthocyanidins, commonly known as condensed tannins, are polymeric flavonoids. Despite well-established or recognized flavonoid biosynthetic pathways, the steps leading to flavonoid condensation or polymerization

(a)

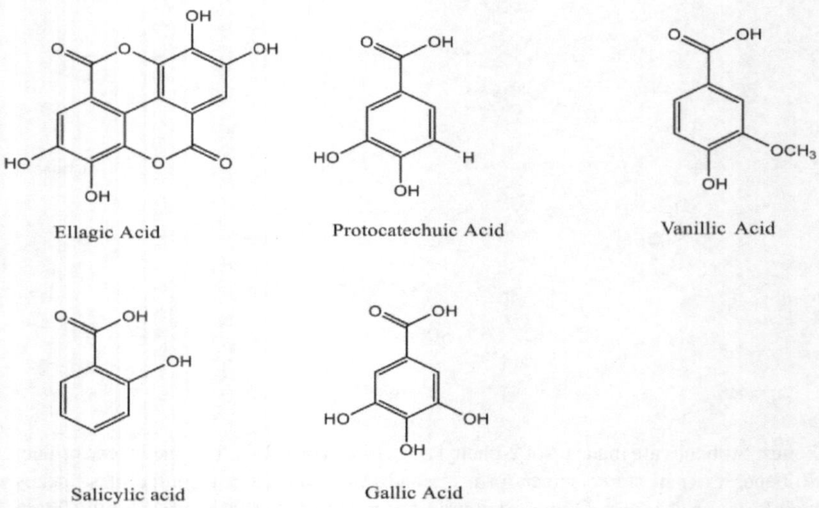
Ellagic Acid
Protocatechuic Acid
Vanillic Acid
Salicylic acid
Gallic Acid

(b)

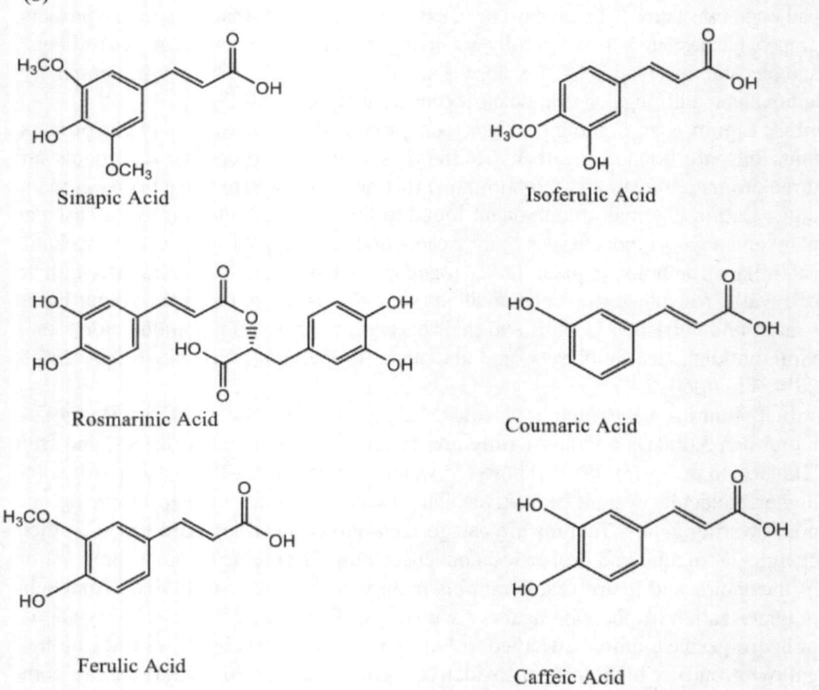
Sinapic Acid
Isoferulic Acid
Rosmarinic Acid
Coumaric Acid
Ferulic Acid
Caffeic Acid

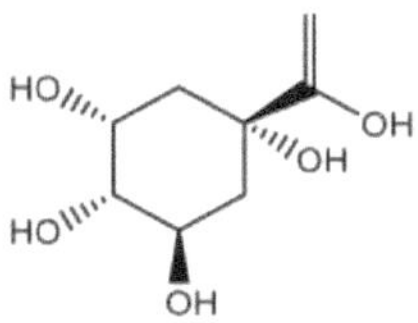

Quinic Acid

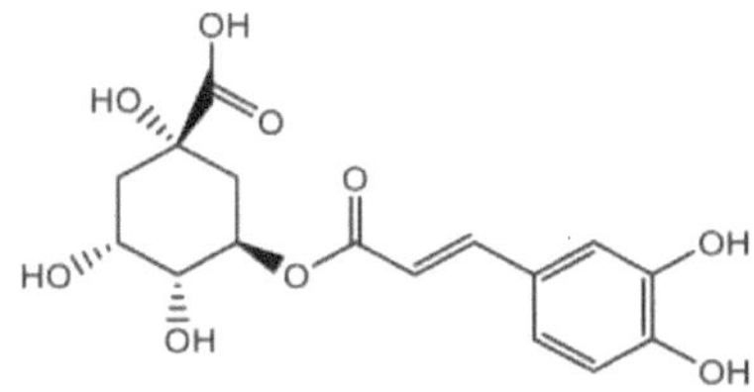

Chlorogenic Acid

(c)

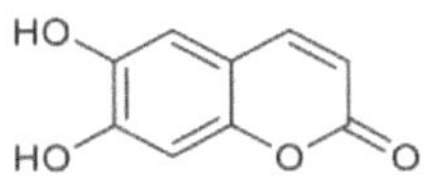

Esculetin

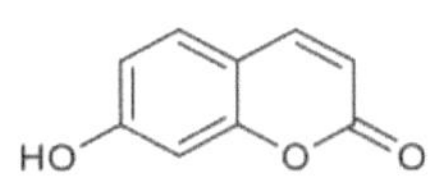

Umbelliferone

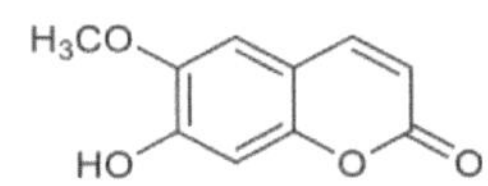

Scopoletin

(d)

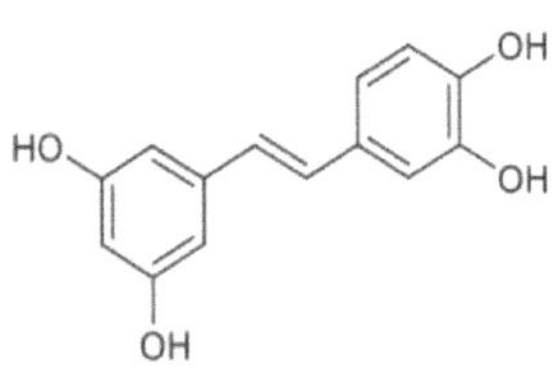

Piceatannol

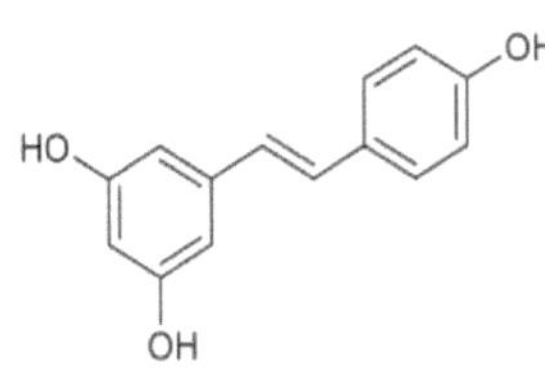

Resveratrol

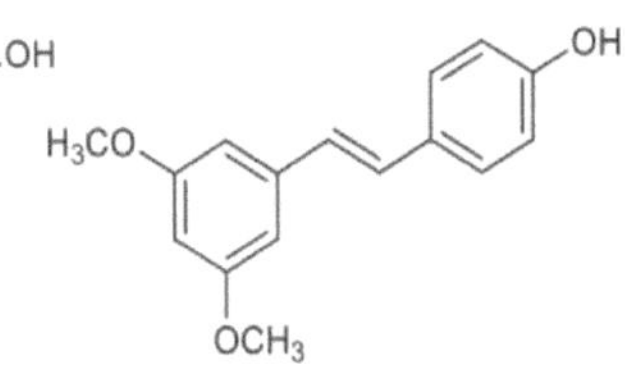

Pterostilbene

(e)

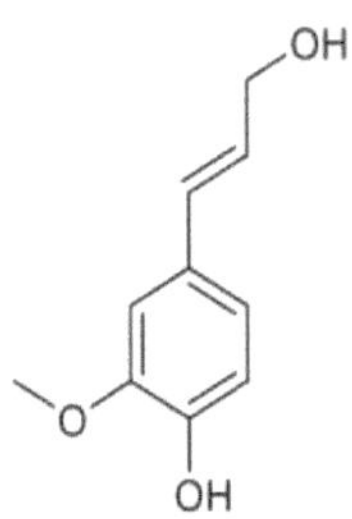

Coniferyl alcohol

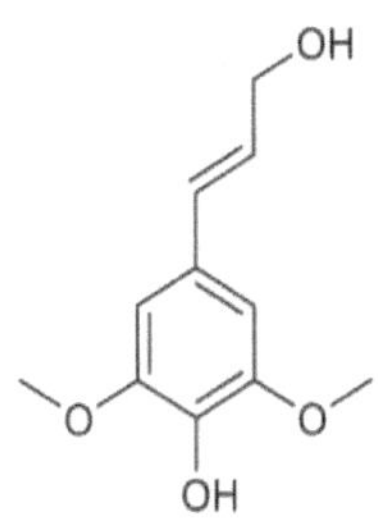

Sinapyl alcohol

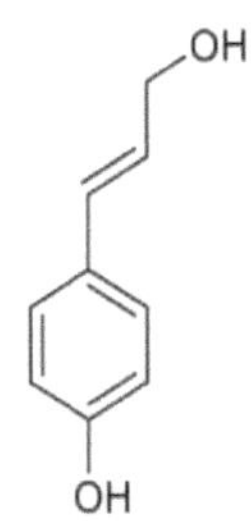

p-Coumaryl alcohol

(f)

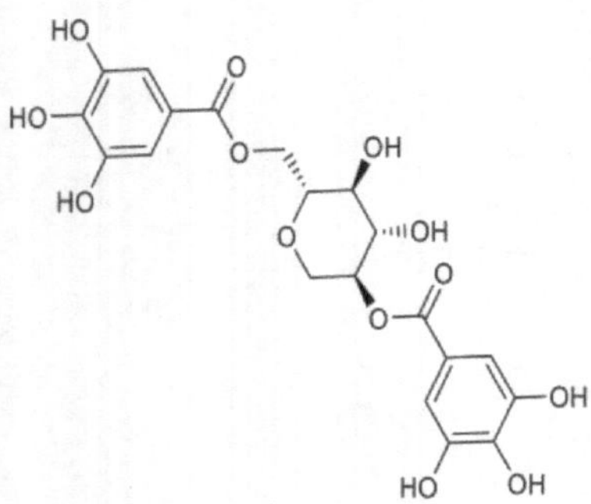

Acertannin

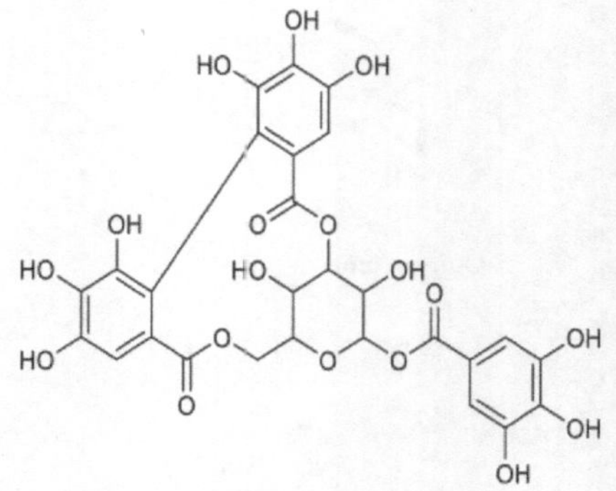

Corilagin

(g)

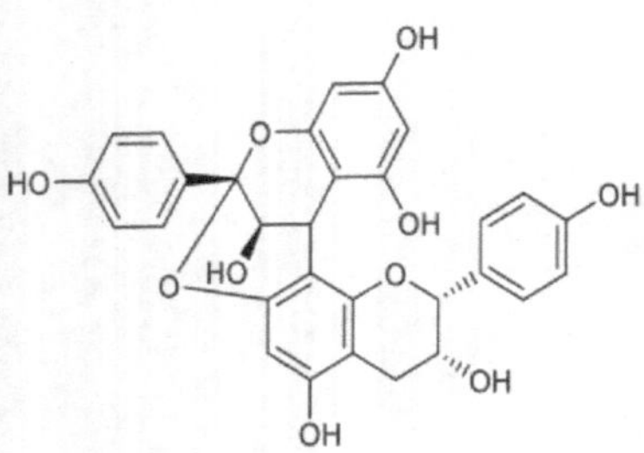

Geranin A

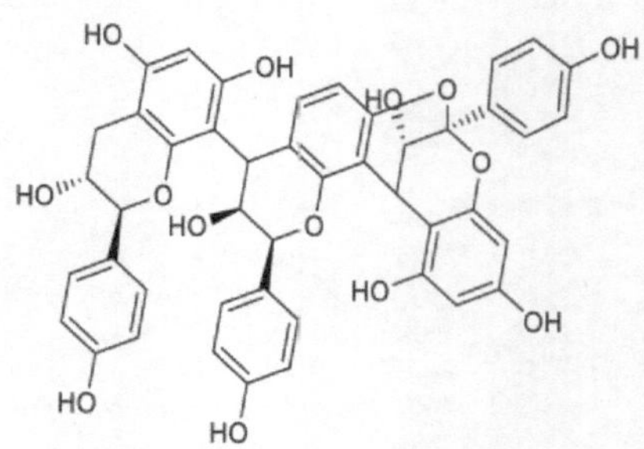

Selligueain A

to form condensed tannins have yet to be unraveled. Flavan-3-ols like (+)-catechin and (–)-epicatechin are used to make the commonly occurring condensed tannins. In biological systems, these can operate as protein precipitators, metal ion chelators, and biological antioxidants. Furthermore, tannins, like other natural PPHs, have antibacterial, anti-inflammatory, cardiovascular, and anticarcinogenic properties. Tannins are found in a wide variety of angiosperms and gymnosperms (Mole, 1993; Chung et al., 2016). Many plant species, including *Sericea lespedeza*, *Acacia* spp., and pasture species including *Lotus* spp., contain tannins (*Journal of American Science, 2011;7(5) Http://Www.Americanscience.Org Effect of Condensed Tannin on Controlling Faecal Protein Excretion in Nematode-Infected Sheep:*, 2011).

Lignans: Lignans are mainly generated from oxidative dimerization of 2-phenylpropane (C_6–C_3) units, while the final product is joined by two particular carbons (C_2–C_{20}). Further reactions of lignans (isomerization/cyclization) result in many different structures. These generally exist in the free form but occur in a minor quantity of their bonded counterparts (glycoside derivatives). Lignans, both naturally occurring and synthetic derivatives, have shown promising roles in chemotherapy (skin, lung, breast, and rectum cancer) (Saleem et al., 2005). Many lignans, such as secoisolariciresinol and matairesinol, are phytoestrogens that are converted into enterodiol and enterolactone by the intestinal microbiota. Some lignans exhibit antiestrogenic and estrogenic effects, similar to isoflavones. Furthermore, these nonflavonoid compounds are also known to exhibit hepatoprotective, antibacterial, antioxidant, antiviral, antifungal, immunosuppressive, and anti-inflammatory properties (Ignat et al., 2011). Linseed, grains, cereals, flax, legumes, algae, vegetables, and some fruits, are the key sources of lignans. Pinoresinol, podophyllotoxin, secoisolariciresinol and matairesinol are some naturally occurring lignans (Sharma et al., 2020).

The abovementioned natural PPHs are most abundant in our diet; however, these are not always the ones with the best bioavailability (Archivio et al., 2010). As a result, the clinical applicability of these natural PPHs is limited. To overcome these disadvantages, nanodelivery and nanoencapsulation methods have been extensively explored these days (Etheridge et al., 2013; Acosta, 2009; Nair et al., 2010).

14.4 TECHNIQUE/INSTRUMENTS USED IN NANOENCAPSULATION OF BIOACTIVE COMPOUNDS/POLYPHENOLICS

Various nutraceutical compounds, such as dietary fibers (inulin), antioxidants (flavonoids, tocopherols, PPH), minerals (Fe^{+2}, Mg^{+2}), essential fatty acids (omega 3), phytosterols (β-sitosterol and stigmasterol), carotenoids (lycopene and carotene), vitamins (thiamin, vitamin D, riboflavin), and bioactive peptides (casein hydrolysates), have been largely consumed worldwide to maintain good health (Jafari, 2017b). When these nutraceutical compounds are incorporated into food structures, most of them are susceptible to breakdown and degradation throughout storage and processing under different environmental conditions. So, to protect these nutraceuticals from breakdown and degradation during processing and storage, nowadays different types of microencapsulation technologies are commonly used to encapsulate these nutraceuticals before their incorporation into foods products (Shimoni, 2009). The different encapsulation technologies are discovered based on physical (e.g., spray drying), chemical (e.g., in situ polymerization), and physicochemical (e.g., hydrophobic interactions, ionic gelation, etc.) principles. A microcapsule is made up of a minute sphere enclosed with the monolithic shell. The substance inside it is attributed to the inner segment (encapsulant, fill/payload segments, core material), while the wall of the shell is known as the coating, wall material, external segment, encapsulating agent, membrane shell, carrier material, or matrix (Jafari et al., 2008; Mahdavi et al., 2014). As described before, microencapsulation has the ability to hide the unpleasant taste of some BACs in food preparations, as well as to convert liquid to solid phases.

14.4.1 Methods of Nanoencapsulation of Nutraceuticals and Food Components

Nanoencapsulation involves the encapsulation of active material on a nanoscale using nanodispersions, films, or layers. Its main advantage is the induced uniformity (homogeneity), which leads to better chemical and physical characteristics (Khare &

Vasisht, 2014; Sanguansri & Augustin, 2006). The nanocapsules have greater advantages than micron-sized capsules, owing to their bigger surface area, which result in higher solubility, targeted liberation, and improved bioavailability (Ezhilarasi et al., 2013; Mcclements et al., 2009). Nanoencapsulation technologies are mainly classified into two groups: "top-down" (emulsification solvent-evaporation and simple emulsification) or "bottom-up" (coacervation, supercritical fluids, nanoprecipitation, and inclusion complexation). In the "top-down" process, particle size is abridged throughout the encapsulation procedure (e.g., by employing various mechanical forces), while in the "bottom-up" method, the particle size is raised through the various technique of self-assembly (Jafari, 2017b; Anandharamakrishnan, 2014). Nanoencapsulation techniques primarily utilized for the synthesis of nanocapsules have been chiefly classified into five main groups as presented explained below:

- lipid formulations (nanoemulsions, nanostructured lipid carriers (NLCs), nanoliposomes);
- natural nanocarriers (cyclodextrins (CDs), caseins, and nanocrystals);
- specialized equipment (electrosprayed Nano particles (NPs), nanospray-dried particles, and electrospun nanofibers);
- functional biopolymeric NPs (single biopolymer NPs and biopolymer nanocomplexes);
- other techniques (amylose nanostructures, protein nanotubes, and carbohydrate nanogels).

14.4.1.1 Lipid-Formulation Nanoencapsulation Technologies

Lipid-based nanoencapsulation methods have been largely used by pharmaceutical and food industries along with various research and development agencies (Assadpour et al., 2016). Despite numerous benefits of biopolymer nanocapsules, mass production is impossible owing to the need for complex chemical and thermal procedures that must be continuously monitored. On the other hand, water-insoluble nanocarriers have the ability to be scaled up, as well as offer increased encapsulation efficiency and reduced toxicity. Most of the BACs, such as fatty acids, preservatives, flavonoids, scents, hydrophobic vitamins, and other beneficial substances are all hydrophobic. So, emulsifiers (usually oil in water) are frequently required to encapsulate these lipophilic substances. Furthermore, the availability of digestible lipids facilitates nutrient absorption by increasing the quantity of mixed micelles required to break down and deliver hydrophobic molecules (Jafari, 2017b; Trevaskis et al., 2008). As a result, various studies on the use of lipid-based nanocarriers have been presently going on.

Encapsulation by nanoemulsions: Nanoemulsions are considered to be the best vehicles for encapsulating BACs. These can improve water solubility and limit the decay of food BACs, making them useful as nanocarriers in a variety of products. Oil-in-water (O/W) nanoemulsions are good for incorporating lipophilic BACs, whereas water-in-oil emulsions (W/O) are utilized to encapsulate natural PPHs hydrophilic substances (Mohammadi et al., 2016). As the oil phase is composed of different nutritious oils, the lipophilic components are generally protected in these emulsions. Recently, various researchers from all around the world have also been working on the utilization of double emulsions methods for the nanoencapsulation of various food BACs, especially natural PPHs (Assadpour et al., 2016; Mohammadi et al., 2016; Jafari, 2017a).

Encapsulation by nanoliposomes: Liposomes are vesicles with an aqueous core surrounded by polar lipid bilayers, primarily phospholipids. The self-assembly of amphiphilic lipids in aqueous solutions is responsible for the production of liposomes. The interior aqueous phase is formed by the concentrated packing of amphipathic lipids, which serves as a basin for the encapsulation of hydrophilic constituents (Briuglia et al., 2015). Furthermore, in comparison to the liposomal hydrophilic core, outer lipid bilayers provide a very lipophilic environment. As a result, they are extremely versatile in encapsulating and stabilizing the aqueous components in their aqueous core and hydrophobic components in their lipid bilayers, respectively (Ghorbanzade et al., 2017). Nanoliposome production generates significantly more fresh surface areas, necessitating a higher energy input. The usual diameter range of nanoliposomes is around 50–150 nm in the majority of current applications (Ghorbanzade et al., 2017).

Encapsulation by nanostructured lipid carriers: NLCs and solid lipid nanoparticles (SLNs) have shown benefits over alternative lipid-based nanocarriers including nanoemulsions and nanoliposomes. SLNs are innovative nanoparticulate vehicles that have been designed to extend the release duration of active compounds by virtue of their solid form, insulating them from undesired chemical reactions such as oxidation. SLNs are lipid droplets that have been entirely crystallized and organized and contain BACs in their lipid matrix (Weiss et al., 2008). SLNs were created as a replacement for outmoded lipid-based carriers by combining the benefits of all prior systems into a single structure. These consist of a single lipid (solid) or a combination of lipids (Müller et al., 2002). The SLN's solid matrix is obtained from O/W emulsions in which a hard solid phase replaces the liquid lipid (oil) (i.e., maintains its solid structure at body temperature) (Müller et al., 2002).

Later, Lipo-SNLs carriers were introduced into the nanoencapsulation field to compensate for the shortcomings of SLNs. The primary goal was to increase the loading capacity while also preventing BACs from being expelled. These are made by dispersing solid and liquid lipids, as well as active substances (in the inner phase), in water with emulsifiers (as the outer phase). Unlike SLNs, the occurrence of lipid in the liquid form in the NLCs (inner phase) allows active compounds that are readily solubilized in liquid lipid to be entrapped. The immobilization of the encapsulated active components in solid lipid also protects them against degradation. The lipid combination in NLCs, on the other hand, has a slower polymorphic transition and a low crystallinity index. When compared to SLNs, the inner phase composition of NLCs can provide better bioavailability, loading volume, and good encapsulation efficacy of nanoencapsulated substances (Müller et al., 2002).

14.4.1.2 Nanoencapsulation Techniques Based on Natural Nanocarriers

The concept of encapsulating BACs with natural nanocarriers came from considering the various features of NPs such as being biocompatible, ecofriendly and cost-effective. The isolation methods, biological source, biochemical and physical modifications all influence the functional qualities of naturally occurring nanovehicles. Nanocrystals, caseins, starch granules, CDs, and amylose nanostructures are examples of natural sources used in the nanoencapsulation of nutraceuticals and BACs (Faridi Esfanjani et al., 2017). The nanoencapsulation by different natural sources has been discussed below:

Nanoencapsulation by cyclodextrins: CDs, which have a nanoscale dimension by nature, are commonly used as encapsulating materials for molecular inclusion. CDs are made up of five or more α-d-glucopyranoside bodies that form cyclic oligosaccharide assemblies and are formed from starch molecules. In general, CDs are divided into two categories: natural and modified (Wang et al., 2014). Till date, about 1,500 CD subclasses have been recorded, which mainly belong to α-, β-, and γ-CDs with 6, 7, and 8 glucopyranose units (Singh et al., 2014). These structural characteristics make CDs an ideal medium for entrapping molecules of lower polarity (such as essential oils) within the nonpolar hole via a water-hydrophobic interaction. The compatibility and assembly of the CD with the integrated BACs are connected to the stability and arrangement of the CD-bioactive chemical complex. As a result, the capacity of different CDs to entrap the same BACs varies. CD molecules have a hydrophilic exterior structure and vice versa (Astray et al., 2009). Till date, several efforts have been made to improve the functional features of CDs; for instance, chemical alteration of parent CDs to yield numerous derivatives with increased physiochemical properties and complexing power (Loftsson & Duchêne, 2007).

Nanoencapsulation by caseins: Casein micelles are widely known as food-grade nanocarriers that entrap and deliver nutraceuticals, which are hydrophobic in nature. Casein protein accounts for about 80% of the total protein content (30–35 g/L) in bovine milk and is often found as spherical micelles (5–500 nm). The natural high digestibility of nanoscale micelles and caseins may help to increase the stability of micelles and bioavailability of encapsulated nutraceuticals throughout the process. Caseins, in fact have distinct functional and structural features that make them ideal nanocarriers for nutraceuticals and BACs. Some of the examples related to the key functional and structural features of caseins are binding to molecules or ions, interactions with proteins or polymers (e.g., noncovalent interactions and covalent conjugations), self-assembly/re-assembly capabilities, surface activity, binding to hydrophobic molecules, and gelation properties (Livney, 2015). Casein micelles self-assemble due to electrostatic interactions and hydrophobic interaction, making them ideal for incorporating hydrophobic bioactive components. These naturally occurring protein complexes are capable of transporting/stabilizing essential nutrients. Milk protein and casein aggregates have been recognized as highly effective nanovehicles by cost, nutrition, and safety (Tavares et al., 2014).

Nanoencapsulation by nanocrystals: Nanoencapsulation based on nanocrystals is proposed to increase oral bioavailability of poorly soluble BACs because poorly soluble BACs generally have low bioavailability which can restrict their efficiency and application in food industries. Nanoparticle with dimension ≤100 nm is referred to as a nanocrystal. Thus, reducing the size of bioactive formulations helps to improve the kinetics of solubility and dissolving velocity of BACs. Pure BACs could be used to make nanocrystals. These can be used in food materials (rutin and hesperidin), functional drinks, nutraceutical components (, tablets and capsules), and the antioxidant CoQ10. Nanocrystals are currently accessible commercially as water-based submicron particle suspensions (available as preserved or unpreserved) that are simply blended during the formulation of products.

14.4.1.3 Nanoencapsulation Techniques via Specialized Equipment

The various specialized techniques and equipment used for nanoencapsulation are discussed below:

Electrospinning: Electrospinning is a new process for continuously fabricating nonwoven nanofibers having diameters of just a few nanometers, which could be used to deliver food-based BACs (Faridi Esfanjani & Jafari, 2016). Electrospinning is a simple and reliable approach for making ultrathin eletrospuns that require the pumping of a biopolymer/polymer solution through a spinneret into a prepared collector. Lastly, the collector plate collects nonwoven ultrafine fibers (Ghorani & Tucker, 2015). The produced nanoscale fiber mat has unique qualities, for instance, significant porosity and a high surface area to volume ratio, which makes them ideal carriers for food BACs and nutraceuticals. Furthermore, the ultrathin fibers' size enhances the usefulness and applications of the inserted molecules (Weiss et al., 2008).

Electrospraying: Electrospraying also termed as electrohydrodynamic atomization is a latest nanoencapsulation technology that is alike to electrospinning in terms of its base, but instead of ultrathin fibers, it generates NPs (Torres-Giner et al., 2010). This approach involves the atomization of liquids into fine droplets at high voltages. On the droplet's stretching toward the ground electrode, the solvent eventually vaporizes, which results in the formation of NPs. The single production process and substantial encapsulation efficiency are two notable advantages of using this approach. Certain parameters, for example, collector distance, volume, voltage, velocity, and solution properties (viscosity, concentration, and density) can affect the nanocapsules and nanospheres produced by this method. Furthermore, the properties of the produced nanocapsules may also be affected by environmental conditions like temperature and pressure (Hao et al., 2013).

Nanospray dryer: A spray-dryer technique is utilized to convert liquid solutions into small particles quickly and efficiently. Even sensitive nutrients can be dried in a short time without causing any damage due to the quick drying time (Jafari et al., 2008). Applications of nanospray drying have been largely used in cosmetics, pharmaceutics, and functional food formulations. In 2009, the nanospray dryer was created allowing spray drying to be extended to the submicron scale (Faridi Esfanjani & Jafari, 2016). This represents a watershed moment in this technology allowing the formation of very fine powders with extremely narrow size distributions and high quantity.

14.4.1.4 Nanoencapsulation via Biopolymer NPs

Biopolymer NPs create functional nanovehicles that are being utilized in food and nutraceutical sectors. Biopolymer NPs are designed with the purpose of delivering nutrients in a regulated and targeted manner, manufacturing value-added food products by increasing their physical features (texture and appearance), and improving the end product's marketing potential. Biopolymers are thought to offer a safe substitute to synthetic polymers, particularly in the field tissue engineering, medication delivery, and wound healing. The biopolymers can encapsulate both hydrophilic and hydrophobic BACs and nutraceuticals (Mukhopadhyay & Kundu, 2015).

Desolvation, thermal gelation, precipitation, and self-assembly are some of the encapsulation processes for creating NPs from biopolymers. Desolvation is of particular relevance for protein NPs. It is a thermodynamically controlled, self-assembly method for biopolymers (proteins) to form NPs. This method basically uses a desolvating agent (another solvent), which causes a reduction in solubility and conformational changes of proteins, which in turn results in the precipitation and phase separation of proteins as NPs (Sahoo et al., 2015). On the other hand, biodegradable polysaccharides are involved in manufacturing nanocarriers. Mild approaches for the manufacturing of carbohydrate nanocarriers should be prioritized to avoid matrix degradation, save money, and permit scaling up and large production. There has been an increasing interest in NPs made of starch as nanodelivery systems among polysaccharides. These NPs could be made from a variety of starch sources that are prevalent in nature. These NPs make it easier to release BACs in a controlled manner without causing unwanted liberation, minimize overdosing, and improve the performance of sensitive ingredients (Simi & Emilia Abraham, 2007).

14.4.1.5 Other Techniques of Nanoencapsulation

Nanoencapsulation can also be done by using protein nanotubes. α-lactalbumin has been identified as the sole dietary protein capable of generating nanotubes. There are various properties of α-lactalbumin nanotubular structures that make them distinctive for utilization in nanotechnological systems. A-Lactalbumin-based gels are strong and reversible in nature. Furthermore, the resulting gel is translucent making it a fascinating nanostructure. Nanotubes may encapsulate enzymes, vitamins, and other bioactive molecules (Ipsen & Otte, 2007). Furthermore, carbohydrates nanogels can also be used for nanoencapsulation. Nanogels are amphiphilic/hydrophilic biopolymers that have been cross-linked physically or chemically and are utilized for a variety of applications such as biocatalysts, biosensors, delivery systems for proteins, and biocidal agents (Yan et al., 2006).

14.5 CONTROLLED DELIVERY OF BIOACTIVE INGREDIENTS AND FUTURE PERSPECTIVES

BACs, especially PPHs, have emerged as a health-promoting healing agent in the 21st century paving the way for the formation of new supplements and functional foods. BAC's stability (typically taken from natural sources) is a prime factor for their efficacy as simulation into various food products (Đorđević et al., 2014). As described above, lipid-based delivery systems (NLCs) which include SLNs, micelles, and liposomes as well as protein- and polysaccharide-based biopolymeric NPs (cationic- and anionic-polyelectrolytes) are the apt NPs for the controlled distribution of BACs (Harde et al., 2011). In comparison to synthetic polymeric NPs, natural polymer-based nanosystems have low cost, nontoxic, biodegradable, and possess bio- and mucoadhesive nature which make them an acceptable BACs carrier. When bioactive phytochemicals are loaded into NPs (phytochemicals encapsulated NPs), they become more productive and help to attain site-specific delivery. Some NPs slowly deliver bioactive phytochemicals inside the cells leading to long-term therapeutic benefits. The study of the cytotoxic profile of NPs in the human neuroblastoma model has been the focus of research on the therapeutic benefits of anthocyanins-loaded NPs against neurotoxicity in vitro (SH-SY5Y cell line – Human Neuroblastoma Cell Line). The administration of BACs orally using edible NPs will constitute a valuable approach. Furthermore, during the assessment of the impact of NPs or nanovesicles on the phytochemical bio-efficacy, specific attention should be given to the various substitutions carried out in NPs structure (Xiao et al., 2017). The pH conditions and enzymatic activities in gastrointestinal fluids could affect nanoparticle stability and their role as bioactive drug carriers. As a result, it is vital to assess nanoparticle functioning in complex matrices and biological systems in order to obtain accurate information on their practical benefit. Overall, the encapsulated phytochemicals have improved bioavailability, improved solubility in aqueous media, increased mucus penetration and intestinal permeation, extended residence time inside the body circulation, controlled release rate at a specific site, facilitated cellular uptake, and are less prone to decay during storage and gastrointestinal digestion (Minekus et al., 2014). In this context, it becomes obvious that the interaction among the phytoconstituents and encapsulating matrix material or molecules in complicated biological systems (such as mammals) needs additional investigation in terms of metabolism and bio-efficacy. Despite the rational usefulness of nanotechnology in enhancing the bioavailability and bioactivity of natural products, NPs are synthesized using physical and chemical approaches which involve the use of expensive and hazardous chemicals particularly in the case of metallic biomaterials limiting their actual application in vivo. NPs synthesis necessitated the use of green technologies that did not involve the use of harmful reagents to prepare metal NPs. In this sense, using environmentally friendly and biocompatible chemicals help to reduce the side effects of these processes and encapsulating materials.

14.6 CONCLUSIONS

Natural PPHs are among the most prime bioactive SMs produced by plants, which display an inimitable combination of physiological, biological, biochemical, and chemical activities. Despite the rising evidences related to the beneficial effects of bioactive PPHs in the cure and prevention of a variety of deteriorating diseases, their low bioavailability embodies a critical constraint for their various clinical applications. The poor bioavailability of these PPHs and BSCs are mainly attributed to their limited solubility and stability. Owing to this, there is the utmost need to resolve these issues in order to enhance their bioavailability and efficacy subsequently to make these most versatile BACs available to fulfill the ever-growing demands for human well-being. Food, pharmacology, and medical fields have lately made some advancement to offer pioneering nanotechnology-based methodologies for the development of efficient cell targeting and drug delivery system. The recent studies revealed that nanoencapsulation of BACs, especially PPHs is a fascinating means to improve their bioavailability and efficacy. The current research studies demonstrated that nanoencapsulation of BACs and PPHs provide a substantial protection against a variety of drastic environments such as thermal degradation and oxidation thus improving their shelf life. Furthermore, noncoagulation also helps to control the release keep the chemical and physical properties of initial constituents intact in the core, and enhance the bioavailability of the BACs and PPHs. The development of these nanoencapsulation techniques for PPHs and BACs has resulted in the improvements of targeted delivery of PPHs, smart packaging, food safety and quality, and sensory properties of food product containing PPHs. Most frequently encapsulated polyphenols are eugenol, catechins, curcumin, EGC, quercetin, and tea polyphenols. Nanoparticles, such as gelatin, zein, whey proteins, casein, CDs and chitosan, are usually used as nanocarriers. Future developments in this area must be aimed to assess possible toxicity, complete validation and characterization of relevant and efficient methods for nanoencapsulation, evaluation of elimination, metabolism, distribution and absorption behavior of these nanocarriers to explore more efficient greener approach for the synthesis of these nanocarriers to access their new therapeutic applications. Overall, there should be a positive attitude and viewpoint to completely introduce and implement the nanoencapsulation techniques in the medical, pharmacology, and food industry which appears to develop in the coming future and improve the health care and living standards of the people.

REFERENCES

Acosta, E. (2009). Bioavailability of nanoparticles in nutrient and nutraceutical delivery. *Current Opinion in Colloid & Interface Science*, *14*(1), 3–15. doi: 10.1016/j.cocis.2008.01.002.

Aderogba, M. A., Ogundaini, A. O., Eloff, J. N. (2006). Isolation of two flavonoids from bauhinia monandra (Kurz) leaves and their antioxidative effects. *African Journal of Traditional, Complementary and Alternative Medicines*, *3*, 59–65.

Agarwal, M., & Kamal, R. (2007). Studies on flavonoid production using in vitro cultures of Momordica charantia L. *Indian Journal of Biotechnology*, *6*, 277–279.

Agati, G., Azzarello, E., Pollastri, S., & Tattini, M. (2012). Plant Science Flavonoids as antioxidants in plants: Location and functional significance. *Plant Science, 196*, 67–76. doi: 10.1016/j.plantsci.2012.07.014.

Anandharamakrishnan, C. (2014). *Techniques for Nanoencapsulation of Food Bioactive.* Springer: Berlin/Heidelberg, Germany. doi: 10.1007/978-1-4614-9387-7.

Ann, M., Hemar, Y., & Ann, M. (2009). Nano- and micro-structured assemblies for encapsulation of food ingredients. *Chemical Society Reviews*. doi: 10.1039/b801739p.

Archivio, M. D., Filesi, C., Varì, R., & Scazzocchio, B. (2010). Bioavailability of the Polyphenols: Status and Controversies. *International Journal of Molecular Sciences*, *11*, 1321–1342. doi: 10.3390/ijms11041321.

Arts, I. C. W., & Hollman, P. C. H. (2005). Polyphenols and disease risk in epidemiologic studies. *The American Journal of Clinical Nutrition, 81*(1 Suppl). doi: 10.1093/ajcn/81.1.317s.

Assadpour, E., Maghsoudlou, Y., Jafari, S., Ghorbani, M., & Aalami, M. (2016). Optimization of folic acid nano-emulsification and encapsulation by maltodextrin-whey protein double emulsions. *International Journal of Biological Macromolecules*. doi: 10.1016/j.ijbiomac.2016.01.064.

Astray, G., Gonzalez-Barreiro, C., Mejuto, J. C., Rial-Otero, R., & Simal-Gándara, J. (2009). A review on the use of cyclodextrins in foods. *Food Hydrocolloids, 23*(7), 1631–1640. doi: 10.1016/j.foodhyd.2009.01.001.

Bajpai, V. K., Kamle, M., Shukla, S., Mahato, D. K., Chandra, P., Hwang, S. K., Kumar, P., Huh, Y. S., & Han, Y. K. (2018). Prospects of using nanotechnology for food preservation, safety, and security. *Journal of Food and Drug Analysis, 26*(4), 1201–1214. doi: 10.1016/j.jfda.2018.06.011.

Bazana, M. T., Codevilla, C. F., & de Menezes, C. R. (2019). Nanoencapsulation of bioactive compounds: Challenges and perspectives. *Current Opinion in Food Science, 26*, 47–56. doi: 10.1016/j.cofs.2019.03.005.

Beckman, C. H. (2000). Phenolic-storing cells : Keys to programmed cell death and periderm formation in wilt disease resistance and in general defence responses in plants? *Physiological and Molecular Plant Pathology*. doi: 10.1006/pmpp.2000.0287.

Belink, A. (2018). Licorice extract and its major polyphenol glabridin protect lipoprotein against lipid peroxidation: In vitro and ex vivo studies in humans and in atherosclerotic apolipoprotein. *American Journal of Clinical Nutrition, 66*(2), 267–275.

Blando, F. (2004). Anthocyanins: More than nature's colours. *Journal of Biomedicine and Biotechnology, 5*, 239–240.

Briuglia, M. L., Rotella, C., McFarlane, A., & Lamprou, D. A. (2015). Influence of cholesterol on liposome stability and on in vitro drug release. *Drug Delivery and Translational Research, 5*(3), 231–242. doi: 10.1007/s13346-015-0220-8.

Carrión-Granda, X., Fernández-Pan, I., & Maté, J. I. (2016). Antimicrobial edible films and coatings. *Edible Films and Coatings: Fundamentals and Applications, 67*(4), 281–295. doi: 10.1201/9781315373713.

Chem, F. (2010). Essential oils, phenolics, and antioxidant activities of different parts of cumin (*Cuminum cyminum* L.). *Journal of Agricultural and Food Chemistry*, 10410–10418. doi: 10.1021/jf102248j.

Cheynier, V. (2018). Polyphenols in foods are more complex than often thought 1–3. *The American Journal of Clinical Nutrition, 81*, 223–229.

Chung, K., Wong, T. Y., Wei, C., Huang, Y., Lin, Y., Chung, K., Wong, T. Y., Wei, C., & Huang, Y. (2016). Tannins and human health : A review. *Critical Reviews in Food Science and Nutrition, 8398*. doi: 10.1080/10408699891274273.

Cook, N. C., & Samman, S. (2006). Dietary sources. *Journal of Nutritional Biochemistry 2863*(Ldl), 66–76.

Craig, W. J. (2018). Health-promoting properties of common herbs. *The American Journal of Clinical Nutrition, 70,* 491S–499S.

Croteau, R., Kutchan, T. M., & Lewis, N. G. (2000). Natural products (secondary metabolites). *Biochemistry and Molecular Biology of Plants*, 24, 1250–1318.

de Pascual-teresa, S., Moreno, D. A., & García-viguera, C. (2010). Flavanols and anthocyanins in cardiovascular health: A review of current evidence. *International Journal of Molecular Sciences*, 1679–1703. doi: 10.3390/ijms11041679.

Đorđević, V., Balanč, B., Belščak-Cvitanović, A., Lević, S., Trifković, K., Kalušević, A., Kostić, I., Komes, D., Bugarski, B., & Nedović, V. (2014). Trends in encapsulation technologies for delivery of food bioactive compounds. In: G. Barbosa-Cánovas (Ed.), *Food Engineering Reviews* (vol. 7, issue 4). Springer Nature: Switzerland. doi: 10.1007/s12393-014-9106-7.

Etheridge, M. L., Campbell, S. A., Erdman, A. G., Haynes, C. L., Wolf, S. M., & Mccullough, J. (2013). The big picture on nanomedicine: The state of investigational and approved nanomedicine products. *Nanomedicine: Nanotechnology, Biology, and Medicine, 9*(1), 1–14. doi: 10.1016/j.nano.2012.05.013.

Ezhilarasi, P. N., Karthik, P., Chhanwal, N., & Anandharamakrishnan, C. (2013). Nanoencapsulation techniques for food bioactive components: A review. *Food and Bioprocess Technology, 6*(3), 628–647. doi: 10.1007/s11947-012-0944-0.

Faridi Esfanjani, A., & Jafari, S. M. (2016). Biopolymer nano-particles and natural nano-carriers for nano-encapsulation of phenolic compounds. *Colloids and Surfaces B: Biointerfaces, 146*, 532–543. doi: 10.1016/j.colsurfb.2016.06.053.

Faridi Esfanjani, A., Jafari, S. M., & Assadpour, E. (2017). Preparation of a multiple emulsion based on pectin-whey protein complex for encapsulation of saffron extract nanodroplets. *Food Chemistry, 221*, 1962–1969. doi: 10.1016/j.foodchem.2016.11.149.

Gaonkar, S. L., & Vignesh, U. N. (2017). Synthesis and pharmacological properties of chalcones: A review. *Research on Chemical Intermediates, 43*(11), 6043–6077. doi: 10.1007/s11164-017-2977-5.

Ghorani, B., & Tucker, N. (2015). Fundamentals of electrospinning as a novel delivery vehicle for bioactive compounds in food nanotechnology. *Food Hydrocolloids, 51*, 227–240. doi: 10.1016/j.foodhyd.2015.05.024.

Ghorbanzade, T., Jafari, S. M., Akhavan, S., & Hadavi, R. (2017). Nano-encapsulation of fish oil in nano-liposomes and its application in fortification of yogurt. *Food Chemistry, 216*, 146–152. doi: 10.1016/j.foodchem.2016.08.022.

Ghoulami, S., Idrissi, A. I., & Fkih-tetouani, S. (2001). Phytochemical study of mentha longifolia of morocco. *Fitoterapia*, 72(5), 596–598.

Gupta, K. K., Taneja, S. C., Dhar, K. L., & Atal, C. K. (1983). Flavonoids of andrographis-paniculata. *Phytochemistry, 22*, 314–315.

Hagerman, A. E., & Butler, L. G. (1981). The specificity of proanthocyanidin-protein interactions. *Journal of Biological Chemistry, 256*(9), 4494–4497. doi: 10.1016/S0021-9258(19)69462-7.

Hao, S., Wang, Y., Wang, B., Deng, J., Liu, X., & Liu, J. (2013). Rapid preparation of pH-sensitive polymeric nanoparticle with high loading capacity using electrospray for oral drug delivery. *Materials Science and Engineering C, 33*(8), 4562–4567. doi: 10.1016/j.msec.2013.07.009.

Harde, H., Das, M., & Jain, S. (2011). Solid lipid nanoparticles: An oral bioavailability enhancer vehicle. *Expert Opinion on Drug Delivery, 8*(11), 1407–1424. doi: 10.1517/17425247.2011.604311.

Herrmann, K. (1976). Flavonols and flavones in food plants, a review. *Journal of food Technology*, 11, 433–448.

Herrmann, K. (1989). Occurrence and content of hydroxycinnamic and hydroxybenzoic acid compounds in foods. *Critical Reviews in Food Science and Nutrition, 28*(4). doi: 10.1080/10408398909527504.

Ignat, I., Volf, I., & Popa, V. I. (2011). A critical review of methods for characterisation of polyphenolic compounds in fruits and vegetables. *Food Chemistry, 126*(4), 1821–1835. doi: 10.1016/j.foodchem.2010.12.026.

Ipsen, R., & Otte, J. (2007). Self-assembly of partially hydrolysed α-lactalbumin. *Biotechnology Advances, 25*(6), 602–605. doi: 10.1016/j.biotechadv.2007.07.006.

Jafari, S. M. (2017a). 1-An overview of nanoencapsulation techniques and their classification. In: *Nanoencapsulation Technologies for the Food and Nutraceutical Industries.* Elsevier Inc. doi: 10.1016/B978-0-12-809436-5/00001-X.

Jafari, S. M. (2017b). Nanoencapsulation technologies for the food and nutraceutical industries. *Acta Alimentaria, 46*(3), 390–394. doi: 10.1556/066.2017.46.3.16.

Jafari, S. M., Assadpoor, E., He, Y., & Bhandari, B. (2008). Encapsulation efficiency of food flavours and oils during spray drying. *Drying Technology, 3937.* doi: 10.1080/07373930802135972.

Khan, M. K., & Dangles, O. (2013). Accept edcrt. *Journal of Food Composition and Analysis.* doi: 10.1016/j.jfca.2013.11.004.

Khare, A. R., & Vasisht, N. (2014). Nanoencapsulation in the food industry : Technology of the future. In: R. Sobel (Ed.), *Microencapsulation in the Food Industry.* Elsevier Inc. doi: 10.1016/B978-0-12-404568-2.00014-5.

King, A., & Young, G. (1999). Characteristics and occurrence of phenolic phytochemicals. *Journal of the American Dietetic Association, 99*(2), 213–218. doi: 10.1016/S0002-8223(99)00051-6.

Klejdus, B., Lojková, L., Plaza, M., Šnóblová, M., & Štěrbová, D. (2010). Hyphenated technique for the extraction and determination of isoflavones in algae : Ultrasound-assisted supercritical fluid extraction followed by fast chromatography with tandem mass spectrometry. *Journal of Chromatography A, 1217,* 7956–7965. doi: 10.1016/j.chroma.2010.07.020.

Koes, R., Verweij, W., & Quattrocchio, F. (2005). Flavonoids : A colorful model for the regulation and evolution of biochemical pathways. *Trends Plant Science,10*(5). doi: 10.1016/j.tplants.2005.03.002.

Kogawa, K., Kazuma, K., Kato, N., Noda, N. & Suzuki, M. (2007). Biosynthesis of malonylated flavonoid glycosides on the basis of malonyl transferase activity in the petals of *Clitoria ternatea. Journal of Plant Physiology, 164,* 886–894. doi: 10.1016/j.jplph.2006.05.006.

Kris-Etherton, P. M., Hecker, K. D., Bonanome, A., Coval, S. M., Binkoski, A. E., Hilpert, K. F., Griel, A. E., & Etherton, T. D. (2002). Bioactive compounds in foods: Their role in the prevention of cardiovascular disease and cancer. *American Journal of Medicine, 113*(9 Suppl. 2), 71–88. doi: 10.1016/s0002-9343(01)00995-0.

Kumar, S., & Pandey, A. K. (2013). Chemistry and biological activities of flavonoids : An overview. *Scientific World Journal, 2013,* 162750.

Kumar, S, Sharma, U. K., Sharma, A. K., & Pandey, A. K. (2012). Protective efficacy of solanum xanthocarpum root extracts against free radical damage: Phytochemical analysis and antioxidant effect. *Cellular and Molecular Biology, 58*(1), 174–181. doi: 10.1170/T938.

Laurichesse, S., & Avérous, L. (2014). Chemical modification of lignins: Towards bio-based polymers. *Progress in Polymer Science, 39*(7), 1266–1290. doi: 10.1016/j.progpolymsci.2013.11.004.

Lewis, C. E., Walker, J. R. L., Lancaster, J. E., & Sutton, K. H. (1998). Determination of anthocyanins, flavonoids and phenolic acids in potatoes. I: Coloured cultivars of Solanum tuberosum L. *Journal of the Science of Food and Agriculture, 77*(1), 45–57. doi: 10.1002/(SICI)1097-0010(199805)77:1<45::AID-JSFA1>3.0.CO;2-S.

Liu, R. H. (2003). Health benefits of fruit and vegetables are from additive and synergistic combinations of phytochemicals. *American Journal of Clinical Nutrition, 78,* 517S–520S.

Livney, Y. D. (2015). Nanostructured delivery systems in food: Latest developments and potential future directions. *Current Opinion in Food Science*, *3*, 125–135. doi: 10.1016/j.cofs.2015.06.010.

Loftsson, T., & Duchêne, D. (2007). Cyclodextrins and their pharmaceutical applications. *International Journal of Pharmaceutics*, *329*(1–2), 1–11. doi: 10.1016/j.ijpharm.2006.10.044.

Lopez, M. (2001). Analysis of phenolic constituents of biological interest in red wines by high-performance liquid chromatography. *Journal of Chromatography A*, 922, 359–363.

Mahdavi, S. A., Jafari, S. M., & Ghorbani, M. (2014). Spray-drying microencapsulation of anthocyanins by natural biopolymers : A review. *Drying Technology*, 509–518. doi: 10.1080/07373937.2013.839562.

Manach, C., Scalbert, A., Morand, C., Rémésy, C., & Jime, L. (2004). Polyphenols: food sources and bioavailability. *The American Journal of Clinical Nutrition*, *79*(5), 727–747.

Marais, J. P. J., Deavours, B., Dixon, R. A., & Ferreira, D. (2006). Chapter 1: The stereochemistry of flavonoids. In: Grotewold, E. (Ed.), *The Science of Flavonoids*. Springer: New York, pp. 1–46.

McClements, D. J., Decker, E. A., Park, Y., & Weiss, J. (2009). Structural design principles for delivery of bioactive components in nutraceuticals and functional foods. *Critical Reviews in Food Science and Nutrition*, *49*(6). doi: 10.1080/10408390902841529.

Mcmanus, J. P., Davis, K. G., Beart, J. E., Gaffney, S. H., Lilley, T. H., & Haslam, E. (1985). Polyphenol interactions, Part 1: Introduction; some observations on the reversible complexation of polyphenols with proteins and polysaccharides. *Journal of The Chemical Society-perkin Transactions*, *28*(2), 1429.

Middleton, E. (1998). Effect of plant flavonoids on immune and inflammatory cell function. *Advances in Experimental Medicine and Biology*, *439*, 175–182.

Minekus, M., Alminger, M., Alvito, P., Ballance, S., Bohn, T., Bourlieu, C., Carrière, F., Boutrou, R., Corredig, M., Dupont, D., Dufour, C., Egger, L., Golding, M., Karakaya, S., Kirkhus, B., Le Feunteun, S., Lesmes, U., MacIerzanka, A., MacKie, A., … Brodkorb, A. (2014). A standardised static in vitro digestion method suitable for food-an international consensus. *Food and Function*, *5*(6), 1113–1124. doi: 10.1039/c3fo60702j.

Mohammadi, A., Jafari, S. M., Esfanjani, A. F., & Akhavan, S. (2016). Application of nanoencapsulated olive leaf extract in controlling the oxidative stability of soybean oil. *Food Chemistry*, *190*, 513–519. doi: 10.1016/j.foodchem.2015.05.115.

Mole, S. (1993). Systematics review paper the systematic distribution of tannins in the leaves of angiosperms: A tool for ecological studies. *Biochemical Systematics and Ecology*, *21*(8), 833–846.

Mondal, S. (2020). Potential of nanotechnology for rural applications. In: B. M. El Ali (Ed.), *Arabian Journal for Science and Engineering*. Springer: Berlin, Heidelberg. doi: 10.1007/s13369-019-04332-5.

Mukhopadhyay, P., & Kundu, P. P. (2015). Chitosan-graft-PAMAM-alginate core-shell nanoparticles: A safe and promising oral insulin carrier in an animal model. *RSC Advances*, *5*(114), 93995–94007. doi: 10.1039/c5ra17729d.

Müller, R. H., Radtke, M., & Wissing, S. A. (2002). Nanostructured lipid matrices for improved microencapsulation of drugs. *International Journal of Pharmaceutics*, *242*(1–2), 121–128. doi: 10.1016/S0378-5173(02)00180-1.

Nair, H. B., Sung, B., Yadav, V. R., Kannappan, R., Chaturvedi, M. M., & Aggarwal, B. B. (2010). Delivery of antiinflammatory nutraceuticals by nanoparticles for the prevention and treatment of cancer. *Biochemical Pharmacology*, *80*(12), 1833–1843. doi: 10.1016/j.bcp.2010.07.021.

Ne, J., Ping, C., Che, Y. B., & Misran, M. (2008). α-Tocopherol nanodispersions: Preparation, characterization and stability evaluation. *Journal of Food Engineering*, *89*, 204–209. doi: 10.1016/j.jfoodeng.2008.04.018.

Pandey, K. B., & Rizvi, S. I. (2009). Plant polyphenols as dietary antioxidants in human health and disease. *Oxidative Medicine and Cellular Longevity*, *2*(5), 270–278.

Pateiro, M., Gómez, B., Munekata, P. E. S., Barba, F. J., Putnik, P., Kovačević, D. B., & Lorenzo, J. M. (2021). Nanoencapsulation of promising bioactive compounds to improve their absorption, stability, functionality and the appearance of the final food products. *Molecules,* 26(6). doi: 10.3390/molecules26061547.

Pinelo, M., Del, P., Manzocco, L., Jose, M., & Cristina, M. (2005). Optimization of continuous phenol extraction from vitis vinifera by products. *Food Chemistry*, *92*, 109–117. doi: 10.1016/j.foodchem.2004.07.015.

Pisoschi, A. M., Pop, A., Cimpeanu, C., & Turcuş, V. (2018). Nanoencapsulation techniques for compounds and products with antioxidant and antimicrobial activity: A critical view. *European Journal of Medicinal Chemistry*. doi: 10.1016/j.ejmech.2018.08.076.

Reinli, K., & Block, G. (1996). Phytoestrogen content of foods: A compendium of literature values. *Nutrition and Cancer*, 26, 37–41. doi: 10.1080/01635589609514470.

Rezaei, A., Fathi, M., & Mahdi, S. (2019). Food hydrocolloids nanoencapsulation of hydrophobic and low-soluble food bioactive compounds within di ff erent nanocarriers. *Food Hydrocolloids*, *88*, 146–162. doi: 10.1016/j.foodhyd.2018.10.003.

Robbins, R. J. (2003). Phenolic acids in foods: An overview of analytical methodology. *Journal of Agricultural and Food Chemistry*, *51*(10), 2866–2887. doi: 10.1021/jf026182t.

Sahoo, N., Sahoo, R. K., Biswas, N., Guha, A., & Kuotsu, K. (2015). Recent advancement of gelatin nanoparticles in drug and vaccine delivery. *International Journal of Biological Macromolecules*, *81*, 317–331. doi: 10.1016/j.ijbiomac.2015.08.006.

Saibabu, V., Fatima, Z., Khan, L. A., & Hameed, S. (2015). Therapeutic potential of dietary phenolic acids. *Advances in Pharmacological and Pharmaceutical Sciences*, 2015, 823539.

Saleem, M., Hyoung, J. K., Ali, M. S., & Yong, S. L. (2005). An update on bioactive plant lignans. *Natural Product Reports*, *22*(6), 696–716. doi: 10.1039/b514045p.

Sanguansri, P., & Augustin, M. A. (2006). Nanoscale materials development a food industry perspective. *Trends in Food Science & Technology*. doi: 10.1016/j.tifs.2006.04.010.

Sankaranarayanan, S., Bama, P., Ramachandran, J., Kalaichelvan, P. T., & Deccaraman, M. (2010). Ethnobotanical study of medicinal plants used by traditional users in Villupuram district of Tamil Nadu, India. *American Journal of Ethnomedicine*, *4*(12), 1089–1101. doi: 10.5897/JMPR09.027.

Sannomiya, M., Fonseca, V. B., Silva, M. A., Rocha, L. R. M., Santos, L. C., & Hiruma-lima, C. A. (2005). Flavonoids and antiulcerogenic activity from Byrsonima crassa leaves extracts. *Journal of Ethnopharmacology*, *97*, 1–6. doi: 10.1016/j.jep.2004.09.053.

Scalbert, A., & Williamson, G. (2018). Dietary intake and bioavailability of polyphenols. *Journal of Nutrition*, *130*(8S Suppl), 2073S–2085S.

Sharma, A., Bhardwaj, G., & Gaba, J. (2020). Natural antioxidants: Assays and extraction methods/solvents used for their isolation. In: G. A. Nayik & A. Gull (Eds), *Antioxidants in Fruits: Properties and Health Benefits*. Springer, Singapore, pp. 1–33.

Shimoni, E. (2009). Chapter 23: Nanotechnology for foods: Delivery systems. In: G. V. Barbosa-Canovas, et al. (Eds.), *Global Issues in Food Science and Technology*. Academic Press: San Diego, pp. 411–424. doi: 10.1016/B978-0-12-374124-0.00023-5.

Shukla, S., Shankar, R., & Singh, S. P. (2014). Food safety regulatory model in India. *Food Control*, *37*(1), 401–413. doi: 10.1016/j.foodcont.2013.08.015.

Silva, H. D., Cerqueira, M. Â., & Vicente, A. A. (2012). Nanoemulsions for food applications: Development and characterization. *Food and Bioprocess Technology*, *5*(3), 854–867. doi: 10.1007/s11947-011-0683-7.

Simi, C. K., & Emilia Abraham, T. (2007). Hydrophobic grafted and cross-linked starch nanoparticles for drug delivery. *Bioprocess and Biosystems Engineering*, *30*(3), 173–180. doi: 10.1007/s00449-007-0112-5.

Singh, M., Kaur, M., & Silakari, O. (2014). Flavones: An important scaffold for medicinal chemistry. *European Journal of Medicinal Chemistry*. doi: 10.1016/j.ejmech.2014.07.013.

Soobrattee, M. A., Neergheen, V. S., & Luximon-ramma, A. (2005). Phenolics as potential antioxidant therapeutic agents: Mechanism and actions. *Mutation Research*, *579*, 200–213. doi: 10.1016/j.mrfmmm.2005.03.023.

Tabatabaei, R. H., Jafari, S. M., Mirzaei, H., Nafchi, A. M., & Dehnad, D. (2018). Preparation and characterization of nano-SiO2 reinforced gelatin-k-carrageenan biocomposites. *International Journal of Biological Macromolecules*. doi: 10.1016/j.ijbiomac.2018.01.116.

Tavares, G. M., Croguennec, T., Carvalho, A. F., & Bouhallab, S. (2014). Milk proteins as encapsulation devices and delivery vehicles: Applications and trends. *Trends in Food Science and Technology*, *37*(1), 5–20. doi: 10.1016/j.tifs.2014.02.008.

Torres-Giner, S., Martinez-Abad, A., Ocio, M. J., & Lagaron, J. M. (2010). Stabilization of a nutraceutical omega-3 fatty acid by encapsulation in ultrathin electrosprayed zein prolamine. *Journal of Food Science*, *75*(6). doi: 10.1111/j.1750-3841.2010.01678.x.

Törrönen, R., Häkkinen, S., Kärenlampi, S., & Mykkänen, H. (1997). Flavonoids and phenolic acids in selected berries. *Cancer Letters*, *114*(1–2), 191–192. doi: 10.1016/S0304-3835(97)04660-0.

Trevaskis, N. L., Charman, W. N., & Porter, C. J. H. (2008). Lipid-based delivery systems and intestinal lymphatic drug transport: A mechanistic update. *Advanced Drug Delivery Reviews*, *60*(6), 702–716. doi: 10.1016/j.addr.2007.09.007.

Tripoli, E., La Guardia, M., Giammanco, S., Di Majo, D., & Giammanco, M. (2007). Citrus flavonoids: Molecular structure, biological activity and nutritional properties: A review. *Food Chemistry*, *104*, 466–479. doi: 10.1016/j.foodchem.2006.11.054.

Van Dam, R. M., & Hu, F. B. (2005). Coffee consumption a systematic review. *Clinician Corner, 294*(1), 97–104.

Venugopala, K. N., Rashmi, V., & Odhav, B. (2013). Review on natural coumarin lead compounds for their pharmacological activity. *BioMed Research International*, *2013*, 963248.

Vinardell, M. P. (2017). Lignins and their derivatives with beneficial effects on human health. *International Journal of Molecular Sciences*. doi: 10.3390/ijms18061219.

Wang, X., Luo, Z., & Xiao, Z. (2014). Preparation, characterization, and thermal stability of β-cyclodextrin/soybean lecithin inclusion complex. *Carbohydrate Polymers*, *101*(1), 1027–1032. doi: 10.1016/j.carbpol.2013.10.042.

Weiss, J., Decker, E. A., McClements, D. J., Kristbergsson, K., Helgason, T., & Awad, T. (2008). Solid lipid nanoparticles as delivery systems for bioactive food components. *Food Biophysics*, *3*(2), 146–154. doi: 10.1007/s11483-008-9065-8.

Xiao, J., Cao, Y., & Huang, Q. (2017). Edible nanoencapsulation vehicles for oral delivery of phytochemicals: A perspective paper. *Journal of Agricultural and Food Chemistry*, *65*(32), 6727–6735. doi: 10.1021/acs.jafc.7b02128.

Yan, M., Ge, J., Liu, Z., & Ouyang, P. (2006). Encapsulation of single enzyme in nanogel with enhanced biocatalytic activity and stability. *Journal of the American Chemical Society*, *128*(34), 11008–11009. doi: 10.1021/ja064126t.

Index

Note: **Bold** page numbers refer to tables and *italic* page numbers refer to figures.

Taylor & Francis Group
an informa business